Microbial Biotechnology

Microbial Biotechnology

Edited by **Elsa Cooper**

SYRAWOOD
PUBLISHING HOUSE

New York

Published by Syrawood Publishing House,
750 Third Avenue, 9th Floor,
New York, NY 10017, USA
www.syrawoodpublishinghouse.com

Microbial Biotechnology
Edited by Elsa Cooper

Contents

Preface   IX

Chapter 1   **Screening and identification of lactic acid bacteria isolated from sorghum silage processes in west Algeria**   1
Chahrour W., Merzouk Y., Henni J. E., Haddaji M. and Kihal M.

Chapter 2   **Antimicrobial activity of *Xylopia aethiopica, Aframomum melegueta* and *Piper guineense* ethanolic extracts and the potential of using *Xylopia aethiopica* to preserve fresh orange juice**   8
Ogbonna Christiana N., Nozaki, K. and Yajima, H.

Chapter 3   ***In vitro* study of antiamoebic activity of methanol extract of fruit of *Pimpinella anisum* on trophozoites of *Entamoeba histolytica* HM1-IMSS**   14
Quiñones-Gutiérrez Y., Verde-Star M. J., Rivas-Morales C., Oranday-Cárdenas A., Mercado-Hernández R., Chávez-Montes A. and Barrón-González M. P.

Chapter 4   **Influence of plant growth regulators on somatic embryogenesis induction from inner teguments of rubber (*Hevea brasiliensis*) seeds**   18
Kouassi Kan Modeste, Koffi Kouablan Edmond, Konkon N'dri Gilles, Gnagne Michel, Koné Mongomaké and Kouakou Tanoh Hilaire

Chapter 5   **Manumycin from a new *Streptomyces* strain shows antagonistic effect against methicillin-resistant *Staphylococcus aureus* (MRSA)/vancomycin-resistant enterococci (VRE) strains from Korean Hospitals**   24
Yun Hee Choi, Seung Sik Cho, Jaya Ram Simkhada, Chi Nam Seong, Hyo Jeong Lee, Hong Seop Moon and Jin Cheol Yoo

Chapter 6   **Evaluation of non-viable biomass of *Laurencia papillosa* for decolorization of dye waste water**   29
Dahlia M. El Maghraby

Chapter 7   **Protective effect of zinc against cadmium toxicity on pregnant rats and their fetuses at morphological, physiological and molecular level**   38
Ashraf El-Sayed, Salem M. Salem, Amany A. El-Garhy, Zeinab A. Rahman and Asmaa M. Kandil

Chapter 8   **Study of reproductive toxicity of *Combretum leprosum* Mart and Eicher in female Wistar rats**   48
Jamylla Mirck Guerra de Oliveira, Denise Barbosa Santos, Francimarne Sousa Cardoso, Márcia de Sousa Silva, Yatta Linhares Boakari, Silvéria Regina de Sousa Lira and Amilton Paulo Raposo Costa

Chapter 9   **Antimicrobial activity of plant phenols from *Chlorophora excelsa* and *Virgilia oroboides***   53
Thiriloshani Padayachee and Bharti Odhav

Chapter 10   **Efficiency evaluation of three fluidised aerobic bioreactor based sewage treatment plants in Kashmir Valley**   61
Dilafroza Jan, Ashok K. Pandit and Azra N. Kamili

Chapter 11   **Demulsification capabilities of a Microbacterium species for breaking water-in-crude oil emulsions**   71
Hossein Salehizadeh, Aida Ranjbar and Kevin Kennedy

Chapter 12   **Xanthine oxidase activity during transition period and its association with occurrence of postpartum infections in Murrah buffalo (*Bubalus bubalis*)**   79
Bhabesh Mili, Sujata Pandita, Bharath kumar B.S., Anil Kumar Singh, Madhu Mohini and Manju Ashutosh

Chapter 13   **Molecular study on extended spectrum β-lactamase-producing Gram negative bacteria isolated from Ahmadi Hospital in Kuwait**   83
Eshaq A. Mohmid, El-Sayed A. El-Sayed and Mahmoud F. Abdel El-Haliem

Chapter 14   **Marine bacterial prodigiosin as dye for rubber latex, polymethyl methacrylate sheets and paper**   97
Jissa G. Krishna, Ansu Jacob, Philip Kurian, Elyas KK and M. Chandrasekaran

Chapter 15   **Anti-ulcer activity of aqueous leaf extract of *Nauclea latifolia* (rubiaceae) on indomethacin-induced gastric ulcer in rats**   101
Balogun M. E., Oji J.O., Besong E. E., Ajah A. A. and Michael E. M.

Chapter 16   **Studies on some active components and antimicrobial activities of the fermentation broth of endophytic fungi DZY16 Isolated from *Eucommia ulmoides* Oliv.**   108
Ding Ting, Sun Wei-Wei,Qi Yong- Xia and Jiang Hai-Yang

Chapter 17   **Characterization of lactic acid bacteria isolated from indigenous dahi samples for potential source of starter culture**   115
Shabana Maqsood, Fariha Hasanand Tariq Masud

Chapter 18   **Screening of verotoxin-producing *Escherichia coli* (VETC) O104-2011 from Egyptian market in 2011**   121
Nashwa A. Ezzeldeen, Khaled F. Al-Amary, Mohamed M. Abdalla and Sherein I Abd El-Moez

Chapter 19   **Effect of light and aeration on the growth of *Sclerotium rolfsii in vitro***   126
Muthukumar A. and Venkatesh, A.

Chapter 20   **Evaluation of antimicrobial and antioxidant properties of leaves of *Emex spinosa* and fruits of *Citrillus colocynthis* from Saudi Arabia**   130
Mona A. Aldamegh, Emad M. Abdallah and Anis Ben Hsouna

Chapter 21   **Emerging *Acinetobacter schindleri* in red eye infection of *Pangasius sutchi***   136
M. Radha Krishna Reddy and S. A. Mastan

Chapter 22   **Determination of heavy metals and genotoxicity of water from an artesian well in the city of Vazante-MG, Brazil**   **141**
Regildo Márcio Gonçalves da Silva, Eni Aparecida do Amaral, Vanessa Marques de Oliveira Moraes and Luciana Pereira Silva

Chapter 23   **Identification of phytochemical components of aloe plantlets by gas chromatography-mass spectrometry**   **147**
Mansoor Saljooghianpour and Taiebeh Askari Javaran

Chapter 24   **Production and characterization of antimicrobial active substance from some macroalgae collected from Abu-Qir bay (Alexandria) Egypt**   **152**
Mohamed E.H. Osman, Atef M. Aboshady and Mostafa E. Elshobary

Chapter 25   **Plant regeneration through indirect organogenesis of chestnut (*Castanea sativa* Mill.)**   **164**
Mehrcedeh Tafazoli, Seyed Mohammad Hosseini Nasr, Hamid Jalilvand and Dariush Bayat

Chapter 26   **Association of plasma protein C levels and coronary artery disease in men**   **171**
Olfat A. Khalil, Kholoud S. Ramadan, Amal H. Hamza and Safinaz E. El-Toukhy

Chapter 27   **Toxicity study of the anti-hypertensive agent perindopril on the entomopathogenic fungus *Metarhizium anisopliae* (Metschnikoff) Sorokin assessed by conidia germination speed parameter**   **177**
Ligia Maria Crubelati Bulla, Daniela Andressa Lino Lourenço, Sandro Augusto Rhoden, Ravely Casarotti Orlandelli and João Alencar Pamphile

Chapter 28   **Isolation, characterization and antimicrobial activity of *Streptomyces* strains from hot spring areas in the northern part of Jordan**   **183**
M. J. Abussaud, L. Alanagreh and K. Abu-Elteen

Chapter 29   **Helminth parasites of *Synodontis nigrita* at lower Niger (IDAH), Nigeria**   **192**
O. O. Toluhi and S. O. Adeyemi

Chapter 30   **In vitro antibacterial activity of alkaloid extracts from green, red and brown macroalgae from western coast of Libya**   **195**
Rabia Alghazeer, Fauzi Whida, Entesar Abduelrhman, Fatiem Gammoudi and Mahboba Naili

Permissions

List of Contributors

Preface

The main aim of this book is to educate learners and enhance their research focus by presenting diverse topics covering this vast field. This is an advanced book which compiles significant studies by distinguished experts. This book addresses successive solutions to the challenges arising in the area of application, along with it; the book provides scope for future developments.

Microbial biotechnology is an interdisciplinary field of study that incorporates concepts and techniques of microbiology and biotechnology to develop useful products. The topics covered in this extensive book deal with the major applications of microbial biotechnology in drug development, food processing, biocatalysis, etc. The ever growing need of advanced technology is the reason that has fueled the research in this field in recent times. The chapters included herein are appropriate for students seeking detailed information in this area as well as for experts.

It was a great honour to edit this book, though there were challenges, as it involved a lot of communication and networking between me and the editorial team. However, the end result was this all-inclusive book covering diverse themes in the field.

Finally, it is important to acknowledge the efforts of the contributors for their excellent chapters, through which a wide variety of issues have been addressed. I would also like to thank my colleagues for their valuable feedback during the making of this book.

Editor

Screening and identification of lactic acid bacteria isolated from sorghum silage processes in west Algeria

Chahrour W., Merzouk Y., Henni J. E., Haddaji M. and Kihal M.*

Laboratory of Applied Microbiology, Department of Biology, Faculty of Science, Oran University, BP 16, Es-Senia, 31100, Oran, Algeria.

The lactic acid bacteria (LAB) isolated from sorghum (*Sorghum bicolor.* L.) silage were identified during different periods of evolution of sorghum silage in west Algeria. Morphological, physiological, biochemical and technological techniques were used to characterize lactic acid bacteria isolates. A total number of 27 representatives of lactic acid bacterial strains were retained and among them four dominant genus were identified as *Lactobacillus* (44%), *Lactococcus* (14.81%), *Weissella* (29.62%) and *Leuconostoc* (11.11%). The representative species identified were *Lactobacillus brevis* (25%), *Lactobacillus pentosus* (3.7%), *Lactobacillus manihotivorans* (11.11%), and *Lactobacillus fermentum* (3.7%). *Lactococcus lactis* subsp. *lactis* biovar. *diacetylactis* (14.81%), *Weissella cibaria* (7.2%), *Weissella minor* (11.11%), *Weissella soli* (3.7%), *Weissella viridescense* (7.2%) and *Leuconostoc mesenteroides* subsp. *mesenteroides* (11.11%). Only two strains of lactic acid bacteria were amylolytic. These results will enable future research on the relationship between LAB species and silage fermentation quality.

Key words: Lactic acid bacteria, identification, silage, sorghum, evolution, amylolytic, technology, species.

INTRODUCTION

Lactic acid fermentation, an ancient preservation method, is nowadays specially preferred as a "natural" process to increase the shelf life of various products (vegetable, dairy and meat). Natural populations of LAB on plant material are responsible for conservation of a crop as silage by converting water-soluble carbohydrates (WSC) into organic acids, mainly lactic acid. As a result, the pH decreases and the forage is preserved (Chen and Weinberg, 2009). The biochemical and microbiological events that occur during ensiling process can be divided into four distinct stages (Ashbell et al., 1997). The bacterial spectrum includes homofermentative species that exclusively produce lactic acid and heterofementative species, which produce a mixture of lactic and acetic acids and/or other by-products like ethanol and carbon dioxide (Vlková et al., 2012). Habitually, heterofermentative LAB are low in number. Thus, the concept of using microbial inoculants for silage involves adding fast-growing homofermentative LAB in order to dominate the fermentation, but is less effective if insufficient fermentable substrate is available. Some of the commonly used homofermentative LAB in silage inoculants include *Lactobacillus plantarum*, *Lactobacillus acidophilus*, *Pediococcus acidilactici* and *Enterococcus faecium*. Commercially available microbial inoculant contains one or more of these bacteria that have been selected for their ability to realize fermentation. The quality of silage could be improved by the addition of inoculants, consequently lactic acid production occurs more quickly and loss of nutrients during ensilage can be reduced (Widyastuti, 2008). The presence of LAB such as *L. plantarum* and *Pediococcus* spp. were able to maintain the silage quality through increasing lactic acid production (Filya et al., 2006). Recently, in order to prevent the aerobic deterioration of silage, heterofermentative LAB species, such as *Lactobacillus brevis*, *Lactobacillus fermentum* and *Lactobacillus reuterin*, were developed as silage additives for antifungal effect. Moreover, homofermentative LAB inoculum can provide substantial benefit by reducing the risk of the

*Corresponding author. E-mail: Kihalm@gmail.com.

growth of other harmful spoilage organisms such as butyric acid bacteria including clostridia by reducing the pH (Saarisalo et al., 2007). Several factors influence the fermentation processes that transform green forage into silage.

From a microbiological point of view, to our knowledge, no information is available on the microbial ecology of sorghum, especially with regard to the indigenous LAB and their effects during the fermentation process. Especially, the sorghum plant is drought adaptive supporting the arid climate and rich food. The aim of this study was to identify and characterize naturally present microbial populations from the sorghum silage, especially dominants lactic acid bacteria on the basis of their important technological properties in order to select potential autochthonous as grass silage inoculants.

MATERIALS AND METHODS

Preparation of silage

Silage sorghum (*Sorghum bicolor*) was obtained from different experimental sites in the north west of Algeria and it was prepared using the method described by Filya et al. (2004). Whereas, the forage of *S. bicolor* was collected in Oran region. This study was done at different periods of silage incubation. Sorghum was harvested and chopped by a precision forage harvester to 3 to 5 cm theoretical length. A serum bottle of 500 ml was used as microsilo and was incubated at two different temperatures, 30°C and ambient temperature. In all the experiments, the serum bottles were sampled on 2, 5, 8, 10, 18 and 90 days in triplicate (Ashbell et al., 1991; Filya et al., 2004).

Analytical methods

Biochemical test

An amount of 10 g of silage was homogenized for 5 min with 90 ml of distilled water and then the pH of the filtered water was measured by pH meter (Xing et al., 2009). Dry matter was determined by oven drying for 48 h at 60°C for fresh material and silages (Weinberg et al., 1995).

Microbiological examination

For microbiological analysis, the samples of silage were prepared as follows: 1 g of sample was homogenized with 9 ml of 0.85% (w/v) sterile physiological saline in a Stomacher lab-blender for 1 min and serially diluted (10^{-1} to 10^{-7}) in the same diluents. One milliliter of these dilutions was pour-plated in the respective media for LAB. LAB were isolated on MRS agar, after incubation under anaerobic conditions at 30°C for 48 to 72 h. Representative strains of LAB were obtained from MRS plates of the highest sample dilutions. Aerobic mesophilic counts were determined using plate count agar incubated at 30°C for 48 to 72 h.

Isolation and identification of dominant lactic acid bacteria

The selection of colonies was randomly isolated from the plate containing between 25 and 250 colonies. Purity of the isolates was checked by streaking in MRS agar plates, followed by microscopic observation.

For the investigation of the fermentation properties of these isolates, inoculation was done in 10% skimmed milk and to which 0.3% yeast extract, 1% glucose, 1% $CaCO_3$ were added and incubated for 48 h at 30 and 42°C. Subsequently, the coagulation of milk was checked which indicates the presence of LAB (Sengun et al., 2009).

Preliminary identification and grouping was done based on cell morphology, gram staining, catalase activity and other phenotypic properties by using CO_2 production from glucose, hydrolysis of arginine in M16BCP plates, growth at different temperatures of 15°C for 14 days and 37 and 45°C for 2 days, and at different pH: 4 and 9.6 for 7 days as well as the ability to grow in different concentrations of NaCl: 3, 4 and 6.5% for 2 days of incubation in MRS broth (Badis et al., 2004; Bendimerad et al., 2012).

The identification at species level was carried out by the fermentation of carbohydrates determined on MRS broth containing bromocresol purple (0.04 g/l) as a pH indicator. The carbon sources were added to the medium at 1% (w/v) as final concentration. The carbohydrates tested were: glucose, fructose, lactose, galactose, sorbitol, arabinose, xylose, trehalose, raffinose, ramnose, maltose, mannitol, sucrose, starch and esculin to ensure anaerobic conditions. Each tube was supplemented with two drops of sterile liquid paraffin before inoculation.

All strains of lactic acid bacteria were stored without appreciable loss of properties in skimmed milk containing 30% (wt/vol) glycerol at -20°C. Working cultures were also kept in MRS agar slant at 4°C and streaked for every 4 weeks (Badis et al., 2004; Guessas and Kihal, 2004).

Technological properties of the isolates

The physiological group of lactic acid bacteria involved in the degradation of macromolecules (protein, cellulose and starch) in the vegetative parts of the silage sorghum (*S. bicolor*) was investigated by the method of Dubos (1928).

Proteolytic activity

For screening of the proteolytic bacteria, strains were grown on Yeast Milk Agar (YMA) and gelatin medium. The plate counts were evaluated by the presence of the hydrolysis zone around the colonies, and by the change in the appearance of gelatin in liquid medium (Huggins and Sandine, 1984).

Amylolytic activity

The amylolytic activity of isolates of sorghum silage was observed in MRS starch medium where glucose in the medium was replaced by 20 g soluble starch (Brabet et al., 1996). The enumeration of amylolytic bacteria was evaluated by the presence of the clear zone around the colonies after using iodine solution (lugol) (Thapa et al., 2006).

Cellulolytic activity

From retained isolates of lactic acid bacteria strain, the pre-cultures were inoculated by spot in Dubois agar supplemented with 5% of cellulose. The plates were then incubated at 37°C for 72 h. The plates were checked for clear zones surrounding the colonies.

RESULTS AND DISCUSSION

The study of the diversity and the dynamics of microbial

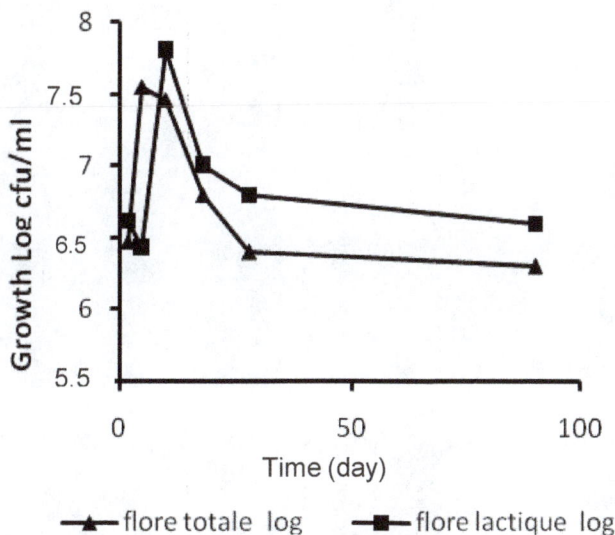

Figure 1. Growth of total flora and lactic flora during the fermentation process of sorghum silage.

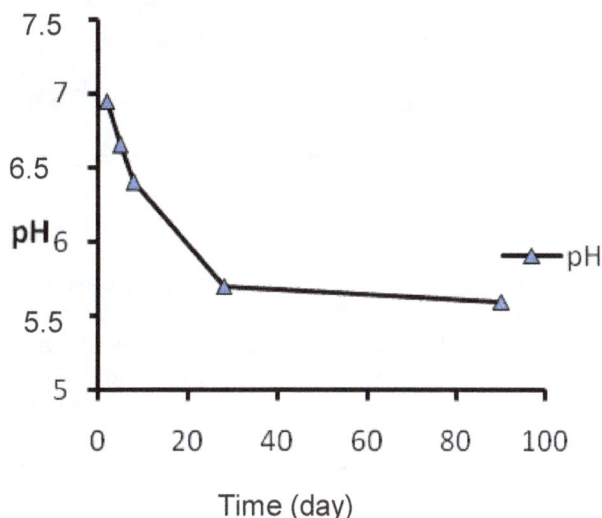

Figure 2. pH variations during the fermentation process of sorghum silage.

populations associated with sorghum silage were investigated for evaluation of the quality of silage which represent a major effect on the feed intake, nutrient utilization and milk production of dairy cows (Saarisalo et al., 2007).

Enumeration of total flora and lactic acid flora

The microbial composition of the sorghum silages is depicted in Figure 1. During the initial time of incubation, the number of total microflora is higher than lactic flora (7.5 and 6.5 log, respectively). But at day 8 of silage

incubation, the number of lactic acid bacteria increased to 7.8 log, whereas, the total microflora remained stable at 7.5 log. The number of total microflora and lactic acid flora decreased gradually during the time of incubation. At 90 days, the number of total microflora is less than lactic acid microflora (6.3 and 6.7 log, respectively). The growth of the total flora during 90 days of silage incubation was found to be maximum on 2nd and 5th day, whereas it was least on the 8th day. The variation in the evolution of pH during sorghum silage was noted and shown in Figure 2. The initial pH was 7 which dropped gradually to 5.6 on the 28th day and after which no change in the pH was observed.

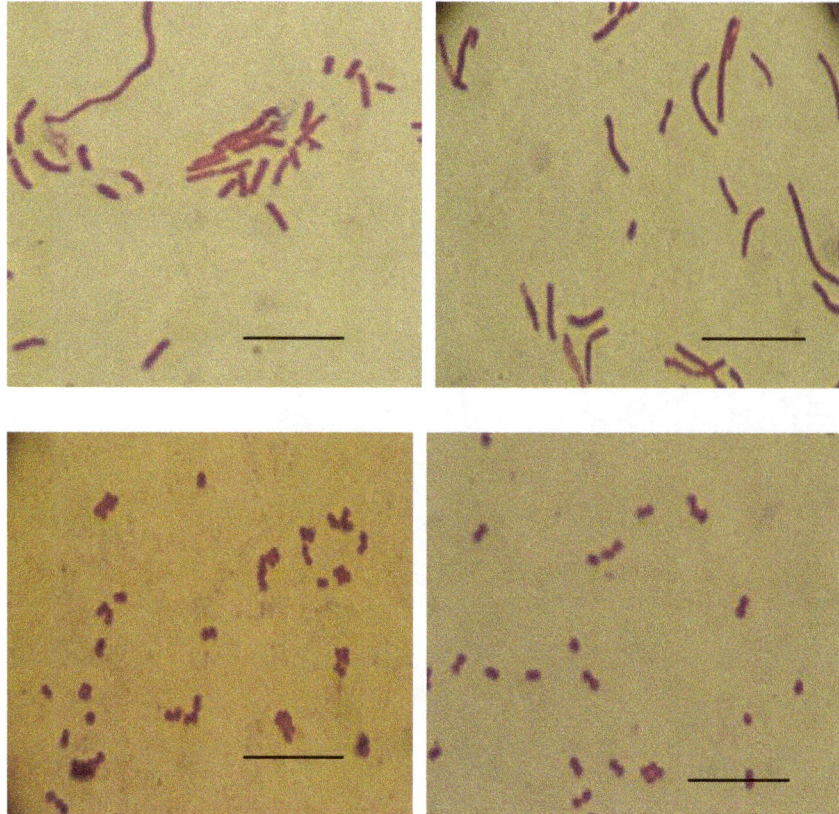

Figure 3. Morphological characters of lactic acid bacteria of silage sorghum observed with microscope (rod form and cocci form).

On the other hand, the growth of lactic flora which is a part of the natural heritage of flora associated with the plant material (Makimattila et al., 2011), acts differently in silage because as shown in Figure 1, the growth is minimal at 2nd and 5th day but in the 8th day it was maximum (ICMSF, 2005; Weinberg et al., 2010). During this period of silage incubation, many factors can present favorable or unfavorable effect on microorganism's growth. The total flora can be inhibited by the absence of O_2 such as yeast, molds and aerobic bacteria (Pahlow et al., 2003), pH sensibility by presence of the organic acid especially lactic acid produced during the growth of the lactic acid bacteria (Paragon, 2004) or acetic acid which has an antifungal effect (Lindsey and Kung, 2010; Keles and Demirci, 2011). The dominance of lactic acid bacteria at the end phase of silage is enhanced by environmental conditions especially gradual absence of oxygen and pH decrease.

Isolation and identification of dominant lactic acid bacteria

From 63 isolates screened, 27 were considered to be LAB by the properties like Gram staining, catalase activity, motility and production of spores or not, as per the method described by Sengun et al. (2009). This lead to the delineation of two principal groups of isolates, each one displaying a distinct carbohydrate fermentation pattern (Table 1 and Figure 3). The various groups presumably represented four different LAB and various percentages of the total genera were found to be *Lactobacillus* (44%), *Lactococcus* (14.81%), *Weissella* (29.62%) and *Leuconostoc* (11.11%).

The LAB isolates were mostly homofermentative and heterofermentative. Most of them showed growth in the pH 4 and at temperature of 45°C.

In the lactobacilli isolates, especially for homofermentative *Lactobacillus* group, the specie of *Lactobacillus manihotivorans* was detected and identified in our silage sample. The *Lactobacillus pentosus* is facultative heterofermentaire species which can produce CO_2 from gluconate and have the same characters of *Lactobacillus manihotivorans*. The later species was also isolated from sour cassava starch fermentation in Colombia (Morlon et al., 1998, 2001).

The eight species of heterofermentative *Lactobacillus* group were found to belong to two species of *L. brevis* (7 isolates) *and L. fermentum* (1 isolate). To differentiate between these two species, xylose and esculin were used. Our results for *Lactobacillus* identification were in agreement with those obtained by Abriouel et al. (2008).

Table 1. Physiological and biochemical characteristics of isolated strains from silage.

Characteristics		Rod group					Cocci group				
		1	2	3	4	5	6	7	8	9	10
		14-13-A5-12-D11-19-A9	50	7*-22-D1	D10	11-A8-A6-20	4-32	7-43-A4	25	A1-42	E1-E2-E3
Number of isolates (27)		7	1	3	1	4	2	3	1	2	3
Form		rods	rods	rods	rods	cocci	cocci	cocci	cocci	cocci	cocci
Gram		+	+	+	+	+	+	+	+	+	+
Catalase		-	-	-	-	-	-	-	-	-	-
CO$_2$ from glucose		+	v	-	+	-	+	+	+	+	+
NH$_3$ from arginine		+	+	+	+	+	+	+	+	+	-
Growth at temperature (°C)	15	+	+	+	+	+	+	+	+	+	+
	45	v	+	+	+	+	+	+	+	+	+
Growth in a medium with NaCl (%)	3	+	+	+	+	+	+	+	+	+	+
	6.5	+	+	+	-	+	+	+	+	+	-
Growth at pH	4	+	+	+	+	+	+	+	+	+	-
	9.6	+	+	+	+	-	+	+	+	+	+
Citrate hydrolysis		+	+	+	+	+	+	+	+	+	+
Heat resistance 63.5°C for 30 min		v	+	+	-	+	+	+	+	v	+
Acid production from											
Trehalose		+	+	+	+	/	+	+	+	+	+
Raffinose		-	-	-	+	+	-	-	+	-	+
Xylose		+	+	+	-	/	+	+	+	+	+
Maltose		+	+	+	+	+	+	+	+	+	+
Galactose		+	+	+	+	+	+	+	+	+	+
Sorbitol		+	+	+	+	+	-	+	-	-	-
Arabinose		v	v	+	+	+	+	-	+	-	+
Mannitol		+	+	+	+	+	+	+	-	-	+
Ramnose		-	-	-	v	-	-	-	-	-	-
Sucrose		+	/	/	+	+	+	+	+	+	+
Fructose		+	+	+	+	+	+	+	+	+	+
Glucose		+	+	+	+	+	+	+	+	+	+
Esculin		+	+	+	-	+	+	+	+	-	+
Starch		-	-	-	-	-	-	-	-	-	-
Pre-identification		*Lactobacillus brevis*	*Lactobacillus pentosus*	*Lactobacillus manihotivorans*	*Lactobacillus fermentum*	*Lactococcus lactis* subsp. *lactis* biovar. *diacetylactis*	*Weissella cibaria*	*Weissella minor*	*Weissella soli*	*Weissella viridescens*	*Leuconostoc mesenteroides* subsp *mesenteroides*

+, Positive; -, negative; v, variable.

Seseña et al. (2005) always considered the predominant LAB, with increased growth of *L. brevis* at the end of fermentation in natural fermentation of green olives. The utilization of ribose, galactose, glucose, fructose, maltose, saccharose, raffinose and esculin was used for the identification of *L. fermentum* by Hammes et al. (1992 and Sawadogo-Lingani et al. (2010). In cocci groups, three isolates were positively identified as *Leuconostoc mesenteroides* subsp. *Mesenteroides* characterized by: Production of CO_2 from glucose, production of dextran from sucrose, inability to use arginine, but the use of arabinose is specific for this species. The properties of *Leuconostoc* species were reported in several works (Carr et al., 2002; Bendimerad et al., 2012). Generally, leuconostocs are found living in association with plant material and dairy products. Other studies have reported leuconostocs as the dominant microbial population on forage crops and silage (Cai et al., 1994). The lower numbers of *Leuconostoc* is probably due to their inability to compete with other LAB in mixed cultures (Teuber and Geis, 1981; Togo et al., 2002).

Eight strains in this group were identified as *Weissella* differing from the *Leuconostoc* by the arginine test and the sugar fermentation profile. The represented species include *Weissella cibaria* (7.2%), *Weissella minor* (11.11%), *Weissella soli* (3.7%) and *Weissella viridescense* (7.2%). The studies of Pang et al. (2011) indicate that perhaps several *Weissella* species could improve silage quality. Some species of *Weissella* have been isolated from a wide range of sources such as soil, fresh vegetables, meat, fish, fermented silage and foods (Björkroth et al., 2002; Sirirat et al., 2008; Valerio et al., 2009).

Four homofermentative strains can hydrolyze arginine and citrate and cannot grow at pH 9, 45°C and 6.5% NaCl. However, the three later physiological characters: pH, temperature and salinity cannot be used as reference for the identification of lactic acid bacteria because there are variations in these characters particularly in this group of bacteria Drici et al. (2010). They later had isolated from camel milk the species of *Lactococcus lactis* subsp. *lactis* biovar. *diacetylactis* which show growth at 50°C. The profile comparison revealed that homofermetaives isolates observed in the present study can be considered as *L. lactis* subsp. *lactis* biovar. *diacetylactis* (14.81%). Further, the research work of Sawadogo-Lingani et al. (2010) has observed the presence of *L. lactis* in sorghum grains fermentation.

Technological properties of the isolated lactic acid bacteria

Fermentation is one of the oldest methods of food preservation technology in the world. The process relies on the biological activity of microorganisms for production of a range of metabolites which can suppress the growth and survival of undesirable microflora in foodstuffs. As a result, fermented products generally have a longer shelf life than their original substrate and present very good safety records. In this study, the absence of proteolytic and cellulolytic activity in isolated lactic acid bacteria was noticed. In contrast, three species of *Lactobacillus* produced an enzyme amylase in culture medium, which was confirmed by the observation of a clear zone of starch hydrolysis when treated with iodine. Three

amylolytic strains identified belonged to the species of *L manihotivorans* which are considered as amylolytic lactic acid bacteria.

Conclusion

On the basis of phenotypic properties, the lactic acid bacterial population in silage sorghum prepared in laboratory consisted of: *Lactobacillus* sp., *Lactobacillus brevis*, *Lactobacillus pentosus*, *Lactobacillus manihotivorans*, *Lactobacillus fermentum*, *Lactococcus lactis* subsp. *lactis* biovar. *diacetylactis Weissella cibaria*, *Weissella minor*, *Weissella soli*, *Weissella viridescense* and *Leuconostoc mesenteroides* subsp. *mesenteroides* which play an important role in fermentation to increase the shelf life and enrichment by degradation of macromolecules. The present investigation revealed the presence of different genus of lactic acid bacteria in sorghum silage. The study of their technological characters might help in the near future for the establishment of a local starter for silage process. To our knowledge, in Algeria, this is the first report on the isolation of amylolytic species of *Lactobacillus manihotivorans* in sorghum silage which can be exploited in the manufacture of fermented foods.

REFERENCES

Abriouel H, Ben Omar N, Pérez Pulido R, López RL, Ortega E, Cañamero MM, Gálvez A (2008). Vegetable Fermentations Molecular Techniques in the Microbial Ecology of Fermented Foods. Chapter 5. Food Microbiol. Food Saf. pp. 145-161, DOI: 10.1007/978-0-387-74520-6_5

Ashbell G, Weinberg ZG, Azrieli A, Hen Y, Horev B (1991). A simple system to study aerobic deterioration of silages. Can. Agric. Eng 33:391-395.

Ashbell G, Weinberg ZG, Brukental I, Tabori K, Sharet N (1997). Wheat silage: effect of cultivar and stage of maturity on yield and degradability *in situ*. J. Agric. Food Chem. 45(3):709-912.

Badis A, Guetarni D, Moussa Boudjema B, Henni DE, Kihal M (2004). Identification and technological properties of lactic acid bacteria isolated from raw goats milk of four Algerian races. Food Microbiol. 21(5):579-588.

Bendimerad N, Kihal M, Berthier F (2012). Isolation, identification and technological characterization of wild leuconostocs and lactococci for traditional raib type milk fermentation. Dairy Sci. Technol. 92(3):249-264.

Björkroth KJ, Schillinger U, Geisen R, Weiss N, Hoste B, Holzapfel WH, Korkeala HJ, Vandamme P (2002). Taxonomic study of *Weissella confusa* and description of *Weissella cibaria* sp. nov, detected in food and clinical samples. Int. J. Syst. Evol. Microbiol. 52:141-148.

Brabet C, Chuzer G, Oufour D, Raimbaultt M, Giraudtt J (1996). Improving cassava sour starch quality in Colombia, Progress in Research and Development, CIRAD chapter 27. ISBN 958-9439-88-8 pp. 242-246.

Cai Y, Ohmomo S, Kumai S (1994). Distribution and lactate fermentation characteristics of lactic acid bacteria on forage crops and grasses. J. Jpn. Soc. Grassl. Sci. 39:420-428.

Carr FJ, Chill D, Maida NR (2002). The lactic acid bacteria: With literature Survey. Curr. Rev. Microbiol. 28(4):281-370.

Chen Y, Weinberg ZG (2009). Changes during aerobic exposure of wheat silages. Anim. Feed Sci. Technol. 154:76-82.

Drici H, Gilbert C, Kihal M, Atlan D (2010). Atypical citrate-fermenting *Lactococcus lactis* strains isolated from dromedary's milk. J. Appl. Microbiol. 108:647-657.

Dubos R (1928). The decomposition of cellulose by aerobic bacteria .J. Bacteriol. 15:223-234.

Filya I, Sucu E, Karabulut A (2004). The effect of *Propionibacterium acidipropionici*, with or without *Lactobacillus plantarum*, on the fermentation and aerobic stability of wheat, sorghum and maize silages. J. Appl. Microbiol. 97:818-826.

Filya I, Sucu E, Karabulut A (2006). The effects of *Propionibacterium acidipropionici* and *Lactobacillus plantarum*, applied at ensiling, on the fermentation and aerobic stability of low dry matter corn and sorghum silages. J. Ind. Microbiol. Biotechnol. 33:353-358.

Guessas B, Kihal M (2004). Characterization of lactic acid bacteria isolated from Algerian arid zone raw goats' milk. Afr. J. Biotechnol. 3:339-342.

Hammes WP, Vogel RF, Wood BJB, Holzapfel WH (1992). The genus *Lactobacillus*. The genera of lactic acid bacteria. The lactic acid bacteria. Blackie Academic and Professional, Scotland, Glasgow. 2:19-54.

Huggins AR, Sandine WE (1984). Differentiation of fast and slow milk-coagulating isolates in strains of *Lactic streptococci*. J. Dairy Sci. 67:1674-1679.

ICMSF, International Commission on Microbiological Specifications for Foods. (2005). Microbial ecology of food commodities, feeds pet and foods, NY.

Keles G, Demirci U (2011). The effect of homofermentative and heterofermentative lactic acid bacteria on conservation characteristics of baled triticale–Hungarian vetch silage and lamb performance. Anim. Feed Sci. Technol. 164:21-28.

Lindsey JR, Limin Kung J (2010). Effects of combining *Lactobacillus buchneri* 40788 with various lactic acid bacteria on the fermentation and aerobic stability of corn silage. Anim. Feed Sci. Technol. 159:105-109.

Makimattila E, Kahala M, Joutsjoki V (2011). Characterization and electrotransformation of *Lactobacillus plantarum* and *Lactobacillus paraplantarum* isolated from fermented vegetables. World J. Microbiol. Biotechnol. 27:371-379.

Morlon GJ, Guyot JP, Pot B, Jacobe de Haut I, Raimbault M (1998). *Lactobacillus manihotivorans* sp. nov, a new starch-hydrolyzing lactic acid bacterium isolated from cassava sour starch fermentation. Int. J. Syst. Bacteriol. 48:1101-1109.

Morlon GJ, Mucciolo RF, Rodriguez SR, Guyot JP (2001). Characterization of the *Lactobacillus manihotivorans* a-amylase gene. DNA-sequence 12:27-37.

Pahlow G, Muck RE, Driehuis F, Oude Elferink SJWH (2003). Microbiology of ensiling. In: Silage Science and Technology. Agronomy Monograph 42 (Eds Buxton DR, Muck RE, Harrison JH), American Society of Agronomy, Crop Science Society of America, Soil Science Society of America, Madison, WI.

Pang H, Qin G, Tan Z, Li Z, Wang Y, Cai Y (2011). Natural populations of lactic acid bacteria associated with silage fermentation as determined by phenotype, 16S ribosomal RNA and recA gene analysis. Syst. Appl. Microbiol. 34:235-241

Paragon BM (2004). Bonnes pratiques de fabrication de l'ensilage pour une meilleure maitrise des risques sanitaires. Agence française de sécurité sanitaire des aliments. Editions afssa, pp. 1-118.

Saarisalo E, Skytta E, Haikara A, Jalava T, Jaakkola S (2007). Screening and selection of lactic acid bacteria strains suitable for ensiling grass. J. Appl. Microbiol. 102:327-336.

Sawadogo-Lingani H, Diawara B, Glover RK, Debrah KT, Traoré AS, Jakobsen M (2010). Predominant lactic acid bacteria associated with the traditional malting of sorghum grains. Afr. J. Microbiol. Res. 4:169-179.

Sengun IY, Nielsen DS, Karapinar M, Jakobsen M (2009). Identification of lactic acid bacteria isolated from Tarhana, a traditional Turkish fermented food. Int. J. Food Microbiol. 135:105-111.

Seseña S, Sánchez I, Palop L (2005). Characterization of *Lactobacillus* strains and monitoring by RAPD-PCR in controlled fermentations of "Almagro" eggplants. Int. J. Food Microbiol. 104:325-335.

Sirirat R, Thosaporn R, Somkiat P (2008). Evaluations of lactic acid bacteria as probiotics for juvenile seabass *Lates calcarifer*. Aquac. Res. 39:134-143.

Teuber M, Geis A (1981). The Family Streptococcaceae (Non-Medical Aspect). In: The Prokaryotes: A Handbook on Habitats, Isolation and Identification of Bacteria, Starr, M.P, Stolp H, Trueper H.G, Balows A.and Schlegel HG. (Eds.). Springer-Verlag, Berlin, pp. 1614-1630.

Thapa N, Pal J, Tamang JP (2006). Phenotypic identification and technological properties of lactic acid bacteria isolated from traditionally processed fish products of the Eastern Himalayas. Int. J. Food Microbiol. 107:33-38.

Togo MA, Feresu SB, Mutukumira AN (2002). Identification of lactic acid bacteria isolated from Opaque beer (Chibuku) for potential use as a starter culture. J. Food Technol. Afr. 7(3):93-97.

Valerio F, Favilla M, De Bellis P, Sisto A, de Candia S, Lavermicocca P (2009). Antifungal activity of strains of lactic acid bacteria isolated from a semolina ecosystem against *Penicillum roqueforti*, *Aspergillus niger*, and *Endomyces fibuliger* contaminating bakery products. Syst. Appl. Microbiol. 32:438-448.

Vlková E, Rada V, Bunešová V, Ročková Š (2012). Growth and survival of lactic acid bacteria in lucerne silage. Folia Microbiol. 57:359-362.

Weinberg ZG, Ashbell G, Hen Y, Azrieli A (1995). The effect of a propionic acid bacterial inoculant applied at ensiling on the aerobic stability of wheat and sorghum silages. J. Ind. Microbiol. 15:493-497.

Weinberg ZG, Khanal P, Yildiz C, Chen Y, Arieli A (2010). Effects of stage of maturity at harvest, wilting and LAB inoculant on aerobic stability of wheat silages. Anim. Feed Sci. Technol. 158:29-35.

Widyastuti Y (2008). Fermentasi silase dan manfaat probiotik silase bagi ruminansia. Med. Pet. 31:225-232.

Xing L, Chen LJ, Han LJ (2009). The effect of an inoculants and enzyme on fermentation and nutritive value of sorghum straw silages. Bioresour. Technol. 100:488-491.

Antimicrobial activity of *Xylopia aethiopica*, *Aframomum melegueta* and *Piper guineense* ethanolic extracts and the potential of using *Xylopia aethiopica* to preserve fresh orange juice

Ogbonna Christiana N.[1,3] , Nozaki, K.[2] and Yajima, H.[2]

[1]South-East Zonal Biotechnology Centre, University of Nigeria, Nsukka, Enugu State, Nigeria.
[2] ASAMA CHEMICAL Co., Ltd. 20-3, Nihonbashi-Kodenmacho, Chuo-ku, Tokyo, 103-0001, Japan.
[3]Department of Plant Science and Biotechnology, University of Nigeria, Nsukka, Enugu state, Nigeria.

Antimicrobial activity of ethanolic extracts of *Xylopia aethiopica*, *Aframomum melegueta* and *Piper guineense* fruits were assayed against fourteen (14) microorganisms commonly associated with food poisoning and/or food spoilage. The microorganisms were *Bacillus subtilis* IAM1069, *Bacillus cereus* IFO 13494, *Staphylococcus aureus* FDA 209p, *Escherichia coli* NRIC 1023, *Salmonella typhimurium* IFO12529, *Lactobacillus plantarum* IAM 1041, *Pediococcus acidilactici*-M, *Leuconostoc mesenteroides*-M, *Lactobacillus casei* TISTR390, *Saccharomyces cerevisiae* OC-2, *Hansenula anomala* IFO 0140 (p), *Pichia memb.*IFO 0128, *Penicillium funiclosum* NBRC 6345 and *Candida* species. All the plant extracts exhibited selective antimicrobial activities on the test organisms. *X. aethiopica* extract exhibited the highest antimicrobial activity on the organisms with a minimum inhibitory concentration (MIC) of 50 ppm on *Bacillus* species and *S. aureus*. *S. cerevisiae* (MIC = 300 ppm), *P. funiclosum* NBRC 6345 and *L. mesenteroides* (MIC = 500 ppm) were also susceptible to *X. aethiopica* fruit extract but the MIC values for the other tested microorganisms were higher than 1000 ppm. This was followed by *A. melegueta* fruit extract with MIC of 100 ppm for *B. cereus* and *S. aureus*. Although *P. guineense* fruit extract inhibited the growth of *B. cereus* and *S. aureus* (MIC = 300 ppm); and *B. subtilis* (MIC = 1000), the MIC for the other microorganisms were higher than 5000 ppm. On the whole, all the plant extracts exhibited the least antimicrobial activities on *Lactobacilli* and fungi species. *X. aethiopica* fruit extract was used to preserve fresh orange juice. The ability of 100 and 1000 ppm extract to preserve the orange juice was significantly greater (p<0.05) than 50 ppm. The microbial concentration in orange juice containing 100 ppm of *X. aethiopica* extract was 4 cfu/mL after 28 days of storage at room temperature.

Key words: Food spoilage, food poisoning, microorganisms, spices, ethanolic extract, natural preservatives, orange juice.

INTRODUCTION

Food poisoning and food borne infections are very common all over the world, especially in tropical countries with elevated temperature and humidity that favour microbial growth (Adebajo, 1993). In most developing countries, food preservation is done mostly by the conventional methods such as drying and salting since refrigeration and freezing facilities are expensive, electricity supply is very unstable and some remote areas do not have electricity supply at all. Although, there are many chemical food preservatives used by man, most of them have adverse side effects on human health. The use of plant materials traditionally used as food spices, condiments and or as medicine would be more beneficial to human health than the use of synthetic chemical food preservatives. In many tropical countries, these medicinal plants and spices are abundant and easily accessible. In Nigeria, for example, it has been estimated that over 40% of known plants serve as food whereas about 30% serve as spices and medicinal plants (Nwobegu, 2002). The spices are used to give aroma and flavour to food and at the same time they can serve as food preservatives because they possess active ingredients which are either microbistatic or microbicidal (Adegoke and Sagua, 1993; Okeke et al., 2001; Okigbo et al., 2005; White, 2006; Okigbo and Igwe, 2007). Among these medicinal plants are *Xylopia aethiopica* (Negro pepper), *Aframomum melegueta* (Aligator pepper) and *Piper gueenese.*

X. aethiopica is a medicinal plant of great repute in West Africa and contains a variety of complex chemical compounds (Adegoke et al., 2003). These active ingredients are extracted in various forms such as crude aqueous or organic extracts or in the form of essential oils. The medical importance of *X. aethiopica* has been extensively reported (Fleischer, 2003; Okigbo et al., 2005; Adewoyin et al., 2006; White, 2006; Okigbo and Igwe, 2007). The fresh and dried fruits, leaf, stem bark and root bark essential oils were reported to have various degrees of activity against some Gram positive and negative bacteria (Fleischer, 2008). *X. aethiopica* (Annonaceae) is widely distributed in the West African rainforest from Senegal to Sudan in Eastern Africa, and down to Angola in Southern Africa (Irvine, 1961; Burkhill, 1985). Almost every part of the plant is used in traditional medicine for managing various ailments including skin infections, candidiasis, dyspepsia, cough and fever (Irvine, 1961; Burkhill, 1985; Mishana et al., 2000).

A. melegueta (Aligator pepper) is another plant popularly used as a food spice and as a traditional medicine for treating various ailments in Nigeria and other parts of the world. There are a lot of scientific work on the activeties of *A. melegueta* such as antinociceptive (Oloke, 1992; Okigbo and Ogbonnaya, 2006; Umukoro and Ashorobi, 2007), antifungal activity (Ejechi et al., 1997; Adejumo and Langenkämper, 2011), for controlling insect pests (Ewete et al., 1996; Ejechi et al., 1997; Oparaeke et al., 2005; Ukeha et al., 2009) and as a flavouring agent in food (Ajaiyeoba and Ekundayo, 1999).

P. guineense commonly referred to as African black pepper or Ashanti pepper belongs to the family Piperaceae. It is known with different vernacular names in Nigeria which include 'Uziza' in Igbo, and 'Iyere' in Yoruba. *P. guineense* has culinary, medicinal, cosmetic and insectcidal uses (Dalziel, 1955; Okwute, 1992). The leaves are

considered aperitive, carminative and eupeptic. They are also used for the treatment of cough and bronchitis, (Martins et al., 1998) intestinal diseases and rheumatism (Sumathykutty et al., 1999; Saganuwana, 2009; Ogbole et al., 2010). The seeds and leaves are used as spices in various African dishes. It has also been used as an insect repellant (Adewoyin et al., 2006). The plant is utilized in different forms, such as whole herbs, powders, extracts and vapours, for a variety of purpo-ses (Martins et al., 1998).

The microorganisms that are mostly involved in food poisoning and spoilage are mainly bacteria and fungi. The Gram negative bacteria such as *Escherichia coli* and *Salmonella typhimurium*, as well as Gram positive bacteria such as *Staphylococcus aureus* are associated with food poisoning and food borne infections. *Bacillus subtilis* and *Bacillus cereus* are known to produce exoenzymes that hydrolyze food materials causing food spoilage and they are also involved in certain types of food poisoning (Pelczar et al., 1993). The lactobacilli are fermentative group of bacteria that are known to be associated with most food materials including dairy products. Although, their activities are required in some cases as probiotics, they need to be controlled in some food materials where they may cause food spoilage by fermenting food mate-rials to lactic acid and other metabolites which may not be desirable in some food products. Another group of microorganisms that are involved in food spoilage are the fungi. Both filamentous and unicellular fungi have been reported to cause food spoilage and/or food poisoning (Pitt and Hocking, 2009).

Majority of the work on African spices have concentrated on their medicinal values with the aim of using them for disease treatment and insect control. There are very few reports on the use of the extract of these plants as natural food preservatives. The aim of this study was to determine the antimicrobial activities of ethanolic extracts of the fruits of *X. aethiopica, A. melegueta* and *P. guineense* on fourteen (14) food spoilage and pathogenic microorganisms with the aim of using them as natural food preservatives.

MATERIALS AND METHODS

Procurement of plant materials

X. aethiopica (Dunal) A. Rich (Negro pepper), *A. melegueta* (Roscoe) K. Schum (Aligator pepper) and *P. guineense* Schumach and Thonn (African black pepper) fruits were bought from Ogbete main market in Enugu, Enugu state, Nigeria. They were authenticated in the Department of Plant Science and Biotechnology, University of Nigeria, Nsukka by Prof. M. Nwosu.

Preparation of plant extracts

The fruits were dried at ambient temperature for five days and milled into powder. Then, 5 to 20 g of each sample was put into a 300 ml Erlenmeyer flask and ten times volume of 50% ethanol (50

Table 1. Percentage ethanolic extract yields of various spices.

Spice	Amount used (g)	Amount of extract (g)	Yield (%)
Aframomum melegueta	20	1.5	7.5
Xylopia aethiopica	5	1.0	20
Piper guineense	20	2.7	13.5

to 200 mL) was added into each flask. The flasks were kept at room temperature with intermittent shaking for three days. Each sample was filtered through No.5C Whatman filter paper. The filtrate was concentrated to about 1/3 the original volume using a rotary evaporator at 40°C. The filtrate was transferred into Petri dishes and dried completely using a drier at 40°C. The amount of extract obtained from each spice was expressed as a percentage of the quantity of powder used for the extraction (Table 1).

Reconstitution of the extract

Each plant extract was reconstituted by dissolving an appropriate quantity in 50% ethanol to give a final concentration of 10% weight per volume. The reconstituted extracts were filter sterilized using 0.45 µM membrane filter. The sterilized extracts were then transferred into Petri dishes and various volumes of sterile agar medium were poured onto the extract to give a final concentration of 50 to 5000 ppm.

Preparation of trypton agar medium

The medium consisted of trypton (1.7 g); peptone (0.3 g); glucose (0.28 g); KH_2PO_4 (0.25 g); NaCl, (0.5 g) and agar powder (1.5 g). All were dissolved in 100 ml of distilled water and the pH was adjusted to 6.0 using tartaric acid. They were then autoclaved at 121°C for 15 min.

Test microorganisms

Pathogenic bacteria, lactic acid bacteria and fungi that are usually associated with food spoilage and/or food poisoning were used. The pathogenic bacteria included *B. subtilis* IAM1069, *B. cereus* IFO 13494, *S. aureus* FDA 209p *E. coli* NRIC 1023 and *S. typhimurium* IFO12529. The lactic acid bacteria were *Lactobacillus plantarum* IAM 1041, *Pediococcus acidilactici*-M, *Leuconostoc mesenteroides*-M, and *Lactobacillus casei* TISTR390, while the fungi species were *Saccharomyces cerevisiae* OC-2, *Hansenula anomala* IFO 0140 (p), *Pichia memb.*IFO 0128, *Penicillium funiclosum* NBRC 6345 and *Candida* species. All the strains were obtained from culture stock of Asama Chemicals Co. Ltd, Tokyo, Japan.

Determination of minimum inhibitory concentration (MIC)

Each microorganism was pre-cultured and diluted with medium to give a concentration of about 10^6 CFU/ml. 10 µL of the diluted cell culture was inoculated into the medium using micro-planter, incubated at 30°C for 72 h and the presence or absence of colonies was recorded. The minimum extract concentration that inhibited the growth of the microorganism was taken to be the minimum inhibitory concentration. The results were confirmed by repeating the experiments two more times.

Preservation of orange juice with ethanolic extract of *X. aethiopica* fruits

Moderately ripped orange fruits were bought from new market in

Enugu, Enugu State, Nigeria. The orange fruits were carefully and thoroughly washed with tap water and aseptically peeled. The juice was squeezed out using an electric juice extractor and filtered through a sterile muslin cloth. The filtered juice was dispensed in 100 mL aliquots into four 250 mL conical flasks. Then, the ethanolic extract of *X. aethiopica* was added into each flask at the concentrations of 0 (control), 50, 100 or 1000 ppm. Each flask was swirled to mix and the content dispensed in 25 mL aliquots into 30 mL bottles and caped. Two sets of four bottles were prepared from each level of the extract concentration. A set was pasteurized at 60°C for 20 min while another set was not pasteurized. All were stored at room temperature (28 ± 2°C) and samples were taken weekly to determine microbial growth by the pour plate method using agar medium composed of sodium chloride (1%), polypetone (1%), yeast extract (1%) and bacteria agar powder (2%). The whole experiments were done three times and the results are the average of the triplicates.

RESULTS

Extract yields from the spice powders

The percentage extract yields of the three different spices are summarized in Table 1. The yield varied among the spices with *X. aethiopica* giving the highest extract yield of 20%, followed by *P. guineense* with 13.5%, while *A. melegueta* gave the lowest yield of 7.5%.

Antimicrobial activities of the spice extracts on pathogenic bacteria

The effects of the three spice extracts on five test pathogenic bacteria are summarized in Figure 1. All the extracts exhibited selective antimicrobial activity on all the organisms. The three extracts showed higher inhibitory effect on the growth of *B. subtilis, B. cereus* and *S. aureus* with the minimum inhibitory concentration ranging from 50 to 1000 ppm. On the other hand, *E. coli* and *S. typhimerium* were less sensitive to the extracts and their MIC values were all above 5000 ppm. *X. aethiopica* extract showed the highest growth inhibitory activity on the test organisms, with MIC of 50 ppm for *B. subtilis, B. cereus* and *S. aureus. A. melegueta* also had a high growth inhibitory effect on the test pathogenic bacteria with a minimum inhibitory concentration of 100 ppm for *B. cereus* and *S. aureus* and 300 ppm for *B. subtilis.* Out of the three spices, *P. guieenese* exhibited the least growth inhibitory effect on the pathogenic bacteria tested with minimum inhibitory concentrations that ranged from 300 to 1000 ppm on *B. subtlis, B. cereus* and *S. aureus.*

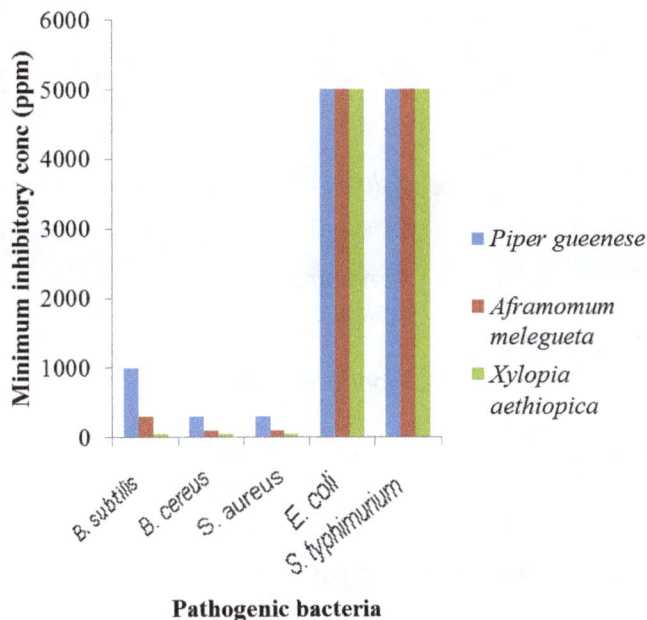

Figure 1. Inhibitory effects of spice fruit extracts on some pathogenic bacteria.

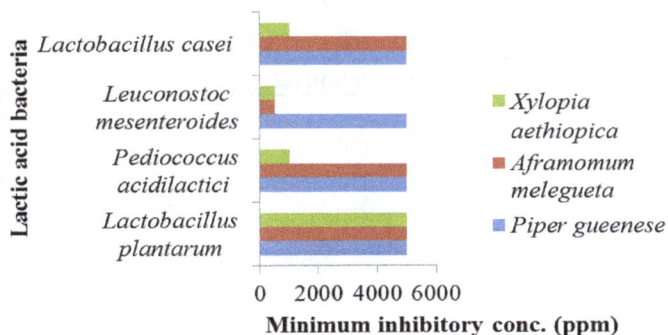

Figure 2. Effects of some spice fruit extracts on the growth of some lactic acid bacteria.

Antimicrobial activities of the spice extracts on lactic acid bacteria

The effects of the spice extracts on the growth of some lactic acid bacteria are shown in Figure 2. *P. acidilactici-*M, *L. mesenteroides*-M and *L. casei* were very sensitive to *X. aethiopica* extracts with a minimum inhibitory concentration of 500 to 1000 ppm. *L. mesenteroides*-M was the most sensitive of the lactic acid bacteria tested with an MIC of 500 ppm for *X. aethiopica* and *A. melegueta*. On the other hand, *L. plantarum, P. acidilactic* and *L. casei* were less sensitive with MIC values higher than 5000 ppm for *A. melegueta and P. guineense* extracts.

Effects of the spice extracts on the growth of fungi

The effects of the spice extracts on the growth of fungi

are shown in Figure 3. *S. cerevisiae* and *P. funiclosum* were the most sensitive to *X. aethiopica* extract with MIC values of 300 and 500 ppm, respectively. *P. memb* was sensitive only to *A. melegueta* but relatively resistant to *X. aethiopica* and *P. gueenese*. *H. anomala* and *Candida* species were the least sensitive to all the extracts.

Potential of using *X. aethiopica* fruit extract to preserve orange juice

Since *X. aethiopica* was the most active of the three spices against all the groups of microorganisms tested, its potential application as food preservative was evaluated, using orange juice as an example. Microbial growth in pasteurized orange juice containing varying concentrations of *X. aethiopica* extract is shown in Table 2. The three different extract concentrations inhibited the growth of microorganisms in the orange juice but the preservative effect of the extract was concentration dependent. However, the preservative abilities of 100 and 1000 ppm (the microbial concentrations in the juice) were not statistically different (p>0.5). The use of 100 ppm for the preservation of orange juice is therefore recommend-ded. In the case of unpasteurized orange juice, microbial growth commenced almost immediately irrespective of the extract concentration, reaching above 10^3 cells/mL within 24 h.

DISCUSSION

The results of this work have shown that all the three spices tested inhibited the growth of all the 14 microorganisms associated with food poisoning and/or food spoilage, though the sensitivities of the microorganisms varied. Antimicrobial activities of these spices, especially *X. aethiopic* on some microorganisms have also been reported (Fleischer, 2008). However, most of the works on them have focused on their medicinal values. The present work seems to be the first comparative study on their antimicrobial activities against a wide range of microorganisms. On the whole, Gram negative bacteria were more resistant to the spice extracts than the Gram positive bacteria. This result was in agreement with that of Agatemor (2009) and Nwinyi et al. (2009) who reported that Gram negative bacteria are more resistant to anti-bacterial agents than the Gram positive species. This may be due to the differences in the cell wall composition and structure, especially the polysaccharide and protein outer membrane in the cell wall of the Gram negative bacteria which limit diffusion of antimicrobial agents into the cell. The antimicrobial activities of these spices can be attributed to the contents of active ingredients such as mono- and ses-quiterpene hydrocarbons in *X. aethiopica* (Karioti et al., 2004), β-pinene, β-caryophyllene, β-elemene, cyclogermacrene and α-humulene in *P. gueenese* (Parmar et al., 1997; Martins et al., 1998).

This work has demonstrated that *X. aethiopica* extract can be used to preserve orange juice. This is the first

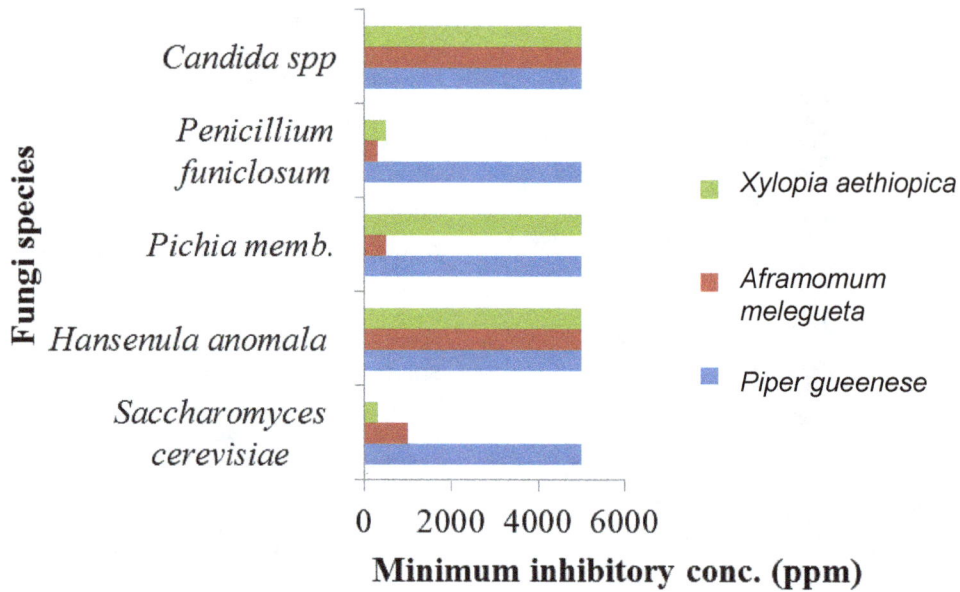

Figure 3. Inhibitory effects of some spice fruit extracts on some fungi species.

Table 2. Effect of addition of *X. aethiopica* fruit extract on microbial growth in orange juice during storage at room temperature.

Extract concentration (ppm)	Incubation period (days)	Cell concentration (cfu/ mL)
	0	-
	7	330
0	14	810
	21	6×10^4
	28	2.5×10^6
	0	-
	7	10±2.4
50	14	8±1.92
	21	7±1.68
	28	10±2.00
	0	-
	7	4±1.3
100	14	3±1.0
	21	2±0.8
	28	3±1.2
	0	-
	7	2±0.8
1000	14	1±0.5
	21	2±0.35
	28	2±0.45

Various concentrations of *X. aethiopica* ethanolic extract were added to freshly squeezed orange juice in 30 mL bottles. They were pasteurized at 60°C for 20 min and stored at room temperature (28 ± 2°C). Samples were taken weekly to determine microbial growth by the pour plate method.

report on the use of *X. aethiopica* fruit extract to preserve orange juice. Even after about one month of storage, the microbial counts in the juice containing 100 ppm was within the range regarded as safe by the FAO. The shelf life of the orange juice can still be extended by combining the spice extract with other safe treatments such as carbonation and high pressure bottling. What can be considered a limitation to the use of *X. aethiopica* extract to preserve food is that it has a strong aroma which may be objectionable for some people in certain food products. Nevertheless, some people like the aroma, hence its wide use as a spice. It is necessary to investigate whether the antimicrobial compounds in the fruits are responsible for the strong aroma. If not, development of a method for extracting the antimicrobial agents that are free of the aroma, or a method of removing or masking the strong aroma can broaden the scope of its applications as a food preservative.

A major advantage of using *X. aethiopica* as a food preservative is that foods preserved by this spice may qualify as a functional food since it has many health benefits such as anti-tumour, anti-asthmatic, anti-inflammatory, antimicrobial (Okigbo et al., 2005; White, 2006; Okigbo and Igwe, 2007), hypotensive and coronary vasodilatory effects (Fleischer, 2003). On the whole, these natural preservatives, especially spices that have very long history of use in many parts of the world, are preferred to chemical preservatives which often have undesirable side effects on the consumers.

REFERENCES

Adebajo LO (1993). The microbial spoilage of "soft" melon ball snack under tropical conditions. Nahrung 37:328-35.

Adegoke GO, Sagua VY (1993). Influence of different spices on the microbial reduction and storability of laboratory-processed tomato ketchup and minced meat .Nahrung 37:352-5.

Adegoke GO, Makinde O, Falade KO, Uzo-Peters PI (2003). Extraction and characterization of antioxidants from *Aframomum melegueta* and *Xylopia aethiopica*. Eur. Food Res. Technol. 216:526-8.

Adejumo TO, Langenkämper G (2011). Evaluation of botanicals as biopesticides on the growth of *Fusarium verticillioides* causing rot diseases and fumonis in production of maize. J. Microbiol. Antimicrobiol. 4(1):23-31.

Adewoyin FB, Odaibo AB, Adewunmi CO (2006). Mosquito repellent activity of *Piper guineense* and *Xylopia aethiopica* fruit oils. Afr. J. Trad. Compl. Alter. Med. 3(2):79-83.

Agatemor C (2009). Antimcrobial activity of aqueous and ethanol extracts of nine Nigerian spices against four food borne bacteria. Elec. J. Environ. Agric. Food Chem. 8(3):195-200.

Ajaiyeoba EO, Ekundayo O (1999). Essential oil constituents of *Aframomum melegueta* (Roscoe) K. Schum. seeds (alligator pepper) from Nigeria. Flav. Frag. J. 14:109-11.

Burkhill HM (1985). Useful Plants of West Africa. 2nd edition. Royal Botanic Gardens, Kew: pp. 130-132.

Dalziel IM (1955). The useful plants in West Tropical Africa hand book. 2nd Edition. Printing Crown Agents.

Ejechi BO, Ojeata A, Oyeleke SB (1997). The effect of extracts of some Nigerian spices on biodeterioration of Okro (*Abelmoschus* (L) Moench) by fungi. J. Phytopath. (Berlin) 145:469-72.

Ewete FK, Arnason JT, Larson J, Philogene BJR (1996). Biological activities of extracts from traditionally used Nigerian plants against the European corn borer, *Ostrinianubilalis*. Entomol. Exp. Appl. 80:531-7.

Fleischer TC (2003). *Xylop70ia aethiopica* A Rich.: A chemical and biological perspective. J. Univ. Sci. Technol. 23:24-31.

Irvine F (1961). Woody Plants of Ghana. London: Crown Agents for Overseas Administration; pp. 23-24.

Martins AP, Salgueiro L, Vila R, Tomi F, Canigueral S, Casanova J, Proenca DA, Cunha A, Adzet T (1998). Essential oils from four *Piper* Species. Phytochem. 49(7):2019-2023.

Mishana NR, Abbiw DK, Addae-Mensah I, Adjanouhoun E, Ahyi MRA, Ekpere JA, Enow-Orock EG, Gbile ZO, Noamesi GK, Odei MA, Odunlami H, Oteng-Yeboah AA, Sarpong K, Sofowora A, Tackie AN (2000). Traditional Medicine and Pharmacopoeia, Contribution to the revision of ethnobotanical and Floristic Studies in Ghana. OAU/STRC Tech. Rep. 67.

Nwinyi OC, Nwodo CS, Olayinka AO, Ikpo CO, Ogunniran KO (2009). Antibacterial effects of extracts of *Ocimum gratissimum* and *Piper guineense* on *Escherichia coli* and *Staphylococcus aureus* Afr. J. Food Sci. 3(3):077-081.

Nwobegu CC (2002). Natural Medicines in the Tropics. Second Edition, Savannah Publishers , Enugu, Nigeria pp. 20-28.

Ogbole OO, Gbolade AA, Ajaiyeoba EO (2010). Ethnobotanical Survey of Plants used in Treatment of Inflammatory Diseases in Ogun State of Nigeria. Eur. J. Sci. Res. 43(2):183-191.

Okeke MI, Iroegbu CU, Jideofor CO, Okoli A, Esimone CO (2001). Anti-Microbial Activity of Ethanol Extracts of Two Indigenous Nigerian Spices. J. Herb, Spic. Med. Pl. 8 (4): 39-48.

Okigbo RN, Mbajiuka CS, Njoku CO (2005). Antimicrobial Potentials of (UDA) Xylopia aethopica and Ocimum gratissimum L. on Some Pathogens of Man. Intern. J. Mol. Med. Adv. Sci. 1(4):392-397.

Okigbo RN, Igwe M (2007). The antimicrobial effects of *Piper guineense* (uziza) and *Phyllantus amarus* (ebe-benizo) on *Candida albicans* and *Streptococcus faecalis*. Act. Microbiol. Immunol. Hung. 54(4):353-366.

Okigbo RN, Ogbonnaya UO (2006). Antifungal effects of two tropical plant leaf extracts (*Ocimum gratissimum* and *Aframomum melegueta*) on postharvest yam (Dioscorea spp.) rot. Afr. J. Biotechnol. 5:727-31.

Okwute SK (1992). Plant derived pesticidal and antimicrobial agents for use in agriculture .A review of phytochemical and biological studies on some Nigerian plants. J. Agric. Sci. Technol. 2 (1): 62-70.

Oloke JK (1992). Fungicidal Effects of the Volatile Oil of *Aframomum Melegueta*. Fitoterapia 63:269-70.

Oparaeke AM, Dike MC, Amatobi CI (2005). Field evaluation of extracts of five Nigerian spices for control of post-flowering insect pests of cowpea, *Vigna unguiculata* (L.) Walp. Pl. Protec. Sci. 41:14-20.

Parmar VS, Jain SC, Bisht KS, Jain R, Taneja P, Jha A, Tyagi OD, Prasad AK, Wengel J, Oslen CE , Boll P (1997). Phytochemistry of the genus *Piper*. Phytochem. 46:597-673.

Pelczar MJ, Chan ECS, Krieg NR (1993) Microbiology: Concept and applications. International edition. McGraw Hill USA.

Pitt JI, Hocking AD (2009) Fungi and Food Spoilage. Third edition. Springer Dordrecht, Heidelberg, London and New York. p 503.

Saganuwana AS (2009). Tropical Plants with Antihypertensive, Antiasthmatic and Antidiabetic Value. J. Herb. Spic. Med. Pl. 15(1):24-44.

Sumathykutty MA, Rao JM, Padmakumari KP, Narayanan CS (1999). Essential oil constituents of some piper species. Flav. Frag. J. 14:279-282.

Ukeha DA, Birkett MA, Pickett JA, Bowman AS, Luntza AJ (2009). Repellent activity of alligator pepper, *Aframomum melegueta* and ginger, *Zingiber officinale*, against the maize weevil, *Sitophilus zeamais*. Phytochem. 70(6):751-758.

Umukoro S, Ashorobi RB (2007). Further studies on the antinociceptive action of aqueous seed extract of *Aframomum melegueta*. J. Ethnopharm. 109:501-504.

White H (2006). Medecine and wart removal, Hemorrhoid Treatment and Herpes Prevention- Without Drugs McGraw-Hill, New York pp. 102-105.

In vitro study of antiamoebic activity of methanol extract of fruit of *Pimpinella anisum* on trophozoites of *Entamoeba histolytica* HM1-IMSS

Quiñones-Gutiérrez Y.*, Verde-Star M. J., Rivas-Morales C., Oranday-Cárdenas A., Mercado-Hernández R., Chávez-Montes A. and Barrón-González M. P.

School of Biological Sciences, UANL Ciudad Universitaria, AP 46-F, CP 66451,
San Nicolas de los Garza, N. L., Mexico.

The aniseed plant *Pimpinella anisum* (Saunf-Hindi) is one of the most ancient medicinal plants used by man. Currently, this plant has several uses in the food industry as spice, whereas in the pharmacopoeia, it is used as an expectorant in digestive disturbances, as mild diuretic, and as insect repellent in external use. In this paper, we evaluated the biological activity of methanolic extract of *P. anisum* on *in vitro* growth of *Entamoeba histolytica* HM1-IMSS under axenic conditions. We observed that the growth inhibition of *E. histolytica* was at CI_{50} = 0.034 μg/mL. Results confirm the antiamoebic activity of the methanolic extract of *P. anisum*.

Key words: *Pimpinella anisum, Entamoeba histolytica,* antiamoebic activity, medicinal plants.

INTRODUCTION

Among parasitic infections, amoebiasis ranks third worldwide among lethal infection, after malaria and schistosomiasis (Walsh, 1988; Petri and Mann, 1993). Amoebiasis is caused by *Entamoeba histolytica,* a protozoan parasite of humans and the causative agent of intestinal amoebiasis. This disease is a major health problem in developing countries (Stanley, 2003). Amoebiasis is acquired by ingestion of the *E. histolytica* cyst in contaminated food or water (Botero and Restrepo, 2003). Although it is asymptomatic in 90% of cases, about 50 million people are estimated to suffer from symptoms associated with amoebiasis, such as hemorrhagic colitis and amoebic liver abscess (Ravdin, 1995). These infections result in 50,000 to 100,000 deaths annually. Amoebiasis contributes to the prevalence of gastrointestinal disease and is defined by the World Health Organization and the Pan American Health Organization as the presence of the parasite with or without clinical manifestations (WHO, 1997).

The Dauphin de France was the first known patient with amoebiasis treated by an extract of the root of ipecacuana plant, and little progress was made during the next 200 years (Laserre, 1966). Many drugs have been used for the treatment of amoebiasis, mainly nitroimidazole derivatives of such, as emetine, metronidazole, and ornidazole. Powell et al. (1966) reported the success of metronidazole in the treatment of amoebic dysentery and liver abscess. Metronidazole and imidazole derivatives are the drugs of choice for the treatment of amoebiasis however, *E. histolytica* has developed resistance mechanisms to these drugs (Samarawickream et al., 1997; Wassmann et al., 1999; Orozco et al., 2002). In addition, there have been reports that these drugs could induce mutagenic (Legator et al., 1975), carcinogenic (Chacko and Bhide, 1986) and neurotoxic activity (Olson et al., 2005). In addition to nutrients, vegetables and fruits may contain several phytochemicals that prevent certain diseases. The classic concept of nutrition has been expanded to include functional nutrition or nutraceuticals, which refers to the potential of certain foods to promote and improve health by reducing the risk for disease (Vaclavik, 2008).

Pimpinella anisum is one of the oldest plants used in food industry, perfumery and medicine. *P. anisum* has

---

*Corresponding author. E-mail: yadiragtz70@hotmail.com.

been used as digestive stimulant, antiparasitic and antifungal (Soliman and Badea, 2002) and antipyretic agent (Afifi et al., 1994). In addition, the plant and especially the essential oil of fruit have been used for the treatment of some diseases including epilepsy and seizures (Avicenna, 1988; Abdul-Ghani et al., 1987); as well for constipation (Chicouri and Chicouri, 2000), and has proven activity as a muscle relaxant (Albuquerque et al., 1995).

Recently, it has been reported that this oil has been used as a substitute for antibiotics in chickens (Mehmet et al., 2005). Nevertheless, there are few reports on antibacterial activity studies of *P. anisum* (Singh et al., 2002; Tabanca et al., 2003). Recently, Akhtar et al. (2008) demonstrated the *P. anisum* bactericidal activity after testing the methanolic extract from seeds against *Staphylococcus aureus*, *Streptococcus pyogenes*, *Escherichia coli* and *Klebsiella pneumoniae*.

In developing countries, medicinal plants are popular because their products are safe and widely available at low cost. Some compounds extracted from medicinal plants already play an important role against infectious disease for example, the quinine extracted from *Cinchona* sp., and the artemisinin from *Artemisia annua*. Both compounds are effective against malaria. Considering their therapeutic potential, it is important to obtain information from the exact doses to become effective. Therefore, the aim of this study was to evaluate the antiamoebic activity of methanolic extract from *P. anisum in vitro* under axenic conditions on *E. histolytica* HM1-IMSS and identified antiamoebic components present in methanolic extract of *P. anisum*.

MATERIALS AND METHODS

Vegetal materials and preparation of methanol extracts

Fruits collection

The fruits of *P. anisum* were purchased from the local market. The material was identified as *P. anisum* by the Department of Botany, Facultad de Ciencias Biológicas, Universidad Autónoma de Nuevo León.

Extraction

The methanolic extract of *P. anisum* was obtained by taking 100 g samples of fruit and individually processed in a Soxhlet unit with methanol (MeOH) at 60°C for 7 h. The extract was then filtered and concentrated in vacuum at 45°C. Methanol extract was dried under room temperature and weighed to calculate the extractability percentage. The extract was stored at 40°C until further use.

Antiamoebic activity of the P. anisum methanolic extract

Trophozoites of *E. histolytica* strain HM1:IMSS at a density of 1 × 10^4 trophozoites per mL were added to TYI-S-33 medium (Diamond et

al., 1978), containing 1% penicillin/streptomycin and 10% heat inactivated bovine serum. *E. histolytica* trophozoites were then incubated in 13 × 100 mm screw-cap tubes (PYREX®) at 37°C for 72 h. Tubes were then placed in iced water for 10 min to detach cells adhering to the base of the tube and were inverted 15 times and then cell density was determined by using a hemocytometer (Neubauer, Hausser Scientific). When the culture of *E. histolytica* was grown logarithmically, trophozoites were harvested and suspended in 50 μL phosphate-buffered saline, pH 7.4 (phosphate buffered saline, PBS) at 1 × 10^4 trophozoites per mL. These suspensions were inoculated separately in 1 mL screw-capped borosilicate vials (Bellco, Vineland, NJ, USA) containing 1 mL of TYI-S-33 medium (Diamond et al., 1978), vials were added containing 50 μL of various concentrations of the *P. anisum* methanolic extract (0.1, 1.0, 3.0 and 5.0 mg per mL) and incubated for six days at 37°C. The solutions were sterilized by filtration through a 0.22 μm pore–size membrane (MercK Millipore).

A control containing 50 μL of DMSO was included in each assay. Preliminary test with dimethyl sulfoxide (DMSO) were performed to ensure that no trophozoite inhibition occurred at the concentrations used. After that, cell density was determined by using a hemocytometer (Barrón-González et al., 2008). The positive controls contained metronidazole (0.124 μg/mL) instead of methanolic extract. Bioassays were done by triplicate. Fifty percent inhibitory concentration was determined by using the Probit test.

The inhibitory effect of the *P. anisum* methanolic extract was estimated as the decreased percentage of trophozoite number with respect to non treated cultures. Corresponding averages and standard deviations (SD) were plotted. The fifty percent inhibitory concentration (IC$_{50}$) of the *P. anisum* methanolic extract was expressed as the concentration that produced a 50% decrease of trophozoite concentration in each of the three protozoa species. Each determination was performed three times, in triplicate. The effect of metronidazole was expressed as a diminution percentage with respect to non treated controls.

Statistical analysis

SPSS software, version 10.0 (SPSS, Inc.) was used for statistical analysis.

Phytochemical tests

Conventional chemical tests were used to identify functional groups in the methanolic extract (Domínguez, 1973).

RESULTS AND DISCUSSION

The yield of the methanolic extraction process for each 100 g of *P. anisum* fruit was 12.13% wt/wt. Methanolic extract from *P. anisum* showed *E. histolytica* trophozoites growth inhibition when tested by *in vitro* culture under axenic conditions. Results showed a dose-reponse behavior, where inhibition rate were significantly different extract doses. At a rate of 0.1 mg/mL, the inhibition ranged from 68 to 74% and it inhibited the dose of 1 mg/mL, at the dose of 3 mg/mL it was 87% and at 5 mg/mL, it was 99.18% (Figure 1).

The inhibition concentration analysis showed that methanolic extract had an IC$_{50}$ of 0.0345 mg/mL on *in*

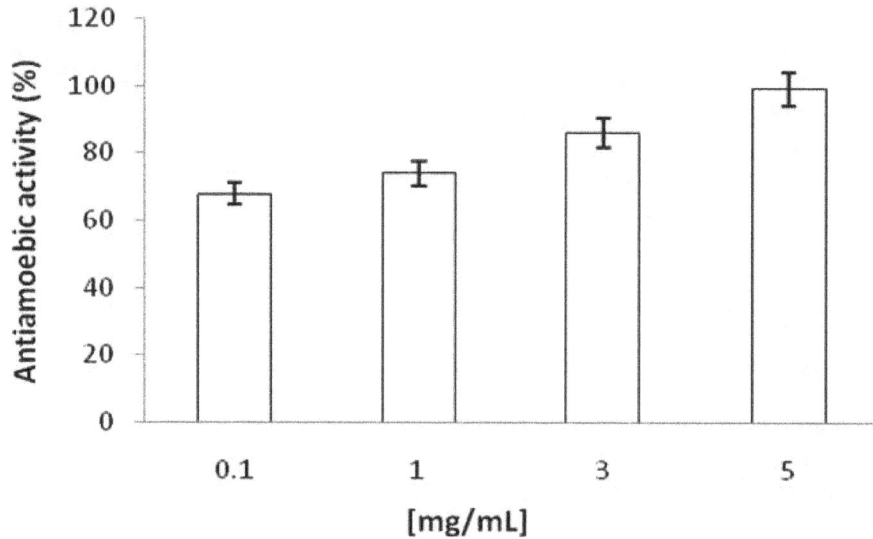

Figure 1. Inhibition of growth *in vitro* axenic of *E. histolytica* in the presence of methanol extract of fruit of *P. anisum*.

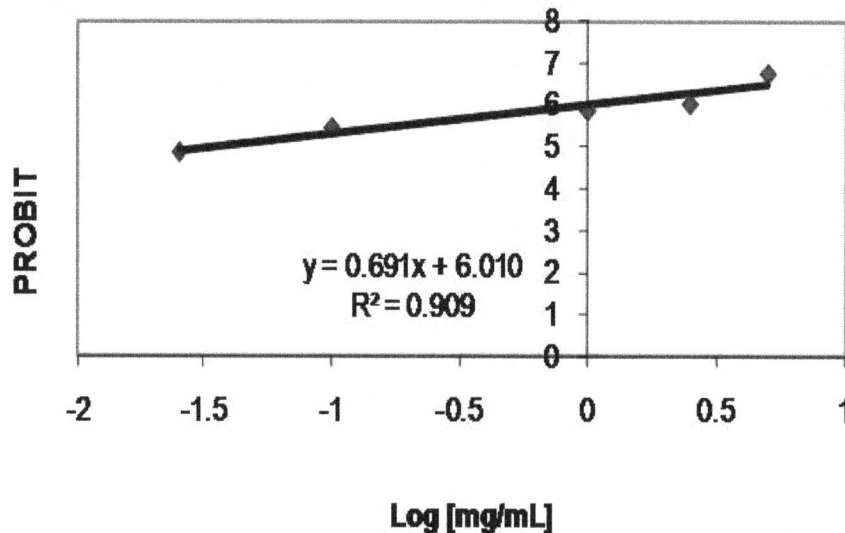

Figure 2. Probit graph of the activity of methanol extract of *P. anisum* against *E. histolytica* HM1-IMSS (50% inhibitory concentration [IC_{50} = 0.0345 mg/mL]).

vitro growth of *E. histolytica* trophozoites (Figure 2), whereas metronidazole had an IC_{50} of 0.1442 µg/mL. Values were expressed as the mean ± standard error (n = 9). The Probit value was expressed relative to the metronidazole effect.

Analysis of the *P. anisum* methanolic extract using silica gel chromatography showed three fractions. The inhibition rate of each fraction was separately evaluated. Results show that regardless of the fraction tested, all presented the highest inhibition growth at 0.47 mg/mL (74% inhibition). In order to identify the compound(s) present in each fraction, the third fraction was analyzed by mass spectrometry, showing a lead compound with a

molecular weight of 175, an aromatic ring, double bonds and a methyl group (methoxy), obtained as the structure of 1-benzopyrylium which is responsible for the antiamoebic activity. 1-benzopyrylium is a widely distributed molecule within plants, and it is partially present as anthocyanins flavones. All together, results provide evidence of *E. histolytica* trophozoites growth inhibition by methanol extract of *P. anisum* fruits. In addition, one out of three fractions obtained from the methanolic extract was positively identified as 1-benzopyrylium. We believe that this support new evidence of metabolites derived from plants with antiamoebic activity. The methanolic extract of *P. anisum* has a great popularity worldwide, as it is used as food which is an advantage in terms of popular acceptance for use.

Other species, such as *Illicium anisatum*, is similar in morphology and chemical composition to *Illicium verum*. In *I. verum*, the veranisatinas A, B and C has been identified, which present low-power neurotoxic, while in *Illicium anisatum*, the presence of anisatin and neoanisatin has been detected, the most toxic therapeutic compounds against pathogenic microorganisms to man, with anti-inflammatory and antifungal activities. It has also been reported that the aqueous extract of the fruit of *I. verum* has antibacterial activity against *Bacillus subtilis*, as well as antiviral activity against herpes virus-type 2, influenza virus, smallpox virus and poliovirus II. The essential oil obtained from the fruit is active against *B. subtilis*, *E. coli* and *Pseudomonas aeruginosa*, as well as antifungal activity against *Candida albicans* and seed essential oil which has activity against *C. lipolytica* (Argueta, 1994).

Methanolic extract of *P. anisum*, as well as three of their fractions, could be evaluated on the process of encystment of *Entamoeba histolytica* and biological activity on other pathogens of man.

Conclusion

The methanolic extract of the fruit *P. anisum* shows growth inhibition activity against *E. histolytica* trophozoites strain HM1:IMSS under axenic conditions *in vitro*.

ACKNOWLEDGEMENT

This work was supported by CONACYT no. 204368.

REFERENCES

Abdul-Ghani AS, El-Lati SG, Sacaan AI, Suleiman MS (1987). *In vitro* Antibacterial activity of *Pimpinella anisum* fruit extracts against some pathogenic bacteria. Int. J. Crude Drug Res. 25:39-43.

Afifi NA, Ramadan A, El-Kashoury EA, El-Banna HA (1994). Some pharmacological activities of essential oils of certain umbelliferous fruits. Vet. Med. J. Giza 42:85-92.

Akhtar Y, Yeoung YR, Isman MB (2008). Comparative bioactivity of selected extracts from Meliaceae and some commercial botanical insecticides against two noctuid caterpillars, *Trichoplusia ni* and *Pseudaletia unipuncta*. Phytochem. Rev. 7:77-88.

Albuquerque A, Sorenson AL, Leal Cardoso JH (1995). Effects of essential oil of Croton zehnteri and of anethole and estragole on skeletal muscles. J. Ethnopharmacol. 49:41-49.

Argueta A (1994). Atlas de las plantas de la medicina tradicional mexicana I., Instituto Nacional Indigenista, Primera Edición. México, D.F.

Avicenna A (1988). Drugs and decoctions used in epilepsy. In: Sharafkandi, A. (Translator), Ghanoon Dar Teb. Soroosh Press, Tehran, pp. 456-459.

Barrón-González MP, Serrano-Vásquez GC, Villarreal-Treviño L, Verduzco-Martínez JA, Morales-Vallarta MR, Mata-Cárdenas BD (2008). Inhibición del crecimiento axénico *in vitro* de *Entamoeba histolytica* por acción de probióticos. Revista Ciencia UANL 11:285-290.

Botero D, Restrepo M, (2003). "Parasitosis Humanas". 4° edición. Corporación para Investigaciones Biológicas. Medellín, Colombia. 14:30-60.

Chacko M, Bhide SV (1986). Carcinogenicity, perinatal carcinogenicity and teratogenicity of low dose metronidazole (MNZ) in Swiss mice. J. Cancer Res. Clin. Oncol. 112:135-140.

Chicouri M. and I. Chicouri (2000). Novel pharmaceutical compositions containing senna with laxative effect. Fr. Demande FR 2791892 A1, Oct 13, 6.

Diamond LS, Harlow DR, Cunnick CC (1978). A new medium for the axenic cultivation of *Entamoeba histolytica* and other Entamoeba. Trans. Royal Soc. Trop. Med. Hyg. 72:431-432.

Domínguez XA (1973). Métodos de investigación fitoquímica. 1ª edición. LIMUSA. México, D.F. 176.

Laserre R (1966). Traitement de l'amibiase-La2-Dehydroemetine. Schweiz Med Wschr. 96:678-701.

Legator MS, Connor TH, Stoeckel M (1975). Detection of mutagenic activity of metronidazole and niridazole in body fluids of human and mice. Science 188:1118-1119.

Mehmet C, Talat G, Bestami D, Nihat EO (2005). The Effect of Anise Oil (*Pimpinella anisum* L) On Broiler Performance. Int. J. Poult. Sci. 4(11):851-855.

Olson EJ, Morales SC, McVey AS, Hayden DW (2005). Putative metronidazole neurotoxicosis in a cat. Vet. Pathol. 42:665-669.

Orozco E.; Lopez, C.; Gomez, C; Perez, D.G.; Marchat, L.; Banuelos, C.; Delgadillo, D.M (2002). Multidrug resistance in the protozoan parasite *Entamoeba histolytica*. Parasitol. Int. 51:353-359.

Petri Jr. WA, Mann BJ (1993). Molecular mechanisms of invasion of *Entamoeba histolytica*. Sem. Cell. Biol. 4(5):305-313.

Powell SJ, Wilmot AJ, MacLeod I, Elsdon-Dew R (1966). The effect of a nitro-thiazole derivative, Ciba 32,644-Ba, in amebic dysentery and amebic liver abscess. Am. J. Trop. Med. Hyg. 15(3):300-302.

Ravdin JI (1995). State of the art clinical article. Clin. Infect. Dis. 20:1453-1466.

Samarawickream NA, Brown DM, Upcroft JA, Thammapalerd N, Upcroft P (1997). Involvement of superoxide dismutase and pyruvate: ferredoxin oxidoreductase in mechanisms of metronidazole resistance in *Entamoeba histolytica*. J. Antimicrob. Hemother. 40:833-840.

Singh Kappoor GIP, Pandey SK, Singh UK, Singh RK (2002). Studies on essential oils: Part 10; antibacterial activity of volatile oils of some species. Phytother. Res.16:680-682.

Soliman KM, Badea RI (2002). Effect of oil extracted from some medicinal plants on different mycotoxigenic fungi. Food Chem. Toxicol. 40:1669-1675.

Stanley SL (2003). Amoebiasis. Lancet 361:1025-1034.

Tabanca NE, Bedir N, Kirmer KH, Baser SI, Khan MR, Jacob, Khan IA (2003). Antimicrobial compounds from *Pimpinella* species growing in Turkey. Planta Med. 69:933-938.

Vaclavik T (2008). Organic retailing development in Europe. Paper at: BioFach Congress, NürnbergMesse, Nuremburg, Germany, February pp. 21-24.

Walsh JA (1988). Prevalence of *Entamoeba histolytica* infection. In: Ravdin JI. ed. Amebiasis. John Wiley and Sons, New York. pp. 93-105.

Wassmann C, Hellberg A, Tannich E, Bruchhaus I (1999). Metronidazole resistance in the protozoan parasite *Entamoeba histolytica* is associated with increased expression of iron-containing superoxide dismutase and peroxiredoxin and decreased expression of ferredoxin 1 and flavin reductase. J. Biol. Chem. 274:26051-26056.

World Health Organization (1997). Amoebiasis-an expert consultation. Wkly. Epidemiol. Rec. No. 14. Ginebra.

Influence of plant growth regulators on somatic embryogenesis induction from inner teguments of rubber (*Hevea brasiliensis*) seeds

KOUASSI Kan Modeste[1], KOFFI Kouablan Edmond[1], KONKON N'dri Gilles[3], GNAGNE Michel[2], KONÉ Mongomaké[4] and KOUAKOU Tanoh Hilaire[4]

[1]Laboratoire Central de Biotechnologies, Centre National de Recherche Agronomique,
01 P. O. Box 1740 Abidjan 01, Côte d'Ivoire.
[2]Station de recherche de Bimbresso, Centre National de Recherche Agronomique,
01 P. O. Box 1740 Abidjan 01, Côte d'Ivoire.
[3]Université de Cocody, UFR Sciences Biosciences, Laboratoire de Botanique, 22 BP 1414 Abidjan 22, Côte d'Ivoire.
[4]Université d'Abobo Adjamé, UFR Sciences de la Nature, Laboratoire de Biologie et Amélioration des Productions Végétales, 02 BP 801 Abidjan 02, Côte d'Ivoire.

Generating somatic embryos from the inner teguments of hevea seeds is difficult. Like other ligneous plants, the rubber-tree is generally considered to be recalcitrant with regard to somatic embryogenesis. In this study, the ability of callus from inner integument explants to develop embryogenic callus lines was highlighted. Combination of 2,4-dichlorophenoxyacetic acid (2,4-D/KT) (9 µM/3.375 µM) revealed the positive effect of the 2,4-D on callogenesis and somatic embryogenesis from the inner integument of the seed of immature fruit. The rate of embryogenic calli of about 50% obtained, suggested that 2,4-D has a similar effect as 3,4-dichlorophenoxy acetic acid (3,4-D). So, although 2,4-D is rarely used as a hormone in biotechnology of rubber, its positive influence on callus induction and somatic embryo development shows that it is an alternative to 3,4-D which is commonly used. Optimal combinations of 2,4-D/thidiazuron (TDZ) (9 µM/34.2 nM) produced abnormal embryos at lower rates (approximately 5%) than the optimal combination of 2,4-D/KT.

Key words: Callus, culture medium, *Hevea brasiliensis*, hormones, rubber-tree, somatic embryogenesis.

INTRODUCTION

Rubber tree (*Hevea brasiliensis* Müll. Arg.) is the South American tropical tree of the spurge family (Euphorbiaceae). Cultivated on plantations in the tropics and subtropics, especially in Southeast Asia and western Africa, it replaced the rubber plant in the early 20th century as the chief source of natural rubber. It has soft wood; high, branching limbs and a large area of bark. The milky liquid (latex) that oozes from any wound to the tree bark contains about 30% rubber, which can be coagulated and processed into solid products, such as tires. Latex can also be concentrated for producing dipped goods, such as surgical gloves.

Nowadays, rubber plantations are established with grafted clones. In the same plantation, half of grafted trees provide 70 to 80% of production (Langlois, 1965). A part of this difference is attributed to the interaction between the rootstock and the scion (Carron et al., 1989). To face these various constraints that are linked to clonal heterogeneity, vigor decline and production related to grafting, somatic embryogenesis was suggested. Since it was used with the inner tegument of rubber's immature seed, somatic embryogenesis recorded some progress regarding the use of growth regulators. The efficiency of various somatic embryogenesis protocols described in rubber depends on the cultivars, as some of them are recalcitrant to *in vitro* culture (El Hadrani et al., 1991). One of the major bottlenecks in somatic embryogenesis procedures is the production of primary calli. Exogenous auxins and cytokinins are main plant growth regulators (PGRs) involved in the control of cell division and differentiation (Féher et al., 2003). The role of these PGRs in the regeneration performance of several plants has been previously described (Zouine et al., 2005; Sané et al., 2006; Zouzou et al., 2008; Ashrafi et al., 2010; Yapo et al., 2011). Different combinations of growth regulators were experimented to obtain more embryogenic calli less subject to browning. In this regard, 9 μM of 2,4-dichlorophenoxyacetic acid (2,4-D) and a combination with 5.7 μM of indolylacetic acid (AIA) were experimented by Carron and Enjarlic (1985). Later, 3,4-dichlorophenoxy acetic acid (3,4-D) was preferred to 2,4-D (Michaux-Ferrière and Carron, 1989). This auxin was first combined with 9 μM of benzyladenine (BA) during the induction of somatic embryogenic phase. In addition, Kinetin (KT) was then preferred to the BA (Montoro et al., 1992). In spite of growth regulators combination research, this protocol is not under full control yet. Carron et al. (1995) reported the fugacity of the embryogenic capacity and the low rate embryos conversion into plantlets. It is therefore of importance to optimize the somatic embryogenesis conditions in rubber which is generally considered to be recalcitrant with regard to somatic embryogenesis (El Hadrami et al., 1991). However, much research input and further refinement considering different growth regulators key factor for devising efficient protocol with particular reference to somatic embryogenesis pathway of rubber is required.

To our knowledge, 2,4-D, KT and thidiazuron (TDZ) combination have never been experimented during the somatic embryogenesis from the inner teguments of rubber. This research is aimed at improving the productions of embryogenic calli. In the present study, the effect of various PGRs, particularly of the auxin, 2,4-D and of the cytokinines, KT and TDZ on the embryogenic calli induction of inner tegument of rubber was explored. The hormonal conditions for the proliferation of calli and the development of somatic embryos were also investigated by combining various concentrations of PGRs.

MATERIALS AND METHODS

Plant material and preparation of explants

Our study was implemented from the inner teguments obtained from seed of immature fruit of rubber. Fruits were harvested after eight to 10 weeks of anthesis in PB 260 clone of rubber. Fruits were harvested in plantations of the Centre National de Recherche Agronomique (CNRA) of Côte d'Ivoire between May and June. Fruits were sterilised with 2.45% aqueous solution of sodium hypochlorite for 30 min and followed by three washings with autoclaved distilled water. Seeds were then extracted by section of the fruit and their inner teguments were aseptically cut into fragments of 5 mm in length. They were then transferred onto Petri dishes containing 30 ml of medium. All the explants were transferred for callus induction with various growth regulators concentrations.

Callus induction

Inner teguments were put in Petri dish containing 30 ml of medium. Mineral basic medium of MB supplemented with 125 mM KH_2PO_4 (MBm) was used. It is a modified Murashige and Skoog (1962) medium that contained 234 mM sucrose, 30 μM $AgNO_3$ and Fossard vitamins (Fossard, 1976) without choline chloride. MBm medium was fortified with different concentrations of 2,4-D (4.5 and 9.0 μM) and KT (1.25, 2.25, 3.375 and 4.5 μM). 2,4-D and KT were used alone or in combination (Table 1). Effect of 2,4-D at 9.0 μM in combination with four concentrations of TDZ (11.40, 22.80, 34.20 and 45.60 nM) were also tested (Table 2). Media were solidified with 2 g/L gelrite (Sigma Chemical Co.), subjected to pH 5.8 before autoclaving (120°C). A total of 13 different hormonal combinations were tested. The effect of the hormonal composition was evaluated by counting the calli obtained after 25 days of culture in the dark at 27 ± 2°C. Percentage of callogenic explants (PCE) were evaluated as follows: PCE = (Number of callus/Total number of explants) x 100.

Somatic embryos induction

The primary calli were chopped with a scalpel and subcultured in embryogenesis induction media for somatic embryos induction. Calli were cultured on 2,4-D and KT combination medium as well as 2,4-D and TDZ combination medium to investigate their effect on somatic embryogenesis. M0 contained 2,4-D (9.0 μM); used as control. Cultures were maintained through subculturing on the same medium condition for three times with intervals of 25 days in the dark at 27 ± 2 °C. At the end of the third subculture (75 days), the percentage of embryogenic calli (PEC) was evaluated [PEC = (Number of embryogenic calli/ Total number of callogenic explants) x100].

Statistical analysis

Data were subjected to analysis of variance (ANOVA) using XLSTAT 7.5.3 program and significant differences among treatments were compared using Duncan test at 5%. Fisher Protected LSD test at P < 0.01 level of significance was used. The means are the result of ten replicates (one replicate containing 30 explants).

Table 1. Combination of 2,4-D and kinetin added in medium for callus induction.

Media code	Concentration of plant growth regulator	
	2,4-D (µM)	KT (µM)
M 1	4.5	1.25
M 2	4.5	2.25
M 3	4.5	3.375
M 4	4.5	4.5
M 5	9.0	1.25
M 6	9.0	2.25
M 7	9.0	3.375
M 8	9.0	4.5

2,4-D, 2,4-dichlorophenoxyacetic acid; **KT,** kinetin.

Table 2. Combination of 2,4-D and TDZ added in medium for callus induction

Media code	Concentration of plant growth regulator	
	2,4-D (µM)	TDZ (nM)
M 0	9.0	-
M 9	9.0	11.4
M 10	9.0	22.8
M 11	9.0	34.2
M 12	9.0	45.6

2,4-D, 2,4-dichlorophenoxyacetic acid; **TDZ,** thidiazuron.

RESULTS

After 25 days of culture, explants were covered with compact yellow calli. After transfer in embryogenic media, the colour of these calli turned to brown at the end of 75 days of incubation with appearance of somatic embryos (Figure 1).

Combination of 2,4-D and KT

Callus were produced in all media containing 2,4-D/KT combination. Percentage of callogenic explants was influenced by the balance of growth regulator. Variables 2,4-D, KT and their interaction explained 91% of the variability observed during callogenesis (data not show). Moreover, they influenced significantly ($P < 0.0001$) the callogenesis. Taken individually, the percentage of callogenic explants increased when the 2,4-D concentration also increased from 4.5 to 9.0 µM.

The difference between the effects of these two concentrations on callogenesis was significant ($P < 0.0001$). The Duncan's test allowed the division of the eight induction media into two different classes on the basis of 2,4-D effect on callogenesis. The first class was constituted by the most callogenic media with a mean of

81.11% callogenic explants (data not show). These media contained 9.0 µM 2,4-D (M5, M6, M7 and M8). The second class was defined by the least callogenic media with an average of 50.56% callogenic explants (data not show). These media contained 4.5 µM of 2,4-D (M1, M2, M3 and M4). Thus, 9.0 µM 2,4-D was more callogenic than 4.5 µM (Table 3). The influence of the kinetin concentration on callogenesis was significant ($P = 0.001$). The percentage of callogenic explants increased when the concentrations of KT were superior to the weakest concentration (1.125 µM) (Table 4). The combined effect of the various levels of the factors 2,4-D and KT on callogenesis revealed three classes. The class of the most callogenic media included media combining 9.0 µM 2,4-D with various concentrations of KT (3.375; 2.25; 4.5 and 1.125 µM). The moderately callogenic media was formed by 4.5 µM 2,4-D combined with KT varying from 2.25 to 4.5 µM. The media combining 4.5 µM 2,4-D to 1.125 KT µM were the least callogenic. 2,4-D, KT and their interaction expressed 87% of the variability observed during somatic embryogenesis induction (data not show). In addition, they have, in general, significantly influenced ($P = 0.0001$) the induction of embryogenic calli. All combinations of 2,4-D/KT were proved to be embryogenic, except M1 medium consisting of 4.5 µM 2,4-D combined with 1.125 µM KT. The most embryogenic

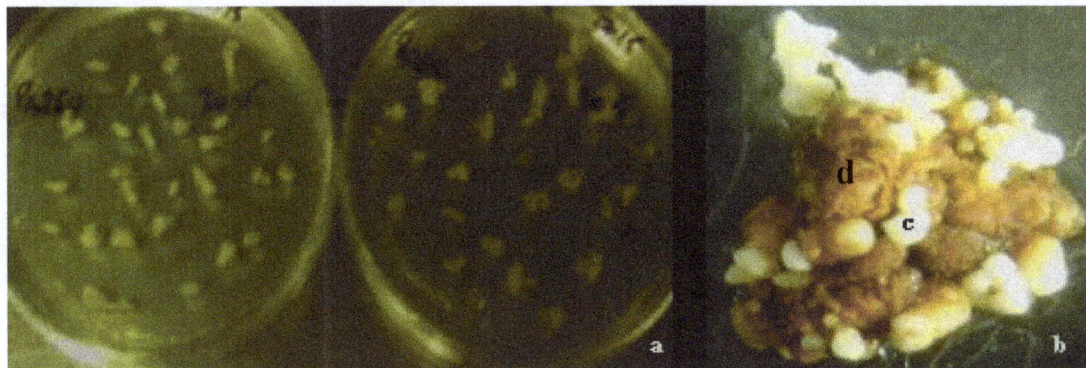

Figure 1. Embryogenic callus induction from inner tegument of *Hevea brasiliensis* fruit in clone PB 260. **a)** Callogenic explants (Gx0.26); **b)** embryogenic callus (Gx12.5) ; **c)** embryogenic cell; **d)** non-embryogenic cell.

Table 3. Effect of 2,4-D and kinetin concentration on callogenesis induction from inner tegument of *Hevea brasiliensis* fruit in clone PB 260.

Hormone	Explants forming callus (%)
2,4-D (µM)	
4.5	50.56 ± 3.30^{b}
9	81.11 ± 2.41^{a}
Kinetin (µM)	
1.125	55.56 ± 9.60^{b}
2.25	65.00 ± 7.32^{c}
3.375	75.00 ± 6.34^{d}
4.5	67.78 ± 5.95^{cd}

Values for each parameter followed by a different letter within each column are significantly different by Duncan test ($P < 0.05$). Each value represents the mean of ten replicates; ± SD, standard deviation; 2,4-D, 2,4-dichlorophenoxyacetic acid.

medium was M7 (9 µM 2,4-D/3.375 µM KT).

However, it is interesting to reveal that the concentrations 2.25 and 3.375 µM KT promoted somatic embryogenesis induction regardless of the concentration of 2,4-D (Table 4).

Combination of 2,4-D/TDZ

The influence of 2,4-D/TDZ combination was consigned in Table 5. Compact callus were observed in media fortified with different concentrations of TDZ (11.4, 22.8, 34.2 and 45.6 nM) in combination with 2,4-D (9.0 µM) alone. The induction media had significantly influenced (P = 0.0001) the percentage of callogenesis because they explain 83% of total variability (data not show). M11 medium (34.2 nM TDZ) was the most callogenic whereas M9 medium (11.4 nM TDZ) was the less callogenic. M10 (22.8 nM) and M11 (34.2 nM) media had callogenesis

rates significantly more elevated than those of the controls. Thus, 2,4-D use alone promotes callogenesis but combination with TDZ improved callus induction. The percentage of embryogenic calli was not significantly different in the four media containing TDZ (0 to 4.96%). Only 21% of the variability observed as regard the percentage of embryogenesis was explained by the induction media. Embryogenic calli were observed on M10 (3.57%) and M11 (4.96%) media.

DISCUSSION

In the present study, 2,4-D alone favored the induction of callogenesis; the emphasis was put on the search for the influence of its association with two cytokinins, KT and TDZ. A factorial influence in the case of the association of 2,4-D/KT showed that the percentage of callogenic explants increased in proportion with the concentration of

Table 4. Effect of 2,4-D and Kinetin combination on callogenesis and somatic embryogenesis from inner tegument of *Hevea brasiliensis* fruit in clone PB 260.

Hormone			Explants forming callus (%)	Embryogenic callus (%)
2,4-D (µM)	Kinetin (µM)	Code media		
4.5	1.25	M1	34.44 ± 3.93^{c}	0^{d}
4.5	2.25	M3	50.00 ± 4.04^{b}	21.67 ± 5.00^{e}
4.5	3.375	M4	61.11 ± 2.00^{b}	22.22 ± 5.07^{e}
4.5	4.5	M5	56.67 ± 2.03^{b}	05.33 ± 1.33^{a}
9.0	1.25	M6	76.67 ± 2.03^{a}	35.19 ± 7.76^{c}
9.0	2.25	M7	80.00 ± 5.13^{a}	16.45 ± 3.67^{ei}
9.0	3.375	M8	88.89 ± 1.00^{a}	46.74 ± 3.65^{b}
9.0	4.5	M9	78.89 ± 7.22^{a}	07.58 ± 2.09^{f}

Values for each parameter followed by a different letter within each column are significantly different by Duncan test (P< 0.05). Each value represents the mean of ten replicates.

Table 5. Effect of 2,4-D (9.0 µM) and TDZ combination on callogenesis and somatic embryogenesis from inner tegument of *Hevea brasiliensis* fruit in clone PB 260

Hormone			Explants forming callus (%)	Embryogenic callus (%)
2,4-D (µM)	TDZ (nM)	Code media		
9.0	0	M0	42.71 ± 5.24^{c}	0^{a}
9.0	11.4	M9	56 ± 8.3^{bc}	0^{a}
9.0	22.8	M10	70 ± 6.12^{ab}	3.57 ± 1.57^{d}
9.0	34.2	M11	76.67 ± 6.01^{a}	4.96 ± 2.37^{d}

Values for each parameter followed by a different letter within each column are significantly different by Duncan test (P< 0.05). Each value represents the mean of ten replicates; ± SD (standard deviation); 2,4-D (2,4-dichlorophenoxyacetic acid), TDZ (thidiazuron).

2,4-D (from 4.5 to 9.0 µM). These results dealing with 2,4-D were similar with those of Carron (1982) which indicated that the level of callus proliferation stemming from epicotyl fragments of *Hevea* increased in the same sense that the concentration of 2,4-D until 9.0 µM and beyond the threshold of toxicity was reached. Kumari et al. (1999) showed from the immature anthers of *Hevea* that the percentage of callogenic explants increased when the concentration of 2,4-D increased from 2.0 m to 3.0 mg/l and decreased when this concentration was 1.0 mg/l. Growth regulators play a key role by intervening in the reactions that lead to a reorientation of the program of gene expression. This expression can lead either to an unorganized growth of the cells (callus) without embryogenesis or to a polarized growth leading to a somatic embryogenesis (Dudits et al., 1995). In the present study, the embryogenesis was induced when the 2,4-D with 4,5 or 9.0 µM was associated to KT (2.25 or 3.375 µM). Montoro et al. (1993) reported with clone PB 260 that percentages values of embryogenic calli varied from 32 to 48% when the concentrations of 3,4-D or KT varied from 2.25 to 9.0 µM. This study registered 47% of embryogenic calli on the induction medium containing 9.0 µM 2,4-D/3.375 µM KT. This rate is comparable to those

obtained in other studies using 3,4-D and KT to induce the embryogenesis from the inner tegument of the seed. The present results evidenced that an inappropriate concentration of 2,4-D combined with an inadequate or a suboptimum concentration of cytokinin may consider the 2,4-D as not very suitable for somatic embryogenesis from inner tegument of rubber's fruit. Actually, 2,4-D is considered as a growth regulator conducive to somatic embryogenesis from the inner tegument of the immature rubber's fruit. It could therefore be used as an alternate solution to 3,4-D.

TDZ has never been used as cytokinin for rubber's callogenesis and embryogenesis. The combination of 2,4-D/TDZ used in this study helped obtain compact calli with a percentage of callogenic induction of more than 70%. Embryogenic calli were induced from 4 to 5% on medium containing 9.0 µM of 2,4-D/34.2 nM of TDZ and 9 µM of 2,4-D/22.80 nM of TDZ. These rates were weaker than those one obtained with the combination of 9 µM of 2,4-D/3.375 µM of KT. Indeed, an optimum association of 2,4-D/cytokinin favorably influences the rate of somatic embryogenesis induction. In addition, all the somatic embryos obtained with TDZ were abnormal. This suggests that the optimal concentrations of this cytokinin associated with 2,4-D (9.0 µM) were not deepened in the present study. Thereby, embryogenic potentialities of TDZ can be improved. With concentrations (being associated with other growth regulators or not) different from these ones used in this study, it could be another alternate to the traditional growth regulators used in rubber's somatic embryogenesis. Moreover, a protocol of secondary somatic embryogenesis from abnormal embryos observed could be envisaged. With regard to many other species, TDZ alone or associated with another growth regulator was used to produce somatic embryos or regenerate shoots with leaves. In case of the cocoa tree, the association of 9.0 µM of 2,4-D/22.70 nM of TDZ proved to be optimum for the induction of somatic embryos (Li et al., 1998). As for lentil, TDZ alone used with 0.25 mg/l regenerated, in eight weeks, young shoots from cotyledons (Kwawar et al., 2003).

ACKNOWLEDGEMENTS

The authors would like to thank the Centre National de Recherche Agronomique (CNRA) of Côte d'Ivoire for the financial and scientific support provided.

REFERENCES

Ashrafi S, Mofid MR, Otroshi M, Ebrahimi M, Khosroshahli M (2010). Effects of plant growth regulators on the callogenesis and taxol production in cell suspension of Taxus baccata L. Trakia. J. Sci. 8(2):36-43.

Carron MP (1982). L'embryogenèse somatique de l'Hevea brasiliensis (Kunth) Müll-Arg : une technique de multiplication au service de l'amélioration génétique. Thèse de troisième cycle en Agronomie, Mention : Phytotechnie. Acad. de Montpellier. Univ. Sci. Tech. Languedoc, France.

Carron MP, Enjarlic F (1985). Somatic embryogenesis from inner integument of the seed of Hevea brasiliensis (Kunth) Müll. Arg. CR Acad. Sci. 300:653-658.

Carron M, Enjarlic L, Deschamps A (1989). Rubber (Hevea brasiliensis Müll. Arg.). In Trees II, YPS Bajaj (Ed.) Biotechnol. Agric. Forest, Berlin, Allemagne, Springer–Verlag. 5:222-245.

Carron MP, Etienne H, Michaux-Ferriere N, Montoro P (1995). Somatic embryogenesis in rubber tree (Hevea brasiliensis Müell.Arg.). In YPS Bajaj (Ed.) Biotechnol. Agric. Forest, Springer Verlag, Berlin-Heidelberg. 30:353-369.

Dudits D, Gyorgvey J, Bogre L, Bako L (1995). Molecular Biology of somatic embryogenesis. In: T.A. Thorpe (ed) In vitro Embryogenesis in Plants. Kluwer Academic Publishers, Dordecht, Boston, London. pp. 267-308.

El Hadrami I, Carron MP, Auzac J (1991). Influence of exogenous hormones on somatic embryogenesis in Hevea brasiliensis. Ann. Bot. 67:511-515.

Féher A, Pasternak TP, Dudits D (2003). Transition of somatic plant cells to an embryogenic state. Review of Plant Biotechnology and Applied Genetics. Plant Cell Tiss. Org. Cult. 74:201-228.

Fossard RA (1976). Tissue Culture for Plant Propagators. Univ. of New England Printery, New South Wales, Australia. Frison, E.A. and Ng, S.Y. 1981.

Kumari JP, Asokan MP, Sobha S, Sankari AL, Rekha K, Kala RG, Jaysree R, Thulaseedharan A (1999). Somatic embryogenesis and plant regeneration from immature anthers of Hevea brasiliensis (Muell.) Arch. Curr. Sci. 76:1242-1245.

Kwawar KM, Sancak C, Uranbey S, Özcan S (2003). Effect of thidiazuron on shoot regeneration from different explants of lentil (Lens culinaris Medik.) via organogenesis. Turk. J. Bot. 28:421-426.

Langlois S (1965). Etude de la production des arbres d'un champ donné. Rapp. Inst. Rech. Caoutch., Cambodge, pp. 56-59.

Li Z, Traore A, Maximova S, Guiltinan MJ (1998). Somatic embryogenesis and plant regeneration from floral explants of Cacao (Theobroma cacao L.) using Thidiazuron. In vitro Cell Dev. Biol. 34:293-299.

Michaux-Ferrière N, Carron MP (1989). Histology of early somatic embryogenesis in Hevea brasiliensis. The importance of timing of subculturing. Plant Cell Tiss. Org. Cult. 19:243-256.

Montoro P, Etienne H, Carron MP (1993). Callus friability and somatic embryogenesis in Hevea brasiliensis. Plant Cell Tiss. Org. Cult. 33:331-338.

Montoro P, Etienne H, Carron MP, Nogarede A (1992). Effect of cytokinins on the induction of embryogenesis and the quality of somatic embryos in Hevea brasiliensis. Müll. Arg. CR. Acad. Sci. 315:567-574.

Murashige T, Skoog F (1962). A revised medium for rapid growth and bioassays with tobacco tissue cultures. Physiol. Plant. 15:473-497.

Sané D, Aberlenc-Bertossi F, Gassama-Dia YK, Sagna M, Duval Y, Borgel A (2006). Histocytological analysis of callogenesis and somatic embryogenesis from cell suspensions of date palm (Phoenix dactylifera L.). Ann. Bot. 98:301-308.

Yapo ES, Kouakou TH, Koné M, Kouadio YJ, Kouamé P, Mérillon JM (2011). Regeneration of pineapple (Ananas comosus L.) plant through somatic embryogenesis. J. Plant Biochem. Biotechnol. 20(2):196-204.

Zouine J, El Bellaj M, Meddich A, Verdeil J-L, El Hadrami I (2005). Proliferation and germination of somatic embryos from embryogenic suspension cultures in Phoenix dactylifera. Plant Cell Tiss. Org. Cult. 82:83-92.

Zouzou M, Kouakou TH , Koné M , Amani NG, Kouadio YJ (2008). Effect of genotype, explants, growth regulators and sugars on callus induction in cotton (Gossypium hirsutum L.). Aust. J. Crop Sci. 2(1):1-9.

5

Manumycin from a new *Streptomyces* strain shows antagonistic effect against methicillin-resistant *Staphylococcus aureus* (MRSA)/vancomycin-resistant enterococci (VRE) strains from Korean Hospitals

Yun Hee Choi[1] , Seung Sik Cho[2] , Jaya Ram Simkhada[1], Chi Nam Seong[3],
Hyo Jeong Lee[4], Hong Seop Moon[2] and Jin Cheol Yoo[1]

[1]Department of Pharmacy, College of Pharmacy, Chosun University, Gwangju 501-759, Korea.
[2]Department of Pharmacy, College of Pharmacy, Mokpo National University, Muan, Jeonnam, 534-729, Korea.
[3]Department of Biology, College of Natural Sciences, Sunchon National University, Sunchon, Jeonnam, 540-742, Korea.
[4]Department of Alternative Medicine, Gwangju University, Gwangju 503-703, Republic of Korea.

An antimicrobial compound, highly effective against multidrug-resistant (MDR) bacteria, purified from a *Streptomyces* strain was identified as manumycin. The minimal inhibitory concentrations (MICs) of manumycin against 8 different strains of methicillin-resistant *Staphylococcus aureus* (MRSA) were ranged 2 to 32 µg/ml. Similarly, MICs of manumycin against 4 vancomycin-resistant enterococci (VRE) strains were ranged 8 to 32 µg/ml while it remained ineffective against 4 other VRE strains. Compared to vancomycin, manumycin provided slightly weaker activity against MRSA strains but stronger activity against 4 VRE strains. This is the first report of antagonistic effect of manumycin against MDR pathogens.

Key words: Manumycin, methicillin-resistant *Staphylococcus aureus* (MRSA), vancomycin-resistant *enterococci* (VRE).

INTRODUCTION

Manumycin is a group of small and discrete class of antibiotics which consist of almost a dozen secondary metabolites produced exclusively by *Streptomyces* (Sattler et al., 1998). Manumycin was first reported by Buzzetti and coworkers in 1963. Its chemical structure has two unsaturated carbon chains, so called m-C7N and C5N unit, which are linked in meta-fashion to unique multifunctional six-membered ring. Manumycin-type compounds derived from the m-C7N unit vary in its stereochemistry and the nature of the oxygen substituent

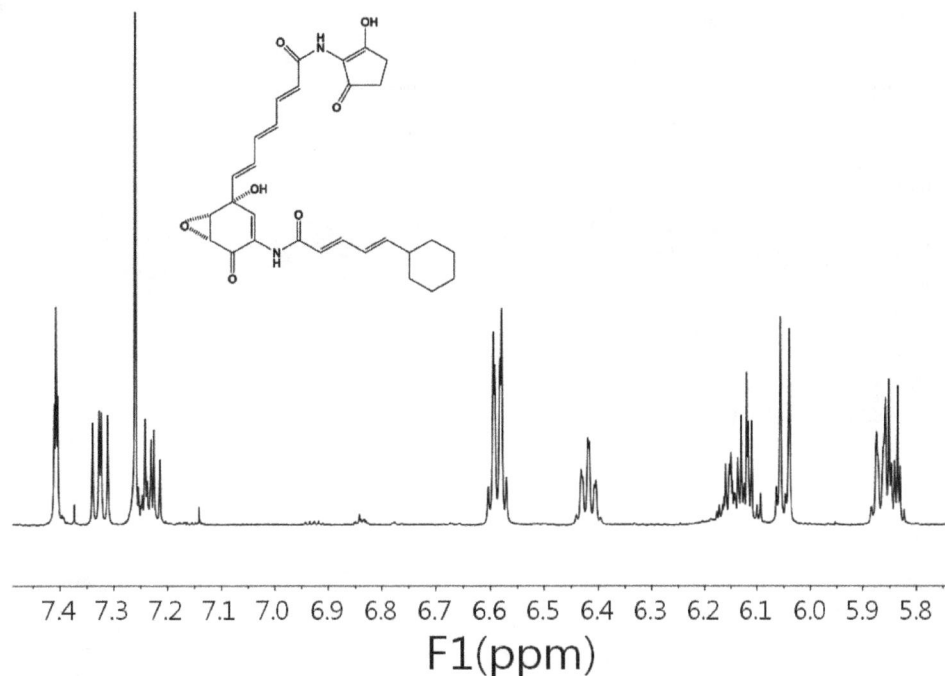

Figure 1. A typical ¹H-NMR spectrum obtained from purified manumycin; *Inset*, structure of manumycin. An active compound (Compound C3) was purified from culture broth of *Streptomyces* sp. CS392 (GenBank accession no. JN128646), according to our recent report (Cho et al., 2012). The compound was identified as manumycin based on NMR along with COSY, TOCSY and HMQC (detailed not shown).

(Hwang et al., 1996; Kohno et al., 1996; Sattler et al., 1998). Manumycin exhibits biological activity against Gram-positive bacteria, fungi and some insects (Hwang et al., 1996; Kohno et al., 1996; Sattler, 1998; Thiericke et al., 1987). In 1993, manumycin A was reported to inhibit Ras farnesyltransferase (FTase) (Hara et al., 1993). Manumycin was also reported to show antitumor activity *in vitro* and *in vivo* in nude mouse xenograft models (Ito et al., 1996; Xu et al., 2001). Manumycin A induces caspase-mediated apoptosis in human hepatoma HepG2 cell line (Zhou et al., 2003). Apart from these broad activities, there is no report dealing with the antimicrobial activity of manumycin against multi drug resistant (MDR) pathogens. The objective of this article is to report the potential of manumycin against hospital-acquired multidrug resistant pathogens such as methicillin resistant *Staphylococcus aureus* (MRSA) and vancomycin resistant enterococci (VRE). The study is also significant being the first report of antimicrobial activity of manumycin against MDR hospital acquired pathogens.

MATERIALS AND METHODS

The active compound (Compound C3) was purified from culture broth of *Streptomyces* sp. CS392 (GenBank accession no. JN128646), according to our recent report (Cho et al., 2012). The compound identification was carried out by using NMR, COSY, TOCSY and HMQC. For cytotoxic effect of the compound (identified as manumycin) against human cell lines, four types of human cancer cell lines (A549 from lung, HepG2 from liver, MCF-7 from breast, and MG-63 from bone) were from Korean Cell Line Bank, Seoul, Korea. Cells were seeded in a 96 well plate at 0.5×10^4 cell/well. Manumycin was added at 1-100 µg/ml. After 24 h incubation with or without manumycin, MTT solution (0.5 mg/mL) was added and the cells were incubated for 4 h at 37°C. After removing the supernatant, DMSO was added and read at 590 nm. The minimal inhibitory concentrations (MIC) of manumycin were determined by agar dilution method according to Mueller-Hinton-agar dilution method (Schreiber and Jacobs, 1995). After inoculation of test organisms in the agar plates containing various concentrations of drugs, results were observed after incubating them at 37°C for 18 h.

RESULTS AND DISCUSSION

In our recent study, we have purified 3 antimicrobial compounds (C1, C2 and C3) from *Streptomyces* sp. CS392 (Cho et al., 2012). In this study, the major compound 'C3' was identified as manumycin according to various structural parameters such as nuclear magnetic resonance (NMR), correlation spectroscopy (COSY), total correlation spectroscopy (TOCSY) and heteronuclear multiple-quantum correlation spectroscopy (HMQC). A typical ¹H-NMR spectrum obtained from purified manumycin with its chemical structure is illustrated in Figure 1. We have investigated the antagonistic effects of

Table 1. MIC of manumycin against various pathogens[1].

Microorganism	Manumycin	Vancomycin
Staphylococcus aureus KCTC 1928	2	0.5
MRSA-693E	2	0.5
MRSA 4-5	2	1
MRSA 5-3	2	1
MRSA-B15	16	4
MRSA-S3	16	4
MRSA-S1	16	4
MRSA-P8	16	4
MRSA-U4	32	4
Enterococcus faecalis ATCC 29212	4	1
VRE-2	>65	>65
VRE-3	>65	>65
VRE-4	>65	>65
VRE-5	>65	>65
VRE-6	32	>65
VRE-82	8	>65
VRE-98	8	>65
VRE-89	16	>65

[1]MIC (µg/mL) value of manumycin were determined by agar dilution method according to Mueller-Hinton-agar dilution method (Schreiber and Jacobs, 1995). After inoculation of test organisms in the agar plates containing various concentrations of drugs, results were observed after incubating them at 37°C for 18 h.

manumycin against *S. aureus* and MRSA strains as well as *Enterococcus faecalis* and VRE strains. Manumycin did not show antimicrobial activity against drug sensitive Gram-negative pathogens such as *Alacligenes faecalis* ATCC 1004, *Salmonella typhimrium* KCTC 1925, *Pseudomonas aeruginosa* KCTC 1637 and *Escherrichia coli* KCTC 1923. Effect of manumycin against various strains in terms of MIC value is illustrated in Table 1. Growth of *S. aureus* KCTC 1928 as well as 3 MRSA strains, namely; MRSA-693E, MRSA 4-5, and MRSA 5 to 3, was inhibited by manumycin at 2 µg/ml. On the other hand, 4 other MRSA strains, namely; MRSA-S1, MRSA-S3, MRSA-B15 and MRSA-P8, were inhibited at 16 µg/ml and rest strain, namely; MRSA-U4, was suppressed at 32 µg/ml of manumycin. Compared to vancomycin with MIC of 4 µg/ml, manumycin remains weaker for 5 tested MRSA strains with MIC values of 16 to 32 µg/ml as mentioned above. However, effect of manumycin against rest 3 MRSA strains with MIC value of 2 µg/ml, namely; MRSA-693E, MRSA 4 to 5, and MRSA 5 to 3, was inferior to vancomycin, only slightly. This discrepancy in the MIC values for the similar types of pathogens may be attributed to the different nature of the strains. Furthermore, although vancomycin showed stronger antimicrobial activity than manumycin against non resistant *E. faecalis*, it did not show any effect against any of the tested VRE strains (Table 1). In contrast, manumycin showed antagonistic effect against 4 VRE strains, namely; VRE-82 and VRE-98 with MIC of 8

µg/ml, VRE-89 with16 µg/ml and VRE-6 with 32 µg/ml. Hence, manumycin shows broader antimicrobial spectra than that offered by vancomycin against MDR bacteria.

So far many attempts have been made to explore effective antimicrobial compounds against resistant bacteria. For example, anti-MRSA compounds such as marinopyrroles A and B from *Streptomyces* sp. CNQ-418 (MIC= ≤ 2 µM) (Hughes et al., 2008), MC21A and MC21B from *Pseudoalteromonas phenolica* (MIC=1-2 to 1-4 µg/mL) (Isnansetyo et al., 2003; Isnansetyo and Kamei, 2009), abyssomicin C from Verrucosispora AB-18-032 (MIC = 4 to 13 µg/mL) (Bister et al., 2004; Keller et al., 2007), lydicamycin from *Streptomyces lydicus* (MIC 6 µg/mL) (Furumai et al., 2002), and so on. Similarly, BE-43472B from *Streptomyces* sp. was reported to exhibit antimicrobial activity against MRSA (MIC = 0.11 to 0.45 µM) and VRE (MIC = 0.24 µM) (Socha et al., 2006) and 2,4-diacetylphloroglucinol (DAPG) isolated from *Pseudomonas* sp. AMSN exhibited antimicrobial activity against MRSA (MIC = 4 µg/mL) and VRE (MIC = 8 µg/mL) (Isnansetyo et al., 2003). Thus, the effect of manumycin against MRSA-693, MRSA 4-5 and MRSA 5-3 (MIC = 2 µg/mL) is stronger than that of DAPG (MIC = 4 µg/mL), lydicamycin (MIC = 6 µg/mL) and abyssomicin C (MIC = 4 to 13 µg/mL) whereas it is relatively weaker than that of BE-43472B (MIC 0.11 to 0.45 µM) and marinopyrroles A and B (MIC = ≤ 2 µM).

Moreover, anti-VRE activity of manumycin (MIC = 8 to 16 µg/mL for 3 VRE strains) was comparable with those

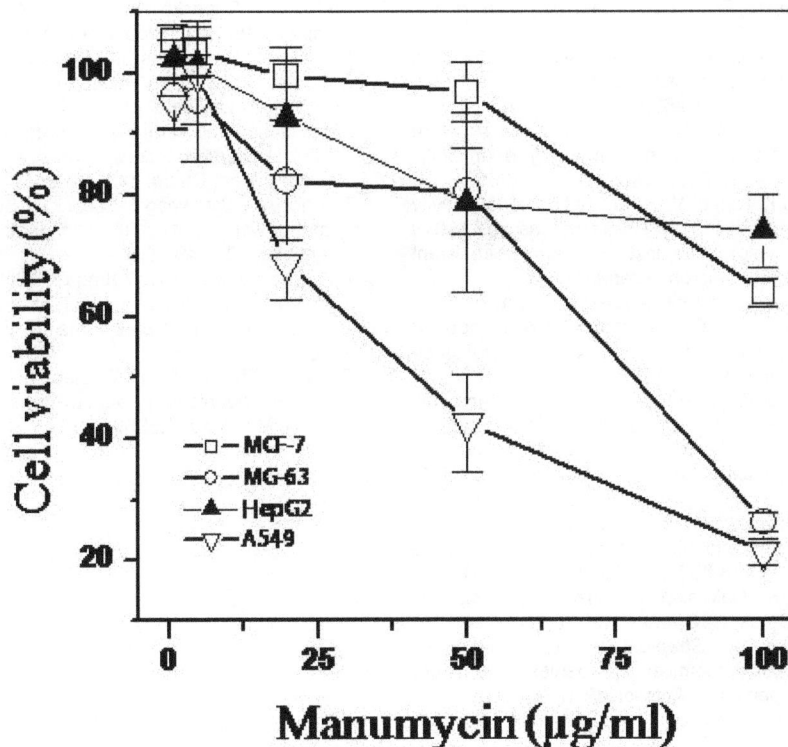

Figure 2. Cytotoxic effect of manumycin against human cell lines. Four types of human cancer cell lines (A549 from lung, HepG2 from liver, MCF-7 from breast, and MG-63 from bone) were from Korean Cell Line Bank, Seoul, Korea. Cells were seeded in a 96 well plate at 0.5×10^4 cell/well. Manumycin was added at 1 to 100 µg/ml. After 24 h incubation with or without manumycin, MTT solution (0.5 mg/mL) was added and the cells were incubated for 4 h at 37°C. After removing the supernatant, DMSO was added and read at 590 nm.

of DAPG (MIC = 8 µg/mL) but weaker than BE-43472B (MIC = 0.24 µM). As illustrated in Figure 2, manumycin with its effective antibacterial concentration (~20 µg/ml, Table 1), when tested for 24 h, did not pose toxic effect against MCF-7 and HepG2 cell lines. It posed ~20 and ~30% of toxicity against MG63 and A549 cell lines, respectively. Although a detailed study is needed, manumycin so far seems safe to use as antimicrobial drug on the basis of cell viability results.

Conclusion

The results show that manumycin isolated from *Streptomyces* sp. CS392 displayed antimicrobial activity against hospital acquired MDR pathogens such as MRSA and VRE strains. Compared to vancomycin manumycin displayed slightly weaker but wider range of antimicrobial spectrum. More detailed study should be done to elucidate its mode of antimicrobial action and antimicrobial effects either as a single antibiotic or in the combination with other commercial antibiotics (synergistic effect), which are our future goals.

ACKNOWLEDGEMENTS

This work was supported by the National Research Foundation of Korea (NRF) grant funded by the Korean government (MEST) (2010-0029178) and a grant from the Next-Generation BioGreen 21 Program (No. PJ009602), Rural Development Administration, Republic of Korea.

REFERENCES

Bister B, Bischoff D, Strobele M, Riedlinger J, Reicke A, Wolter F, Bull AT, Zahner H, Fiedler HP, Sussmuth RD (2004). Abyssomicin C-A polycyclic antibiotic from a marine Verrucosispora strain as an inhibitor of the p-aminobenzoic acid/tetrahydrofolate biosynthesis pathway. Angew. Chem. Int. Ed. Engl. 43:2574-2576.

Cho SS, Choi YH, Simkhada JR, Mander P, Park DJ, Yoo JC (2012). A newly isolated *Streptomyces* sp. CS392 producing three antimicrobial compounds. Bioprocess Biosyst. Eng. 35:247-254.

Furumai T, Eto K, Sasaki T, Higuchi H, Onaka H, Saito N, Fujita T, Naoki H, Igarashi Y (2002). TPU-0037-A, B, C and D, novel lydicamycin congeners with anti-MRSA activity from *Streptomyces platensis* TP-A0598. J. Antibiot. 55:873-880.

Hara M, Akinaga S, Okabe M, Nakano H, Gomez R, Wood D, Uh M, Tamanoi F (1993). Identification of Ras farnesyltransferase inhibitors by microbial screening. Proc. Natl. Acad. Sci. U. S. A. 90: 2281-2285.

Hughes CC, Prieto-Davo A, Jensen PR, Fenical W (2008). The marinopyrroles, antibiotics of an unprecedented structure class from a marine *Streptomyces* sp. Org. Lett. 10:629-631.

Hwang BK, Lee JY, Kim BS, Moon SS (1996). Isolation, structure elucidation, and antifungal activity of a manumycin-type antibiotic from *Streptomyces* flaveus. J. Agric. Food Chem. 44:3653-3657.

Isnansetyo A, Cui L, Hiramatsu K, Kamei Y (2003). Antibacterial activity of 2,4-diacetylphloroglucinol produced by *Pseudomonas* sp. AMSN isolated from a marine alga, against vancomycin-resistant *Staphylococcus aureus*. Int. J. Antimicrob. Agents 22:545-547.

Isnansetyo A, Kamei Y (2009). Anti-methicillin-resistant *Staphylococcus aureus* (MRSA) activity of MC21-B, an antibacterial compound produced by the marine bacterium *Pseudoalteromonas phenolica* O-BC30T. Int. J. Antimicrob. Agents 34:131-135.

Ito T, Kawata S, Tamura S, Igura T, Nagase T, Miyagawa JI, Yamazaki E, Ishiguro H, Matasuzawa Y (1996). Suppression of human pancreatic cancer growth in BALB/c nude mice by manumycin, a farnesyl:protein transferase inhibitor. Jpn. J. Cancer Res. 87:113-116.

Keller S, Nicholson G, Drahl C, Sorensen E, Fiedler HP, Sussmuth RD (2007). Abyssomicins G and H and atrop-abyssomicin C from the marine *Verrucosispora* strain AB-18-032. J. Antibiot. 60:391-394.

Kohno J, Nishio M, Kawano K, Nakanishi N, Suzuki S, Uchida T, Komatsubara S (1996). TMC-1 A, B, C and D, new antibiotics of the manumycin group produced by *Streptomyces* sp. Taxonomy, production, isolation, physico-chemical properties, structure elucidation and biological properties. J. Antibiot. 49:1212-1220

Sattler I, Thiericke R, Zeeck A (1998). The manumycin-group metabolites. Nat. Prod. Rep. 15:221-240.

Schreiber JR, Jacobs MR (1995). Antibiotic-resistant pneumococci. Pediatr. Clin. North Am. 42:519-537.

Socha AM, LaPlante KL, Rowley DC (2006). New bisanthraquinone antibiotics and semi-synthetic derivatives with potent activity against clinical *Staphylococcus aureus* and *Enterococcus faecium* isolates. Bioorg. Med. Chem. 14:8446-8454.

Thiericke R, Stellwaag M, Zeeck A, Snatzke G (1987). The structure of manumycin. III. Absolute configuration and conformational studies. J. Antibiot. 40:1549-1554.

Xu G, Pan J, Martin C, Yeung SC (2001). Angiogenesis inhibition in the in vivo antineoplastic effect of manumycin and paclitaxel against anaplastic thyroid carcinoma. J. Clin. Endocrinol. Metab. 86:1769-1777.

Zhou JM, Zhu XF, Pan QC, Liao DF, Li ZM, Liu ZC (2003). Manumycin induces apoptosis in human hepatocellular carcinoma HepG2 cells. Int. J. Mol. Med. 12:955-959.

6

Evaluation of non-viable biomass of *Laurencia papillosa* for decolorization of dye waste water

Dahlia M. El Maghraby

Department of Botany and Microbiology, Faculty of Science, Alexandria University, 21511 Alexandria, Egypt.

The uptake of fast orange dye by the red seaweed *Laurencia papillosa* has been demonstrated in order to explore its potential use as low-cost adsorbent. The adsorption kinetics of fast orange dye on the alga with respect to initial dye concentration, contact time, particle size and pH were investigated. The dye removal percentage increased from 25.92 to 67.08% and the equilibrium states were attained at almost 60 min within the experimental concentration range. The adsorption kinetic was analyzed using pseudo-first-order and pseudo-second-order models. The pseudo-second-order model was more appropriate to describe the sorption kinetics based on the relatively high values of the linear squared regression correlation coefficient. The nature of the possible adsorbent and fast orange interactions was examined by the Fourier transform infrared technique. This technique confirmed that hydroxyl, carboxyl, amine, sulfonyl, carbonyl and alkyl groups are responsible for the dye binding process. Significant increase in dye adsorption was observed with the decrease in sorbent particle size coupled with its large surface area. Maximum removal efficiency was determined to be 65.7% at a solution pH of 5. However, *Laurencia papillosa* proved to be a promising material for removing fast orange dye from aqueous solutions.

Key words: Dye adsorption, Macroalga, *Laurencia papillosa,* kinetics.

INTRODUCTION

Textile processing operations are considered as important part of the industrial sector in both developing and undeveloped countries (Mahmoud et al., 2007). Several types of textile dyes are available for usage with various types of textile materials (Marungruenga and Pavasant, 2006). Over 7×10^5 tons and about 10,000 different types of dyes are produced in the world. Unfortunately, about 10 to 15% of the total produced dyes is released into the aquatic ecosystems without being removed from the effluents and large volumes of highly polluted wastewater are produced (Sheng and Chi, 2003; Hoda et al., 2006; Senthilkumaar et al., 2006; Bukallah et al., 2007). The presence of these pollutants in water reduces light penetration and photosynthesis (Chen et al., 2003). In addition, dyes in the water bodies

undergo chemical and biological changes that consume dissolved oxygen resulting in fish kills and the destruction of other aquatic organisms (Muthuraman and Palanivelu, 2006). Some dyes have been also reported to cause allergy irritation, cancer and even mutation in humans (Bhattacharyya and Sharma, 2004). Therefore, removal of dyes from the effluents of textile industries is of vital importance for the proper maintenance of the ecosystem health (Cengiz and Cavas, 2008).

Dye molecules comprise of two components: the chromophores, responsible for producing the color, and the auxochromes, which cannot only supplement the chromophore but also render the molecule soluble in water and give enhanced affinity toward the fibers (Gupta et al., 2003). Some of the techniques used in treatment of

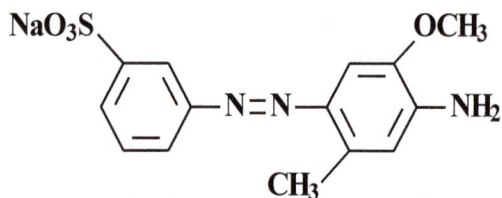

Figure 1. Molecular structure of fast orange dye.

wastewater containing dyes are flocculation, coagulation, precipitation, adsorption, membrane filtration, electrochemical techniques and ozonation (Dabrowski, 2001). Nevertheless these processes are not always effective and economic. This has prompted the use of various materials as adsorbents in order to develop cheaper alternatives by utilizing many types of biosorbents as fungi (Kaushik and Malik, 2009; Mishra et al., 2011), bacteria (Banat et al., 1996; Yang et al., 2011) and yeasts (Ertugrul et al., 2009; Phugare et al., 2010). As well, one of the growing interest and promising biosorbents is "algae" (Veglio and Beolchini, 1997; Pengthamkeerati et al., 2008; Kousha et al., 2012) due to its high sorption capacity and its availability in almost unlimited amounts (Klimmek et al., 2001). Both viable and non-viable algae have been used in color removal from dyes and wastewater. This is may be achieved via bioconversion and biosorption. Through bioconversion, some algae can break down the dyes to more simple compounds (Lim et al., 2010). On the other hand, Biosorption is known as a promising technique concerned with the uptake of undesired ions from aqueous solutions using biological materials. Biosorption in algae has mainly been attributed to the cell wall properties where both electrostatic attraction and complexation can play a role (Davis et al., 2003). In many cases, algal cell walls frequently consisting of proteins and carbohydrates provide functional groups for binding various metals and dyes (Volesky, 1990; Srinivasan and Viraraghavan, 2010). Research in the field of biosorption has mostly concerned itself with brown algae (Matheickal and Yu, 1999; Matheickal et al., 1999; Yu et al., 1999), green algae (Dönmez et al., 1999; Aksu et al., 1997, 1999) and red algae (Holan and Volesky, 1994). The cell walls of most red algae include a rigid inner part composed of micro fibrils and a mucilaginous matrix. The matrix is composed of sulfated polymers of galactose such as agar and carrageenan, which are responsible for flexible, slippery texture of the red algae (Prescott et al., 2002).

Laurencia papillosa is a red alga (Rhodophyta) notable for its importance as an agarophyte. Its cylindrical thallus may reach 15 cm tall and having pale-brown, sometimes yellowish color. The main focus of this study was to discuss the adsorption behavior of fast orange dye using *L. papillosa* as low cost and renewable biosorbent material. Additionally, the equilibrium and kinetics of dye adsorption from aqueous solutions were investigated.

The effects of initial dye concentration, contact time particle size and pH on the adsorption capacity were also considered.

MATERIALS AND METHODS

Adsorbent material

The raw biomass of *L. papillosa* (Forskal) Greville was harvested from Abou-kir, Alexandria coastline of the Mediterranean Sea during the spring season. The wet algal material was carried to the laboratory in an aquarium. Samples were preliminary visual cleaned of impurities followed by several washes with copious quantities of deionized water to remove extraneous materials and common ions (e.g. Na^+ and Ca^+) present in seawater. The washed biomass was sun-dried then oven dried at 60°C for 8 h, crushed to a fine powder sieved and preserved for further use.

Dye solutions preparation

Fast orange 37 was supplied by a local manufacturer (Dyestuffs and Chemicals Company at Kafr El-Dawar, Egypt) and used in commercial purity as received. The molecular structure of the dye is represented in Figure 1. The dye stock solution was prepared by dissolving accurately weighed dye in distilled water to the concentration of 1000 mg/l. The experimental solutions were obtained by diluting the dye stock solution in accurate proportions to different initial concentrations (10, 20, 30, 40 and 50 mg/l). Dye concentration determination was performed calorimetrically using a Perkin Elmer Lambda ultraviolet and visible (UV-Vis) spectrophotometer. The absorbance of the colors was read at 428 nm (λ_{max}).

Adsorption procedure

Adsorption experiments were carried out in batch conditions. A series of 250 ml Erlenmeyer flasks containing 120 ml dye solution of known initial concentrations in the range of 10 to 50 mg/l were prepared at room temperature (25 ± 2°C). Weighed amounts (2 g) of dry algal biomass were added to each flask and stirred. The pH of the mixtures was kept without measurement. Equilibrium process is directly correlated with time. Samples were drawn at suitable time intervals 10, 20, 30, 40, 50, 60, 90, 120 and 150 min and then centrifuged for 15 min at 5000 rpm. The left out concentration of dye in the supernatants were analyzed using the spectrophotometer by monitoring the absorbance changes at a wavelength of 428 nm. The removal percentage of the dye was calculated by using the following equation:

$$Removal \% = \frac{C_i - C_e}{C_i} \times 100 \qquad (1)$$

Where, C_i and C_e are the initial and equilibrium dye concentrations, respectively. The adsorbed dye quantity per gram of biomass at any time (q) can also be calculated from the difference between the initial and the equilibrium concentrations as shown in the following equation:

$$q = \frac{(C_i - C_e)V}{M} \qquad (2)$$

Where, q is the dye uptake capacity (mg g^{-1}), M the adsorbent dosage (g), and V the solution volume.

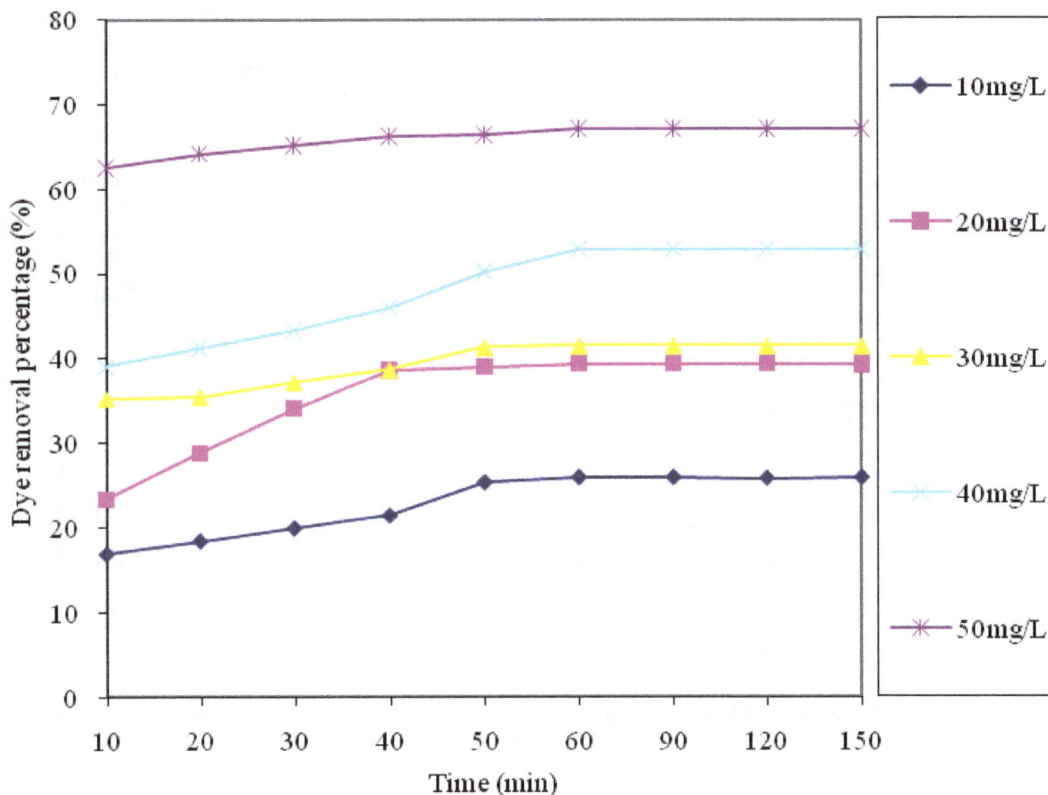

Figure 2. Dye removal percentage as a function of contact time and initial dye concentration (adsorbent dose = 2 g, temperature = 25 ± 2°C).

Kinetic models

Several kinetic models were available to understand the behavior of the adsorbent, to examine the controlling mechanism of the adsorption process and to test the experimental data. In this investigation, the kinetic data obtained were analyzed by using pseudo-first order and pseudo-second order models. The first-order rate expression of Lagergren (1898) is given as:

$$\log(qe - q) = \log(qe) - \frac{K1}{2.303}t \qquad (3)$$

Where, q_e and q are the amounts of dye adsorbed on adsorbent at equilibrium and at time t, respectively (mg/g) and K_1 is the rate constant (min^{-1}).

In many cases, the first-order equation of Lagergren does not fit well to the whole range of contact time and is generally applicable over the initial stage of the adsorption processes (Lagergren, 1898; McKay and Ho, 1999). The linear form of pseudo-second order equation expressing the chemisorption behavior of the reaction (Ho and McKay, 1999; Marungrueng and Pavasant, 2006; Ncibi et al., 2007) was calculated as follows:

$$\frac{t}{q} = \frac{1}{K2qe^2} + \frac{t}{qe} \qquad (4)$$

Where, K_2 is the pseudo-second order rate constant (g/mg.min).

The best-fit model was selected based on the linear regression correlation coefficient, R^2, values.

Fourier transform infrared spectroscopy analysis

After incorporating of an algal sample into a KBr pellet, detection of functional groups located on algal surface after and before adsorption process was specified. FTIR analyses within the range of 500 to 4000 cm^{-1} were recorded with Perkin Elmer Fourier transform infrared spectrophotometer (RXIFT-IR system).

RESULTS AND DISCUSSION

Effect of contact time and different dye concentrations

To design effective and user friendly adsorption model, it was considered necessary to carry out adsorption with a kinetic view-point as a function of contact time and initial dye concentration. The dye removal percentage (%) was represented in Figure 2. The results show that the equilibrium states were attained at almost 60 min within the experimental concentration range. Furthermore, raising the dye concentration from 10 to 50 mg/l allows the dry alga to increase their adsorption capacities from 25.92 to 67.08%, respectively. The curve of contact time is smooth and continuous leading to saturation due to intra particle diffusion process. These data indicate the possible monolayer coverage of dye on the surface of dry algal biomass (Doğan and Alkan, 2003).

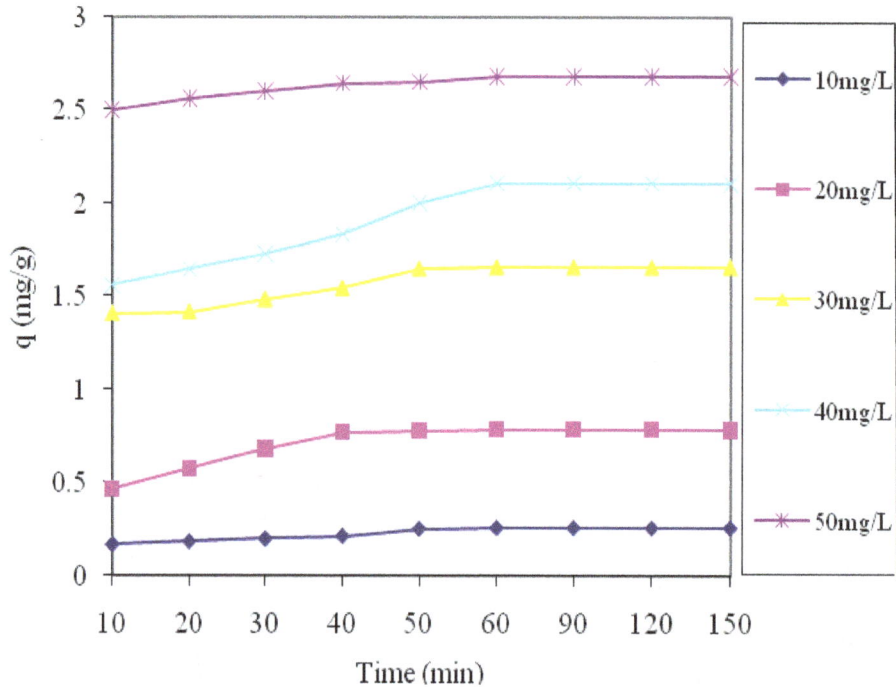

Figure 3. Kinetics of fast orange dye uptake by *Laurencia papillosa* at different initial dye concentrations (adsorbent dose = 2 g, temperature = 25 ± 2°C).

Equilibrium adsorption

It is obvious from Figure 3 that most of the dye was adsorbed to achieve adsorption equilibrium in about 60 min although the data were measured in 150 min. In the view of this result, the dye uptake capacity (q) increased with time and at certain time period; it reached a constant value indicated that no dye was further removed from the solution. It is interesting to note that the surface of adsorbent may contain a large number of active sites and the uptake of dye can be linked to these active sites on equilibrium time. The higher sorption rate at the initial period may be due to an augmented number of vacant sites available. Also, it indicates the strong electrostatic force of attraction between dye molecules and the sorbent binding-sites (Kaewsarn and Yu, 2001). As time proceeds this sorption rate is reduced due to the accumulation of dye particles in the vacant sites (Uddin et al., 2009). This result suggests that there is a high affinity between fast orange dye and functional groups on the wall surface of *L. papillosa*. A similar finding was reported by Marungrueng and Pavasant (2007) and Cengiz and Cavas (2008).

Sorption kinetics

Adsorption involves the mass transfer of a solute (adsorbate) from the fluid phase to the adsorbent surface for evaluating the applicability of sorption process as a unit operation. In order to characterize the kinetic behavior of a reaction, it is desirable to determine how the rate of reaction varies as the reaction progresses. In the present investigation, the validity of the pseudo-first order model can be checked by linearized plot of log (q_e-q) versus t (Figure 4). It was evident that the linear dependency was not obtained between log (q_e-q) and t. Therefore, first-order Lagergren rate kinetics is not convenient for the adsorption of the dye onto the alga. On the other hand, the linear plots of t/q versus t for the pseudo-second order were illustrated in Figure 5. Regressing the observed values of t/q on t afforded with coefficients of correlation allowing estimation of the amount of dye adsorbed at equilibrium and the rate constant. It is clearly found from the model parameters q_e and K_2 given in Table 1 that pseudo-second order model data fall on straight lines. This indicates that this model is in good agreement with the experimental data. Besides it is more appropriate to describe the sorption kinetics of fast orange dye onto the alga based on the relatively high values of the linear squared regression correlation coefficient R^2. This finding supports the assumption that the sorption process was due to chemisorption which required exchange or sharing of electrons between dye cations and functional groups of adsorbent (Ho, 2003; Marungrueng and Pavasant, 2007).

Fourier transform infrared spectroscopy analysis

The FTIR spectroscopy has been frequently used to detect vibrational frequency changes in seaweeds (Park et al., 2004; Sheng et al., 2004; Figueira et al., 1999). It offers excellent information on the nature of the bonds

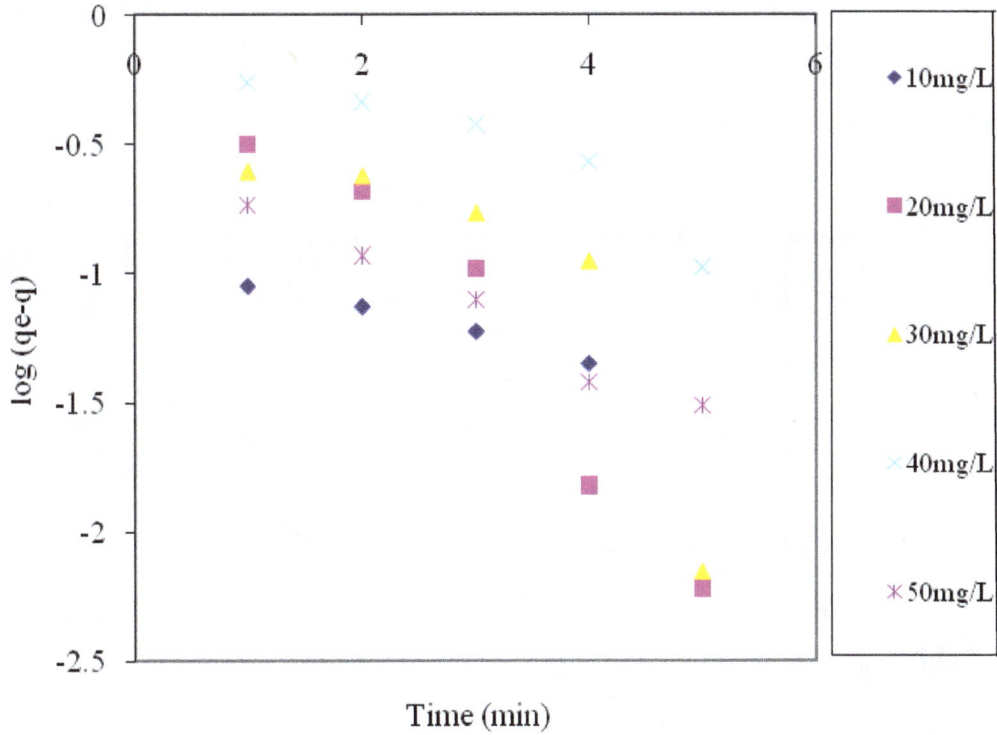

Figure 4. Pseudo-first-order sorption kinetics of fast orange dye by *Laurencia papillosa* at different initial dye concentrations (adsorbent dose = 2 g, temperature=25 ± 2°C).

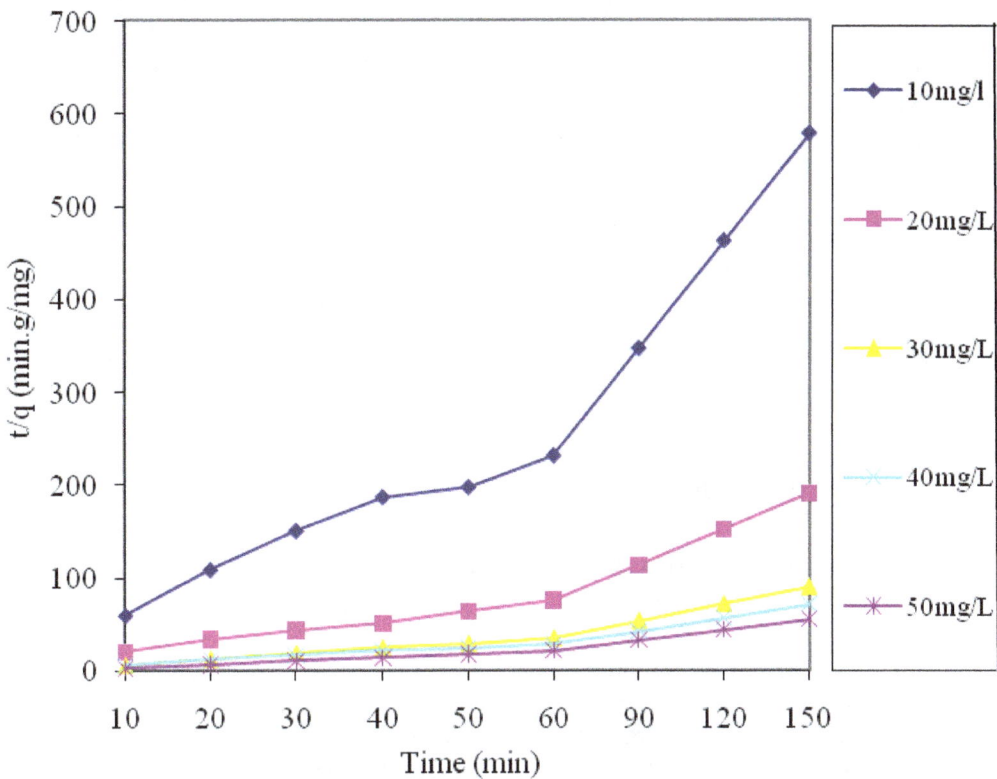

Figure 5. Pseudo-second-order sorption kinetics of fast orange dye by *Laurencia papillosa* at different initial dye concentrations (adsorbent dose = 2 g, temperature=25 ± 2°C).

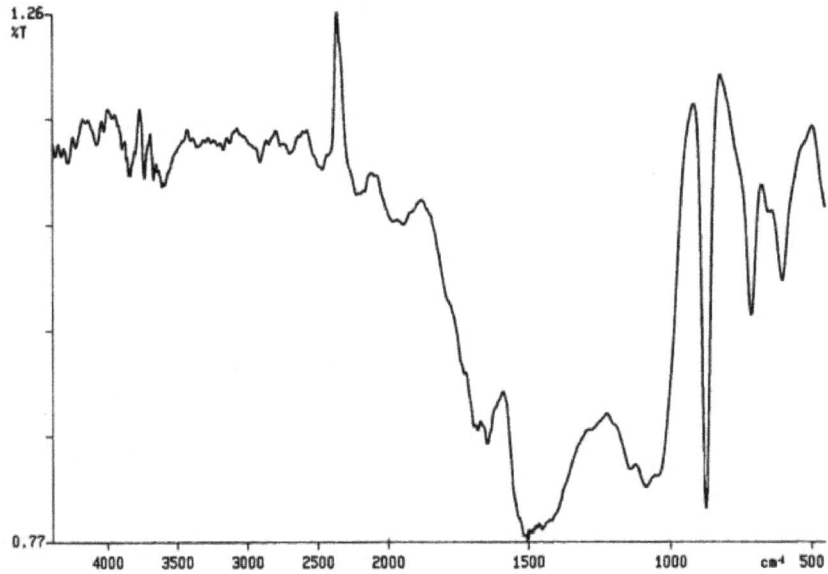

(a)

(b)

Figure 6. Changes in the spectra of the dried *Laurencia papillosa* before (a) and after (b) fast orange adsorption.

present and allows identification of different functionalities on the cell surface. The assignment of FTIR bands and detailed spectroscopy for the dried pure and treated alga are summarized in Figure 6. The functional groups on algal surface exhibits adsorption bands that ranged between 3250 to 3700, 2400 to 3300, 3300 to 3500, 1050 to 1300, 1040 to 1200, 1670 to 1780, 550 to 650 and 2500 to 3100 which indicate the presence of O-H, COOH, NH_2, C-O, S=O, C=O, S-O and C-H groups respectively. In comparison between pure and treated algal biomass, it was observed that there was a shift in wave number of dominant peaks associated with the loaded dye. This shift in the wavelength showed that there was a dye binding process taking place on the surface of the alga

Table 1. Pseudo second order rate constants at various initial dye concentrations.

Dye concentration(mg/L)	Qe (mg/g)	k2 (g/mg min)	R2
10	0.259	0.0644	1.022
20	0.786	0.0212	1.122
30	1.659	0.01	0.904
40	2.111	0.0078	1.002
50	2.684	0.0062	1.018

Table 2. The effect of particle size on the adsorption rate of fast orange onto Laurencia papillosa.

Particle size (μm)	50 - 100	100 - 150	150 - 200	200 - 250	250 - 300
Dye removal percentage (%)	65.3	63.7	59.5	43.3	39.2

(Matheickal, 1998). The extent of band shifting gives an indication of the degree of interaction of functional groups with dye and once equilibrium had been achieved, no further band shifting was observed in the FTIR spectra. Definitely, all the dominant functional groups seemed to play an important role for dye sorption as a shift in the wavelengths was always found.

Effect of particle size on adsorption

The effect of sorbent particle size on the rate of fast orange adsorption was studied in the range of 50 to 100, 100 to 150, 150 to 200, 200 to 250 and 250 to 300 μm keeping the other parameters as constant [dye concentration (20 mg/l); adsorbent dose (2 g); contact time (60 min) and temperature (25±2°C)]. Significant increase in dye adsorption was observed with the decrease in sorbent particle size (Table 2). For larger sorbent particle size (200 to 300 μm), the internal surface area of the particles may not be utilized for adsorption. However, for smaller sorbent particle size (50 to 200 μm), the increase in the dye adsorption is associated with large surface area of the particles. This showed that the grind dried biomass more rapidly adsorbed the dye ions, and the equilibrium was reached faster than those achieved with the whole algal thallus. This was because particles with smaller size allowed a faster contact between the dye molecules and the binding sites (Doğan et al., 2009).

Effect of pH on adsorption

The effect of pH on adsorption of dye was investigated over a range of pH values from 1 to 10 under constant parameters [dye concentration (20 mg/l); adsorbent dose (2 g); contact time (60 min) and temperature (25±2°C)]. The pH was adjusted using 0.1 N HCl and 0.1 N NaOH. Magnetic stirrer was used to agitate the solution continuously. The removal percentages of fast orange dye at different chosen pH are shown in Figure 7. The results show that the adsorption of dye on the biomass surface is controlled by ionic attraction. When pH value was raised from 1 to 5, the adsorption capacity was enhanced significantly from 28.7 to 65.7% and then the dye removal percentages were not significantly altered beyond pH 5. As pH decreased, the number of negatively charged adsorbent sites decreased and the number of positively charged surface sites increased, which did not favor the adsorption of positively charged dye cations due to electrostatic repulsion. Also, lower adsorption of fast orange at acidic pH is due to the presence of excess H+ ions competing with dye cations for the adsorption sites. Similar findings were reported by many researchers (Doğan et al., 2004; Wang et al., 2005; Vadivelan and Vasanth Kumar, 2005; Bhattacharyya and Sharma, 2005; Hamdaoui, 2006). At higher pH, the surface of biomass gets negatively charged, which enhances the positively charged dye cations through electrostatic force of attraction. Higher uptakes obtained at lower pH values may be due to the electrostatic attractions between these negatively charged dye anions and positively charged cell surface. Hydrogen ion also acts as a bridging ligand between the algal cell wall and the dye molecule (Srinivasan and Viraraghavan, 2010)

Conclusion

Dead algal biomass L. papillosa was considered to serve potentially as sorbent material. The maximum removal percentage was 67.08% and the adsorption equilibrium took place within 60 min. A relatively high correlation coefficient (R=1.002) implied that the pseudo-second-order kinetic model was encouraging for the fast orange dye adsorption on L. papillosa. FTIR results in the current work offered the possibility of the coupling between the dye species and the functional groups on the algal surface. Small particle size was recommended for optimum adsorption process. At a pH of 5, the maximum

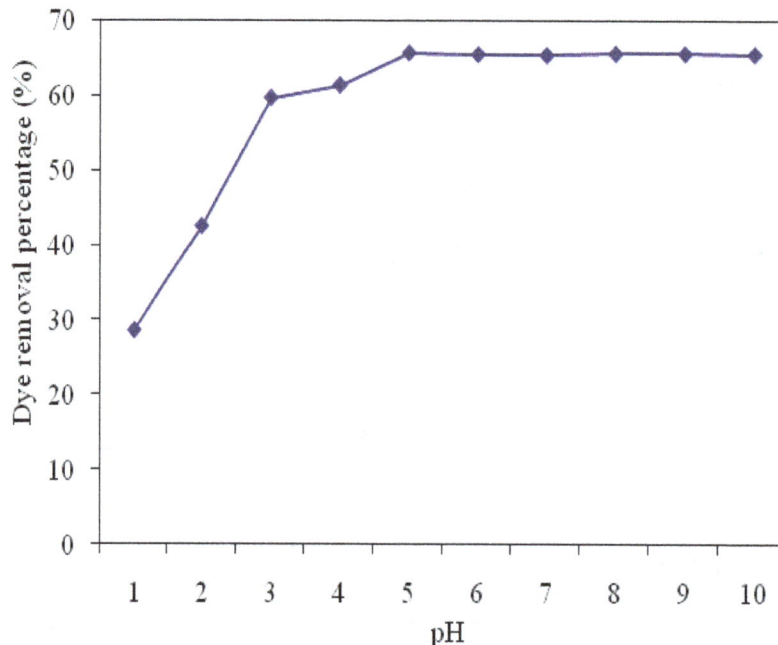

Figure 7. Effect of pH on adsorption of dye by non-living biomass *Laurencia papillosa* (dye concentration=20 mg/l, contact time = 60 min adsorbent dose = 2 g, temperature= 25±2°C).

removal capacity of the biomass was 65.7%. Owing to lower equilibrium time and considerable adsorption capacity, the dried red marine alga *L. papillosa* could be used as an alternative low-cost material for dye adsorption.

REFERENCES

Aksu Z, Acikel Ű, Kutsal T (1997). Application of multicomponent adsorption isotherms to simultaneous biosorption of iron (III) and chromium (VI) on *C. vulgaris*. J. Chem. Tech. Biotechnol. 70:368-378.

Aksu Z, Acikel Ű, Kutsal T (1999). Investigation of simultaneous biosorption of copper (II) and chromium (VI) on dried *Chlorella vulgaris* from binary metal mixtures: application of multicomponent adsorption isotherms. Separ. Sci. Technol. 34:501-524.

Banat IM, Nigam P, Singh D, Marchant R (1996). Microbial decolorization of textile-dye containing effluents: a review. Bioresour. Technol. 58:217-227.

Bhattacharyya KG, Sharma A (2004). Azadirachta indica leaf powder as an effective biosorbent for dyes: a case study with aqueous Congo Red solutions. J. Environ. Manage. 71(3):217-229.

Bhattacharyya KG, Sharma A (2005). Kinetics and thermodynamics of methylene blue adsorption on Neem (*Azadirachta indica*) leaf powder. Dyes Pigments. 65:51-59.

Bukallah SB, Rauf MA, Al Ali SS (2007). Removal of methylene blue from aqueous solution by adsorption on sand. Dyes Pigments 74:85-87.

Cengiz S, Cavas L (2008). Removal of methylene blue by invasive marine seaweed: *Caulerpa racemosa* var. cylindracea. Bioresour. Technol. 99:2357-2363.

Chen KC, Wua JY, Huang CC, Liang YM, Hwang SCJ (2003). Decolorization of azo dye using PVA-immobilized microorganisms. J. Biotechnol. 101:241-252.

Dabrowski A (2001). Adsorption from theory to practice. Adv. Colloid Interface 93:135-224.

Davis TA, Volesky B, Mucci A (2003). A review of the biochemistry o' heavy metal biosorption by brown algae. Water Res. 37:4311-4330.

Döğan M, Abak H, Alkan M (2009). Adsorption of methylene blue onto hazelnut shell: Kinetics, mechanism and activation parameters. J. Hazard. Mater. 164:172-181.

Doğan M, Alkan M (2003). Removal of methyl violet from aqueous solutions by perlite. J. Colloid Interface Sci. 267:32-41.

Doğan M, Alkan M, Turkyilmaz A, Ozdemir Y (2004). Kinetics and mechanism of removal of methylene blue by adsorption onto perlite. J. Hazard. Mater. B. 109:141-148.

Dönmez G, Aksu Z, Özturk A, Kutsal T (1999). A comparative study or heavy metal biosorption characteristics of some algae. Process Biochem. 34:885-892.

Ertugrul S, Sam NO, Dönmez G (2009). Treatment of dye (Remazo Blue) and heavy metals using yeast cells with the purpose o' managing polluted textile wastewaters. Ecol. Eng. 35:128-134.

Figueira MM, Volesky B, Mathieu HJ (1999). Instrumental analysis study of iron species biosorption by *Sargassum* biomass. Environ. Sci. Technol. 33(11):1840-1846.

Gupta VK, Ali I, Suhas Mohan D (2003). Equilibrium uptake and sorption dynamics for the removal of a basic dye (basic red) using low-cost adsorbents. J. Colloid Interface Sci. 265:257-264.

Hamdaoui O (2006). Batch study of liquid-phase adsorption o' methylene blue using cedar sawdust and crushed brick. J. Hazard. Mater. B. 135:264-273.

Ho YS (2003). Removal of copper ions from aqueous solution by tree fern. Water Res. 37(10):2323-2330.

Ho YS, McKay G (1999). Pseudo second-order model for sorption processes. Process Biochem. 34:451-465.

Hoda N, Bayram E, Ayranci E (2006). Kinetic and equilibrium studies or the removal of acid dyes from aqueous solutions by adsorption onto activated carbon cloth. J. Hazard. Mater. B. 137:344-351.

Holan ZR, Volesky B (1994). Biosorption of lead and nickel by biomass of marine algae. Biotech. Bioeng. 43:1001-1009.

Kaewsarn P, Yu Q (2001). Cadmium removal from aqueous solutions by retreated biomass of marine algae *Padina* sp.. Environ. Pollut. 112:209-213.

Kaushik P, Malik A (2009). Fungal dye decolorization: recent advances and future potential. Environ. Int. 35:127-141.

Klimmek S, Stan HJ, Wilke A, Bunke G, Buchholz R (2001). Comparative analysis of the biosorption of cadmium, lead, nickel and zinc by algae. Environ. Sci. Technol. 35:4283-4288.

Kousha M, Daneshvar E, Sohrabi MS, Koutahzadeh N, Khataee AR (2012). Optimization of C.I. Acid black 1 biosorption by *Cystoseira indica* and *Gracilaria persica* biomasses from aqueous solutions. Int. Biodeterior. Biodegrad. 67:56-63.

Lagergren S (1898). Zur theorie der sogenannten adsorption geloster stoffe, Kungliga Svenska Vetenskapsakademiens, Handlingar. 24:1-39.

Lim SL, Chu WL, Phang SM (2010). Use of *Chlorella vulgaris* for bioremediation of textile wastewater. Bioresour. Technol. 101:7314-7322.

Mahmoud A S, Ghaly AE, Brooks SL (2007). Influence of Temperature and pH on the Stability and Colorimetric Measurement of Textile Dyes. Am. J. Biotechnol. Biochem. 3(1):33-41.

Marungrueng K, Pavasant P (2006). Removal of basic dye (Astrazon Blue FGRL) using macroalga *Caulerpa lentillifera*. J. Environ. Manage. 78:268-274.

Marungrueng K, Pavasant P (2007). High performance biosorbent (*Caulerpa lentillifera*) for basic dye removal. Bioresour. Technol. 98:1567-1572.

Matheickal JT, Yu Q (1999). Biosorption of lead (II) and copper (II) from aqueous solutions by pre-treated biomass of Australian marine algae. Bioresour. Technol. 69:223-229.

Matheickal JT, Yu Q, Woodburn GM (1999). Biosorption of cadmium (II) from aqueous solutions by pre-treated biomass of marine alga *Durvillaea potatorum*. Water Res. 33: 335-342.

Matheickal TJ (1998). Biosorption of heavy metals from waste water using macro-algae *Durvillaea potatorum* and *Ecklonia radiata*. Ph.D. dissertation, Environmental Engineering, Griffith University, Queensland.

McKay G, Ho YS (1999). The sorption of lead (II) on peat. Water Res. 33:578-584.

Mishra A, Kumar S, Kumar Pandey A (2011). Laccase production and simultaneous decolorization of synthetic dyes in unique inexpensive medium by new isolates of white rot fungus. Int. Biodeterior. Biodegrad. 65:487-493.

Muthuraman G, Palanivelu K (2006).Transport of textile dye in vegetable oils based supported liquid membrane. Dyes Pigments 70:99-104.

Ncibi MC, Mahjoub B, Seffen M (2007). Kinetic and equilibrium studies of methylene blue biosorption by *Posidonia oceanica* (L.) fibres. J. Hazard. Mater. 139:280-285.

Park D, Yun YS, Cho HY, Park JM (2004). Chromium biosorption by thermally treated biomass of the brown seaweed, *Ecklonia* sp. Ind. Eng. Chem. Res. 43(26):8226-8232.

Pengthamkeerati P, Satapanajaru T, Singchan O (2008). Sorption of reactive dye from aqueous solution on biomass fly ash. J. Hazard. Mater. 153:1149-1156.

Phugare S, Patil P, Govindwar S, Jadhav J (2010). Exploitation of yeast biomass generated as a waste product of distillery industry for remediation of textile industry effluent. Int. Biodeterior. Biodegrad. 64:716-726.

Prescott LM, Harley JP, Klein DA (2002). Microbiology. McGraw-Hill Science/Engineering/Math. ed Fifth.

Senthilkumaar S, Kalaamani P, Porkodi K, Varadarajan PR, Subburaam CV (2006). Adsorption of dissolved Reactive red dye from aqueous phase onto activated carbon prepared from agricultural waste. Bioresour. Technol. 97:1618-1625.

Sheng HL, Chi ML (2003). Treatment of textile waste effluents by ozonation and chemical coagulation. Water Res. 27:1743-1748.

Sheng PX. Ting YP, Chen JP, Hong L (2004). Sorption of lead, copper, cadmium, zinc, and nickel by marine algal biomass: characterization of biosorptive capacity and investigation of mechanisms. J. Colloid Interface Sci. 275(1):131-141.

Srinivasan A, Viraraghavan T (2010). Decolorization of dye wastewaters by biosorbents: a review. J. Environ. Manage. 91:1915-1929.

Uddin T, Islam A, Mahmud S, Rukanuzzaman (2009). Adsorptive removal of methylene blue by tea waste. J. Hazard. Mater. 164:53-60.

Vadivelan V, Vasanth Kumar K (2005). Equilibrium, kinetics, mechanism and process design for the sorption of methylene blue onto rice husk. J. Colloid Interface Sci. 286:90-100.

Veglio F, Beolchini F (1997). Removal of metals by biosorption: a review. Hydrometallurgy 44:301-316.

Volesky B (1990). in: Volesky B (Ed.), Biosorption of Heavy Metals, CRC Press, Boca Raton, FL, p. 3.

Wang S, Li L, Wu H, Zhu ZH (2005). Unburned carbon as a low-cost adsorbent for treatment of methylene blue-containing wastewater. J. Colloid Interface Sci. 292:336-343.

Yang Y, Hu H, Wang G, Li Z,Wang B, Jia X, Zhao Y (2011). Removal of malachite green from aqueous solution by immobilized Pseudomonas sp. DY1 with *Aspergillus oryzae*. Int. Biodeterior. Biodegrad. 65:429-434.

Yu Q, Matheickal JT, Yin P, Kaewsarn P (1999). Heavy metal uptake capacities of common marine macro algal biomass. Water Res. 33:1534-1537.

7

Protective effect of zinc against cadmium toxicity on pregnant rats and their fetuses at morphological, physiological and molecular level

Ashraf El-Sayed[1,2]*, Salem M. Salem[2], Amany A. El-Garhy[3], Zeinab A. Rahman[3] and Asmaa M. Kandil[3]

[1]Cairo University Research Park (CURP), Faculty of Agriculture, Cairo University, 12613 Giza, Egypt.
[2]Department of Animal Production, Faculty of Agriculture Cairo University, 12613 Giza, Egypt.
[3]Department of Pharmacology, National Organization for Drug Control and Research, Giza, Egypt.

Cadmium is a potent teratogen in laboratory animals, causing exencephaly when administered at early stages of development. Due to its heterogenicity with respect to molecular targets, the mechanisms behind cadmium toxicity are not well understood. In the present study, 40 pregnant rats (Sprague-Dawley) were divided into four groups (10 each); first group served as the control (G1), the second group (G2) received 61.3 mg/kg cadmium chloride daily from 7th to 16th day of gestation (organogenesis period) by oral tube. Group 3 (G3) was administrated a solution of 25 mg/kg zinc chloride orally from the 1st day to 20th day of pregnancy. Group 4 were administrated a solution of cadmium chloride (61.3 mg/kg) and zinc chloride (25 mg /kg) daily from the 7th to16th day of gestation. Maternal body weights were measured on gestational day 0, 6, 9, 12, 15 and 20. At the 20th day of gestation, blood samples were collected from the eye, using orbital sinus technique. Serum aspartate transaminase (AST) and alanine transaminase (ALT) were determined calorimetrically and serum, urea and creatinine were determined. All of the pregnant rats were sacrificed by ether anaesthesia at the 20th day of gestation and foetuses were removed from the uterus. The implantation sites, corpora lutea, living, dead and reabsorbed foetuses were counted and recorded. Liver of pregnant rats and their fetuses were used to isolate a total RNA for quantification of *Msx1*, *Cx43*, *Bcl2* and *Bax* genes. The results show the toxic effect of Cd on the pregnant rats and their fetuses, at morphological, physiological and molecular level but, zinc has a very effective protection against cadmium-induced developmental toxicity.

Key words: Cadmium, zinc, rat, organogenesis, gene expression.

INTRODUCTION

Industrial development has brought man into contact with several persistent chemicals, including heavy metals, such as lead, mercury and cadmium (Cd). Cd has been reported to produce several toxic effects in animals and man, while peculiar accumulation kinetics in the kidney cortex mammals has been reported by many authors. Absorption and accumulation of cadmium in tissues is determined by a wide range of factors, like nutritional and vitamin status, age and sex (Salvatori et al., 2004). Derived from natural and anthropogenic sources, widespread environmental exposure to arsenic (As) and cadmium (Cd) remains of public health concern due to their potential to cause adverse effects in the human population. In animal models, Cd is a well characterized teratogens inducing embryo-toxicity, including growth effects, mortality and a range of congenital malformations (Salvatori et al., 2004).

Cadmium (Cd) is particularly important as it is the 7th highest priority hazardous substance according to the agency for toxic substances and disease registry. Key sources of Cd in the environment include industrial production of pigments, plastic stabilizers, alloys, nickel-cadmium batteries as well improper discharge of many manufactured products (IARC, 1993). In the Earth's crust, Cd has an average concentration of about 0.1 to 0.2 mg kg^{-1} (Ursínyová and Hladíková, 2000; Lalor, 2008), contaminating the air, food and water, and so increasing the routes of exposure to animals such as ingestion and inhalation (IARC, 1993).

Exposure to this metal can occur in the workplace and in the natural environment because it is utilized in a number of industrial practices and is a contaminant of the environment and dietary products (World Health Organization, 1992). Cd toxicity in humans (Jarup et al., 1998) and experimental animals (Sharma et al., 1991; Brzóska and Moniuszko-Jakoniuk, 2001) has been widely studied and reported. Dietary Cd exposure of mice and rats during pregnancy results in anemia and reduced body weight of pups at birth (Webster, 1976) and distinct changes in trace metal metabolism in pups (Kuriwaki et al., 2005). Cadmium produces oxidative modifications of DNA, such as the formation of 8-hydroxydeoxy-guanosine, and the generation of strand breaks in different cell types, for example, liver and kidney cells (Forrester et al., 2000; Littlefield and Hass, 1995). Oxidative DNA damage produced by cadmium has been associated with an increased production of reactive oxygen species (ROS) (Ochi et al., 1987), and interactions between this metal and DNA repair enzymes (Assmus et al., 2000; Waalkes, 2000). In human lymphoma cells, cadmium has been shown to cause apoptosis by two independent pathways: the Ca^{2+}-cal-pain and the caspase-mitochondria pathways (Li et al., 2000), indicating that apoptosis could play an important role in acute and chronic toxicity from this metal.

It is well known that the metabolism and toxicity of Cd may be modified by many factors, including substances essential for life (Berglund et al., 1994; Brzóska and Moniuszko-Jakoniuk, 1998) as well as very toxic chemical compounds (Brus et al., 1995; Moniuszko-Jakoniuk et al., 2001). One of these substances is zinc. Microelements such as zinc (Zn) play an important role in metabolic pathways affected by Cd. Disturbances in metabolism of these metals in humans and experimental animals were observed after chronic Cd intoxication (Oishi et al., 2000).

Zinc supplementation prior to cadmium administration prevents several of the effects observed when cadmium is added alone (Dreosti, 2001). Thus, it has been shown that zinc inhibits the apoptotic protease caspase-3 (Truong-Tran et al., 2001), stabilizes the structure of p53 and DNA repair proteins (Chai et al., 1999), acts as an antioxidant by decreasing ROS production in cell cultures (Dally and Hartwig, 1997; Szuster-Ciesielska et al., 2000), and prevents the gross teratogenic effects of cadmium by restoring normal development (Warner et al., 1984). One approach to investigating complex developmental processes and coordinating them with genetic regulation has been to disrupt morphogenesis with specific teratogens and study their consequences at the molecular and morphological levels.

In the present paper, we studied the abnormal morphology in dam and foetus induced by cadmium, in addition, alteration in the expression levels of selected genes by using the Real-PCR technique. The genes under study (MSX1, CX43, Bcl-2 and Bax) were selected according to their implication in organogenesis and in the apoptotic pathway.

MATERIALS AND METHODS

Adult male and female Sprague Dawley rats were used in this study, with weight of 150 to 200 g, obtained from the animal house of National Organization for drug control and research (NODCAR). Animals were kept under standard conditions and allowed free access to food and water.

The reproductive cycles of every rat which were kept in cages at room temperature, were followed for 15 days. Every female rat determined to be in oestrus or pro-oestrus phase of their cycles were mated with male rats in the same cage for one day and those which had sperm in their vaginal smears were considered to be zero day of pregnancies.

40 pregnant rats (Sprague-Dawley) were divided into four groups (10 each); the first group served as the control (G1), the second group (G2) received 61.3 mg/kg cadmium chloride (Cadmium chloride as monohydrate obtained from LOBA Chemise) dissolved in distilled water daily from 7th to 16th day of gestation (organogenesis period) by oral tube. Group 3 (G3) was administrated a solution of 25 mg/kg zinc chloride (zinc chloride purified obtained from LOBA Chemise) dissolved in distilled water orally from 1st day to 20th day of pregnancy. Group 4 were administrated a solution of cadmium chloride (61.3 mg/kg) and zinc chloride (25 mg /kg) daily from 7th to16th day of gestation.

All pregnant females were observed daily throughout gestation for mortality and general appearance. Maternal body weights were measured on gestational day 0, 6, 9, 12, 15 and 20. At the 20th day of gestation, blood samples were collected from the eye, using orbital sinus technique (Sanford, 1954). The blood was collected and allowed to clot, then serum was separated by centrifugation at 3000 rpm for 20 min and the clear non haemolysed serum was collected, divided into several aliquots and stored at -20°C until assayed. Serum aspartate transaminase (AST) and alanine transaminase (ALT) were determined calorimetric according to the method of Reitman and Frankel (1957). Serum, urea and creatinine were determined according to Bartel (1972). All of the pregnant rats were sacrificed by ether anaesthesia at the 20th day of gestation and foetuses were removed from the uterus. The implantation sites, corpora lutea, living, dead and reabsorbed foetuses were counted

Table 1. Oligonucleotide primers used for real-time RT-PCR analysis.

Gene	Primer	Product size (bp)
GAPDH	F: GGCTCTCTGCTCCTCCCTGTTCTA R: TGCCGTTGAACTTGCCGTGG	242
Msx1	F: GCCTGCACCCTACGCAAGCA R: AGCAGGCGGCAACATTGGCT	261
Cx43	F: TCCTTTGACTTCAGCCTCCAAGGAG R: GCAGACGTTTTCGCAGCCAGG	279
Bcl2	F: CTG GTG GAC AAC ATC GCT CTG R: GGT CTG CTG ACC TCA CTT GTG	228
Bax	F: TTCATC CAGGAT CGA GCA GA R: GCA AAG TAG AAG GCA ACG	263

and recorded. All living foetuses were weighed and evaluated for externally visible abnormalities. 50% of the foetuses were fixed in 96% ethanol and their soft tissues were removed in 1.0% KOH solution. After staining with Alcian blue- Alizarin Red-S combined technique, skeletal system was examined (Mcleod, 1980). Data obtained from the treated and the control groups were compared statistically by analysis of variance (ANOVA) test.

RNA isolation and real-time reverse transcription polymerase chain reaction

Total RNA was extracted from liver of mothers and their fetuses using analytic jena bio solution (innuPREP RNA Mini Kit. Germany). For reverse transcription (RT), first strand complementary DNA (cDNA) was synthesized from RNA by using a cDNA synthesis kit (RevertAidTm First Strand cDNA Synthesis Kit, Fermentas,) according to the manufacturer's instructions. After RT at 42°C for 60 min, polymerase chain reaction (PCR) was performed using a Jena Bioscience PCR-101 Taq Master Mix (Jena bioscience, Germany) according to the manufacturer's protocol. The specific primer pairs used in this study are listed in Table 1. Serially, diluted cDNA samples were used as standards. After an initial denaturation step of 5 min at 95°C, 35 cycles of amplification for Msx1 and Cx43, Bcl2 and Bax primer pair, were carried out. Each cycle included a denaturation step, 30 s at 95°C; an annealing step, 30 s at 56°C; and an elongation step, 30 s at 72°C. Final elongation temperature was 72°C for 5 min. Relative levels of gene expression were measured by QuantiTect SYBR Green PCR kit (Qiagen, Clinilab, Egypt) according to the manufacturer's instructions using Mx3000 instrument (Stratagene). The expression levels of Msx1, Cx43, Bcl2 and Bax genes were normalized to the level of *GAPDH* gene expression in each sample.

RESULTS

Morphological parameters

Among the experimental groups, food intake decreased in group treated with cadmium (G2) than that of the other groups. The average maternal body weight showed a steady increase during the gestation period, while the rate of increase during the gestation period was found to be relatively less in groups G2 and G4 compared to control and G3 group (Table 2). No abortion was recorded among mothers of the control and G3 groups. However, the rate of abortion in G2 was high than that of G4. The effect of cadmium on the pregnant rats was indicated by reduction in the uterine weight of pregnant rats and increase in the resorption rates of fetuses (Table 2, Figures 1a, b and c). The uterine weight of G2 was very low compared to that of G4. Cd resulted in abortion and resorption. By co-administration of zinc, the toxicity of Cd in pregnant rats decreased. Body weight of pregnant rats was reduced by Cd while relative weight of liver and kidney was increased (Table 5).

In the present study, Cd reduced growth parameters of the offspring (Table 3) and increased percentage of fetal malformation (paralysis of forelimbs and external hematomas) by co-administration of zinc; no significant decrease in growth parameters were observed meanwhile percentage of malformation decreased in fetuses

Skeletal examination

Fetal skeletal abnormalities were most obvious during embryogenesis period (7th to 14th day of gestation) (Figures 2a, b, c and d; Figures 3a and b).

The skeletal defects observed in fetuses included incomplete ossification of skull bones, sternum, ribs, vertebrae, forelimbs bones, pelvic girdle and hind limbs bones (Figure 2c).

The major skeletal defects were observed mainly in pelvic girdle and hind limbs bones in the form of shortness and partial ossification of ilium ischium, pubis, femur, tibia and fibula (Figure 3b). Shortness of the 13th

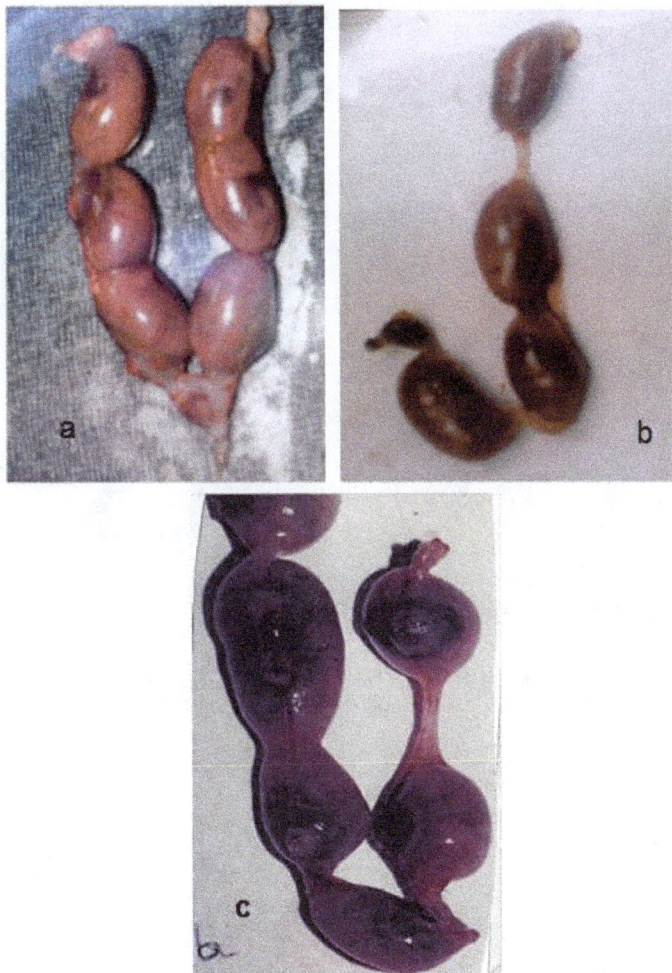

Figure 1. Uterus of control pregnant rat (a), uterus of pregnant rat treated with Cd (b) and uterus of pregnant rat treated with Cd and Zinc (c).

rib was observed in some fetuses (Figure 2c). Incomplete ossifications of vertebral column were observed mainly in sacral and caudal vertebrae (Figure 3b). Forelimbs bones were less affected than hind limbs bones. Abnormalities detected in forelimbs bones were in the form of partial ossifications of hummers, ulna, radius and missed ossifications of metacarpals and phalanges. However, the group treated with Cd and zinc showed that zinc improved the defects of skeletal system.

Liver and kidney function

A significant increase in blood urea and serum creatinine was found in pregnant rats administrated cadmium (G2). By co-administration of zinc, a significant decrease was observed (Table 4). The same trend was observed in the liver enzymes in cadmium treated group (G2) which was slightly decreased by co-administration of zinc (G4) (Table 4).

Quantitative real-time PCR confirmation of selected transcripts

Four genes (*Msx1*, *Cx43*, *Bcl2* and *Bax*) were quantified in independent samples from the four groups (G1, G2, G3 and G4) by quantitative real-time PCR (Figures 4 and 5).

Regarding the expression level in the cadmium treated pregnant rats (Figure 4), the lowest relative expression of Cx43 (0.03) was found in group G2, followed by Bcl2 and Msx1 (0.07, 0.23) respectively. On the opposite direction, Bax recorded the highest level (0.23) in this group compared to the control (G1). The effect of zinc (G3) was very clear in our experiment, where it resulted in a high expression for Msx1, Cx43 and Bcl2 and reduced the expression for Bax to minimum level compared to control group. The protective effect of zinc against cadmium (G4) was greater in Cx43 (3.6 fold), followed by Msx1 (2 fold) and Bcl2 (1.8 fold) and finally Bax (0.65 fold) compared to cadmium group (G2).

Figure 2. Skeletal system of fetus at the 20th day of gestation. **a.** Control. **b.** Maternally treated with zinc during organogenesis period. **c,** Maternally treated with Cd. **d.** Maternally treated with Cd and zinc.

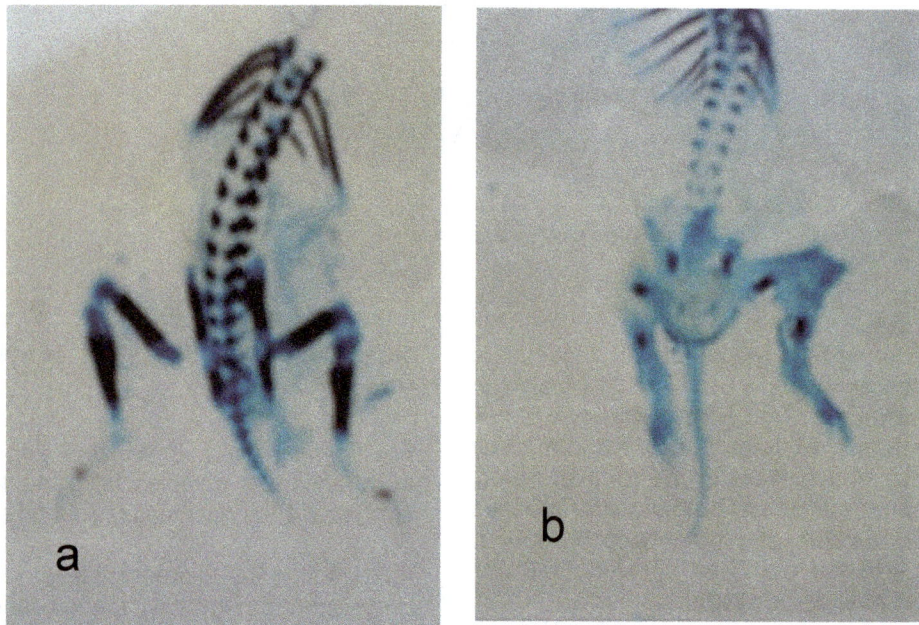

Figure 3. Hind limb and vertebral column of control fetus (a) and of fetus maternally treated with Cd (b).

In the foetuses samples (Figure 5), cadmium has resulted in the same trend as found in mothers, where the lowest relative expression of Cx43 (0.08) was found in group G2, followed by Bcl2 and Msx1 (0.16 and 0.42), respectively. On the opposite direction, Bax recorded the highest level (0.70) in this group compared to the control (G1). Regarding the effect of zinc (G3), it has shown also the same trend as a high expression for Msx1, Cx43 and Bcl2 was observed and reduction in the expression for Bax to minimum level was observed compared to control group. The protective effect of zinc against cadmium (G4) showed a slightly different trend compared to mothers,

Table 2. The effects of Cadmium and /or zinc on the pregnant rats.

Group	No. of pregnant	No. of aborted (%)	No. of sacrificed (%)	Average wt. of pregnant rats			Uteri			
				At 1st day	At 20th day	Average increase (%)	Total No. of uteri	Without resorption (%)	Partially resorbed (%)	Average wt. of uteri ±SE
Control G1	10	-	10 (100)	150.2	208	45.8±0.59 (30.5)	10	10 (100)	-	35.53±2.84
Cadmium G2	10	2 (20)	8 (80)	149.8	179.7	29.9±0.43 (20)	8	4 (50)	4 (50)	20±2.81
Zinc G3	10	-	10 (100)	139.8	189.7	49.9±0.42 (35.7)	10	9 (90)	1 (10)	39.5±1.53
Zinc+ cadmium G4	10	1 (10)	9 (90)	169.5	201.3	31.8±0.23 (18.76)	9	6 (66.67)	3 (33.33)	30.04±1.09

Table 3. The effects of zinc and /or cadmium on the fetuses.

Group	Number of sacrified	Number of implantation/mother	% of living fetuses	% of resorbed fetuses	Average fetal body weight (g)	Average fetal length (mm)	% of malformed fetuses	% of hematomas
Control	10	74 (7.4)	74 (100)	-	3.97±0.012	4.1±0.002	-	-
Zinc	10	77 (7.7)	75 (97.4)	2 (2.6)	3.89±0.003	3.95± 0.002	9 (12)	6
Cadmium	8	53 (6.625)	40 (75.47)	13 (24.53)	3.01±0.007	3.23±0.004	21 (52)	24
Cadmium + Zinc	9	62 (6.88)	56 (90.32)	6 (9.68)	3.51±0.02	3.53±0.01	11 (19.64)	13

where the expression was greater in Cx43 (2.5 fold), followed by Bcl2 (2.25 fold) and Bax (0.51 fold) and finally Msx1 (1.6 fold) compared to cadmium group (G2).

We can conclude that, the effect of cadmium and the protective effect of zinc at molecular level are transferred from treated mothers to their foetuses and follow the same trend.

DISCUSSION

Cd induced nephrotoxicity in pregnant rats which was reduced by co-administration of zinc. In rabbits, Cd treatment resulted in increase of blood urea. Simultaneous administration of zinc prevented Cd induced uremia (Stowe et al., 1972). In uterus, exposure of cadmium leads to toxic renal effects in adult offspring (Jacquillet et al., 2007). Chronic Cd exposure can cause renal proximal tubular dysfunction resulting from the release of Cd metallothionein from the liver and its accumulation and degradation in the renal tubular epithetlial cells. Pretreatment with zinc can protect against acute Cd nephrotoxicity (Tang et al., 1998). Messaoudi et al. (2009) demonstrated the beneficial effects of selenium and zinc combination in the treatment of Cd nephrotoxicity. In Wistar rats, cadmium reaches the placenta or embryo at an organogenetically sensitive time (day 9 of gestation), and zinc may protect the embryo by decreasing the exposure to cadmium this time (García and González, 2010). Cadmium administrated to pregnant mice increased primary DNA damage and activated the apoptotic pathway. These effects could be ameliorated by zinc pretreatment so the mechanism of cadmium teratogenicity could be related to zinc metabolism (Fernández et al., 2003).

Salvatori et al. (2004) showed that Cd treatment during organogenesis was not able to induce maternal toxicity; induced external malformations; increased significantly fetus anomalies and malformations, with reduced metacarpus ossification and cleft palate. Haldsrud and Krokje (2009) found that exposure to high concentration of combination of Copper and cadmium produced a significant increase in the occurrence of DNA strand break and addition of low zinc to the mixture of Cadmium and copper restored DNA damage level back to that of control.

The effects of Cd on prenatal hepatocytes was studied by Bruscalupi et al. (2009) which found that fetal hepatocyte are less sensitive to Cd toxicity and the adverse effects of the metal are always better counteracted by fetal cells. Abrahim et al. (2010) added more proof that Cd exposure has a genotoxic effect and early detection of Cd

Table 4. Effects of zinc and /or cadmium on liver and kidney functions of pregnant rats (mean ±SE).

Group	Liver	Kidney	Spleen
Control	3.64 ± 0.065	0.621 ± 0.012	0.426 ± 0.076
Cadmium	4.95 ± 0.061	0.701 ± 0.021	0.367 ± 0.043
Zinc	3.73 ± 0.041	0.602 ± 0.02	0.432 ± 0.034
Cadmium+ zinc	3.95 ± 0.053	0.671 ± 0.031	0.0421 ± 0.012

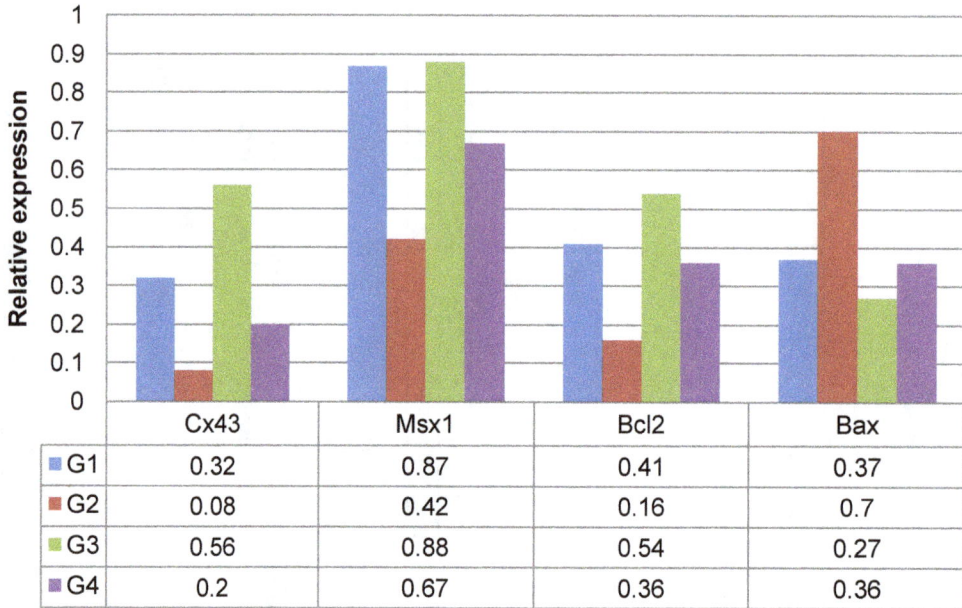

	Cx43	Msx1	Bcl2	Bax
G1	0.32	0.87	0.41	0.37
G2	0.08	0.42	0.16	0.7
G3	0.56	0.88	0.54	0.27
G4	0.2	0.67	0.36	0.36

Figure 4. Expression level of selected genes in pregnant rats at different groups; control (G1), zinc treated (G2), cadmium treated (G3) and both cadmium and zinc treated (G4).

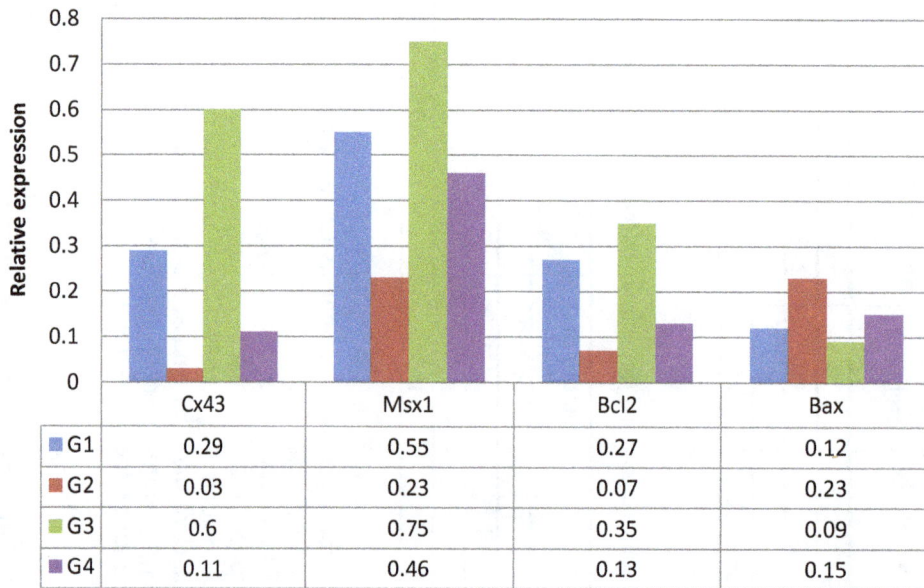

	Cx43	Msx1	Bcl2	Bax
G1	0.29	0.55	0.27	0.12
G2	0.03	0.23	0.07	0.23
G3	0.6	0.75	0.35	0.09
G4	0.11	0.46	0.13	0.15

Figure 5. Expression level of selected genes in rat fetuses of treated mothers at different groups; control (G1), zinc treated (G2), cadmium treated (G3) and both cadmium and zinc treated (G4).

induced mutagenecity is required.

Hepatic dysfunction has been observed after cadmium poisoning and liver function tests may be effective for diagnosis of cadmium effects. Increased activities of serum GOT and serum GPT have been found, and this in in accordance with the results of Nomiyama (1980) and Weigel et al. (1984) who studied the effects of 7.15 ppm dietary cadmium in male Wister rats for 40 and 60 days and found that that the activities of serum GOT and GPT were increased indicating disturbed hepatic functions. On the same side, Jihan et al. (2009) concluded that selenium and zinc protect the liver against oxidative stress induced by Cd.

Cd studies (GD8.0, Cd 4mg/kg) suggest that peak cellular morphological changes (increased pyknotic nuclei) that occur 10 to 12 h correlated with subsequent alterations in neural tube development (Webster and Messerle, 1980). Under these conditions, As and Cd exposures can affect DNA-damage, cell cycle arrest, oxidative stress and cell death pathways (Pulido and Parrish, 2003), yet the means by which these metals interfere with signaling pathways and subsequent morphogenesis remains unresolved. Investigations suggest additional effects such as glucose impairment and disruption in ion (Zn) or folic acid regulation also may be potential mechanisms of metal-induced embryo toxicity (Fernandez et al., 2007; Hill et al., 2009; Robinson et al., 2010).

Our data demonstrates the importance of Cx43 for limb development, as confirmed before using antisense oligonucleotides inhibition of Cx43 expression in the chick embryo which resulted in limb malformations, including truncation of the limb bud, fragmentation into two or more domains, or complete splitting of the limb bud into two or three branches (Becker et al., 1999; Green et al., 1994). These data implicate that Cx43 plays a very important role in osteogenic function and stated that, the genetic deletion of Cx43 resulted in skeletal ossification abnormalities. Therefore, the lack of Cx43 causes a generalized osteoblast dysfunction and this is in accordance with what was reported by Lecanda et al. (2000). Other role was found also for Msx1 as an impor-tant gene for embryo development as reported by Tesfaye et al. (2010) and El-Sayed et al. (2011). In ad-dition, Msx genes may modulate the regulation of type I collagen possibly affecting the formation of extracellular matrix (ECM) development (Dodig et al., 1996; Alappat et al., 2003). Moreover, mice with homozygous mutations in both Msx1 and Msx2 die in late gestation with severe craniofacial malformations, including exencephaly, cleft palate, agenesis of teeth, and unossified calvarial bones (Bei and Maas, 1998; Satokata et al., 1994). Our results concerning expression level of both Cx43 and Msx1 are in accordance with what was stated before about the importance of these genes in organogenesis and embryo development.

Bcl-2 and Bax produce mitochondrial-related proteins with antagonistic effects; the former having an anti-apoptotic activity and the latter a pro-apoptotic activity. Bcl-2 was significantly down regulated, more than three fold, as opposed to Bax, the expression levels of which were significantly up regulated in G2 compared to the control (G1). Interactions between Bcl-2 and Bax regulate cytochrome c release from mitochondria and establish baseline sensitivity to apoptotic stimuli. In this study we observed a significant downregulation of Bcl-2 and a concurrent upregulation of Bax at the transcription level. It is thus likely that a decreased Bcl-2/Bax ratio promotes apoptosis signaling activation (Oltvai et al., 1993).

The injection of zinc before cadmium treatment can keep the expression levels of the genes under study at basal levels. Zinc supplementation thus maintained the expression Bcl-2/Bax ratio (Fernández et al., 2003).

In general toxicological studies, four common predominant mechanisms have been proposed regarding metal toxicity: 1) alterations in cell proliferation pathways; 2) inhibition of major DNA repair systems; 3) interference with cellular redox regulation/generation of oxidative stress, and 4) induction of apoptosis or cell death (Pulido and Parrish, 2003).

Conclusion

From the results obtained in this study, it appears that Cd has toxic effects on pregnant rats by reduction of body weight, increased percentage of abortion and resorption and produced elevation of both liver and kidney functions. As regards to its toxic effects on the fetuses, Cd caused fetal growth retardation and congenital abnormalities. Moreover, from our findings, it could be suggested that, the supplementation with zinc during gestation in conditions of exposure of Cd, is an effective therapy which could reduce Cd induced toxicity.

In addition, we have clearly shown that, cadmium can be a potent inducer of primary DNA damage in embryonic cells, and that it can activate several transcription factors implicated in the apoptotic pathway, (Bcl-2, Bax) and organogenesis and embryo development (Cx43, Msx1). We have also shown that, treatment with zinc could ameliorate the effects induced by cadmium, supporting previous data implicating that the mechanisms of cadmium teratogenicity are in some way related to zinc homeostasis. However, future studies may be able to answer many such unanswered questions in the mechanism involved in the beneficial effect of Zinc against Cd deleterious effects.

REFERENCES

Abrahim KS, Abdel-Gawad NB, Mahmoud AM, El-Gowaily MM, Emara AM, Hwaihy MM (2010). Genotoxic effect of occupational exposure to cadmium. Toxicol Ind Health 27 (2):173-179.

Alappat S, Zhang ZY, Chen YP (2003). Msx homeobox gene family and craniofacial development. Cell Res. 13:429-442.

Asmuss M, Mullenders LH, Eker A, Hartwig A (2000). Differential effects of toxic metal compounds on the activities of Fpg and XPA, two zinc finger proteins involved in DNA repair. Carcinogenesis 21:2097-2104.

Bartel SH (1972). Colourimetric method for determination of creatinine. Clin. Chem. Acta 1972:37:193.

Becker DL, McGonnell I, Makarenkova HP, Patel K, Tickle C, Lorimer J, Green CR (1999). Roles for alpha 1 connexin in morphogenesis of chick embryos revealed using a novel antisense approach. Dev. Genet. 24:33.

Bei M, Maas R (1998). FGFs and BMP4 induce both Msx1-independent and Msx1-dependent signaling pathways in early tooth development. Development 125:4325-4333.

Berglund M, Akesson A, Nermell B, Vahter M (1994). Intestinal absorption of dietary cadmium in women depends on body iron stores and fibre intake. Environ. Health Persp. 102:1058-1066.

Brus R, Kostrzewa R M, Felińska W, Plech A, Szkilnik R and Frydrych J (1995). Ethanol inhibits cadmium accumulation in brains of offspring of pregnant rats that consume cadmium. Toxicol. Lett. 76:57-62.

Bruscalupi G, Massimi M, Devirgiliis LC, Leoni S (2009). Multiple parameters are involved in the effects of cadmium on prenatal hepatocytes. Toxicol In Vitro 23(7):1311-1318.

Brzóska MM, Moniuszko-Jakoniuk J (1998).The influence of calcium content in diet on cumulation and toxicity of cadmium in the organism. Archiv. Toxicol. 72:63-73.

Brzóska MM, Moniuszko-Jakoniuk J (2001). Interactions between cadmium and zinc in the organism. Food. Chem. Toxicol 39:967-980. cadmium-induced damage during pregnancy and lactation in rats' pups. J. Food Sci. 75 (1):T18-23.

Chai F, Truong-Tran AQ, Ho LH, Zalewski PD (1999). Regulation of caspase activation and apoptosis by cellular zinc fluxes and zinc deprivation: A review. Immunol. Cell Biol. 77:272-278.

Dally H, Hartwig A (1997). Induction and repair inhibition of oxidative DNA damage by nickel (II) and cadmium (II) in mammalian cells. Carcinogenesis 18:1021-1026.

Dodig M, Kronenberg MS, Bedalov A, Kream BE, Gronowicz G, Clark SH, Mack K, Liu YH, Maxon R, Pan ZZ, Upholt WB, Rowe DW, Lichtler AC (1996). Identification of a TAAT-containing motif required for high level expression of the COL1A1 promoter in differentiated osteoblasts of transgenic mice. J. Biol. Chem. 271:16422-6429.

Dreosti IE (2001). Zinc and the gene. Mutat. Res. 475:161-167.

El-sayed A, Badr H S, Yahia R, Salem M S, Kandil A M (2011). Effects of thirty minute mobile phone irradiation on morphological and physiological parameters and gene expression in pregnant rats and their fetuses. Afr. J. Biotechnol. 10(26):19670-19680.

Fernandez EL, Dencker L, Tallkvist J (2007). Expression of ZnT-1 (Slc30a1) and MT-1 (Mt1) in the conceptus of cadmium treated mice. Reprod. Toxicol. 24:353-358.

Fernández EL, Gustafson AL, Andersson M, Hellman B, Dencker L (2003). Cadmium-Induced Changes in Apoptotic Gene Expression Levels and DNA Damage in Mouse Embryos Are Blocked by Zinc. Toxicol. Sci. 76:162-170.

Forrester LW, Latinwo LM, Fasanya-Odewumi C, Ikediobi C, Abazinge MD, Mbuya O, Nwoga J (2000). Comparative studies of cadmium-induced single strand breaks in female and male rats and the ameliorative effects of selenium. Int. J. Mol. Med. 6: 449-452.

García MT, González EL (2010). Natural antioxidants protect against

Green CR, Bowles L, Crawley A, Tickle C (1994). Expression of the connexin43 gap junctional protein in tissues at the tip of the chick limb bud is related to the epithelial-mesenchymal interactions that mediate morphogenesis. Dev. Biol. 161: 12.

Haldsrud R, Krokje A (2009). Induction of DNA doublestrand breaks in the H4IIE cell line exposed to environmentally relevant concentrations of copper, cadmium, and zinc, singly and in combinations. J Toxicol Env Health-Part A—Current Issues 2009; 72 (3-4):155-163.

Hill DS, Wlodarczyk BJ, Mitchell LE, Finnell RH (2009). Arsenate-induced maternal glucose intolerance and neural tube defects in a mouse model. Toxicol appl pharmacol 239:29-36.

IARC (1993). Beryllium, Cd, mercury and exposures in the glass manufacturing industry. IARC Monographs on the valuation of Carcinogenic Risks to Humans. IARC, Lyon, S. 58, 119-237.

Jacquillet G, Barbier O, Rubera I, Tauc M, Borderie A, Namorado MC,

Martin D, Sierra G, Reyes JL, Poujeol P, Cougnon M (2007) Cadmium causes delayed effects on renal function in theoffspring o cadmium-contaminated pregnant female rats. Am J Physiol Rena Physiol. 293 (5):F1450-1460.

Jarup L, Berglund M, Elinder CG, Nordberg G, Vahter M (1998). Healt effects of cadmium exposure - a review of the literature and a ris estimate.Scandinavian. J. Work. Environ. Health 24(Suppl. 1):1-51.

Jihen el H, Imed M, Fatima H, Abdelhamid K (2009). Protective effect of selenium (Se) and zinc (Zn) on cadmium (Cd) toxicity in the liver o the rat: Effects on the oxidative stress Ecotoxicol. Environ. Saf. 7 (5):1559-1564.

Kuriwaki J, Nishijo M, Honda R, Tawara K, Nakagawa H, Hori E, Nishij H (2005). Effects of cadmium exposure during pregnancy on trace elements in fetal rat liver and kidney. Toxicol. Lett. 156:369-76.

Lalor GC (2008). Review of cadmium transfers from soil to humans and its health effects in the Jamaican environment. Sci. Total Environ 400:162-172.

Lecanda F, Warlow PM, Sheikh Sh, Furlan F, Steinberg TH, Civitelli F (2000). Connexin43 Deficiency Causes Delayed Ossification Craniofacial Abnormalities, and Osteoblast Dysfunction. J. Cell Biol 151:931-943.

Li M, Kondo T, Zhao QL, Li FJ, Tanabe K, Arai Y, Zhou ZC, Kasuya M (2000). Apoptosis induced by cadmium in human lymphoma U937 cells through Ca2-calpain and caspase-mitochondria-dependen pathways. J. Biol. Chem. 275:39702-39709.

Littlefield NA, Hass BS (1995). Damage to DNA by cadmium or nickel ir the presence of ascorbate. Ann. Clin. Lab. Sci. 25:485-492.

Mcleod MJ (1980). Defferential staining of cartilage and bone in whole mouse etuses by Alcian blue and Alizarin red S. Teratology 22:255 301.

Messaoudi I,EL Heni J , Hammouda F ,Said K, Kerkeni A (2009).Protective effects of selenium ,zinc, or their combination or cadmium induced oxidative stress in rat kidney. Biol. Trace Elem Res. 130(20):152-161.

Moniuszko-Jakoniuk J, Gałażyn-Sidorczuk M, Brzóska MM, Jurczuk M Kowalczyk M (2001). Effect of short-term ethanol administration or cadmium excretion in rats. Bull. Environ. Contam. Toxicol. 66:125-131.

Nomiyama K (1980). Recent progress and perspectives in cadmium health effect studies .Science Total Environment 14:199-232.

Ochi T, Takahash K, Ohsawa M (1987). Indirect evidence for the induction of a prooxidant state by cadmium chloride in culturec mammalian cells and a possible mechanism for the induction. Mutat Res.180:257-266.

Oishi S, Nakagawa J, Ando M (2000). Effects of cadmium administration on the endogenous metal balance in rats. Biol. Trace Elem. Res. 76:257-278.

Oltvai ZN, Milliman CL, Korsmeyer SJ (1993). Bcl-2 heterodimerizes ir vivo with a conserved homolog, Bax, that accelerates programmec cell death. Cell 74:609-619.

Patton CJ, Crouch SR (1977). Spectrophotometric and Kinetic investigation of the Berthelot reaction for the determination o ammonia. Anal. Chem. 49:464-469.

Pulido MD, Parrish AR (2003). Metal-induced apoptosis: Mechanisms Mutat. Res. 533:227-241.

Reitman S, Frankel S, Amer J (1957). A Colorimetric method for determination of serum glutamic pyruvic transaminases .Am.J.Clin,Pathol 28:56.

Robinson JF, Yu X, Hong S, Zhou C, Kim N, DeMasi D, Faustman EM (2010). Embryonic toxicokinetic and dynamic differences underlying strain sensitivity to cadmium during neurulation. Reprod. Toxicol 29:279-285.

Salvatori F, Talassi CB, Salzgeber SA, Spinosa HS, Bernardi MM (2004). Embryotoxic and long-term effects of cadmium exposure during embryogenesis in rats. Neurotoxicol. Teratol. 26: 673-680.

Sanford, H. S. (1954). Method of obtaining venous blood from orbita sinus of the rat or mouse. Science 199: 100.

Satokata I, Maas R. (1994). Msx1 deficient mice exhibit cleft palate an abnormalities of craniofacial and tooth development. Nat Genet. 6:348-356.

Sharma G, Sandhir R, Nath R, Gill K (1991). Effect of ethanol on cadmium uptake and metabolism of zinc and copper in rats exposed to cadmium. J Nutrition 121: 87-91.

Stowe HD, Wilson M, Goyer RA (1972). Clinical and morphological effects of oral cadmium toxicity in rabbits. Arch. Pathol. 94:389-405.

Szuster-Ciesielska A, Stachura A, Slotwinska M, Kaminska T, Sniezko R, Paduch R, Abramczyk D, Filar J, Kandefer-Szerszen M (2000). The inhibitory effect of zinc on cadmium-induced cell apoptosis and reactive oxygen species (ROS) production in cell culture. Toxicology 145:159-171.

Tang W Sadovic S, Sheikh ZA (1998). nephrotoxicity of cadmium - metallothionein:protection by zinc and role of glutathione. Toxicol. Appl. Pharmacol. 151(2):276-82.

Tesfaye D, Regassa A, Rings F, Ghanem N, Phatsara C,Tholen E, Herwig R, Un C, Schellander K, Hoelker M (2010). Suppression of the transcription factor MSX1 gene delays bovine preimplantation embryo development in vitro. Reproduction 139:857-870.

Truong-Tran AQ, Carter J, Ruffin RE, Zalewski PD (2001). The role of zinc in caspase activation and apoptotic cell death. BioMetals 14:315-320.

Ursínyová M, Hladíková V (2000). Trace elements—their distribution and effects in the environment. In: Cadmium in the Environment of Central Europe. 3:87-107.

Waalkes MP (2000). Cadmium carcinogenesis in review. J. Inorg. Biochem. 79:241-244.

Warner CW, Sadler TW, Tulis SA, Smith MK (1984). Zinc amelioration of cadmium-induced teratogenesis in vitro. Teratology 30: 47-53.

Webster WS, Messerle K (1980). Changes in the mouse neuroepithelium associated with cadmium-induced neural tube defects. Teratology 1980(21)79-88.

Weigel HJ, Jager HJ, Elmadfa L (1984). Cadmium accumulation in rats organs after extended oral administration with low concentration of cadmium oxide. Arch. Environ.Contam.Toxicol.13:279-287.

World Health Organization (1992). Environmental Health Criteria, 134 Cadmium. IPCS, Geneva.

Study of reproductive toxicity of *Combretum leprosum* Mart and Eicher in female Wistar rats

Jamylla Mirck Guerra de Oliveira[1], Denise Barbosa Santos[1], Francimarne Sousa Cardoso[2], Márcia de Sousa Silva[2], Yatta Linhares Boakari[2], Silvéria Regina de Sousa Lira[2] and Amilton Paulo Raposo Costa[1,2*]

[1]Medicinal Plants Research Center, Federal University of Piauí, Teresina, Piauí, Brazil.
[2]Departament of Veterinary Morphophysiology, Federal University of Piauí, School of Agrarians Sciences, Teresina, Piauí, Brazil.

Most plants culturally used in Brazil for medicine do not have pre-clinical studies of reproductive toxicity, therefore risks of using such products on the reproductive system are unknown. The aim of the study was to evaluate possible reproductive toxicity of ethanolic extract of *Combretum leprosum* Mart and Eicher (EECL) in female Wistar rats. The animals, weighing between 180 to 250 g, were maintained in controlled environmental conditions of temperature, humidity, light/dark cycle of 12/12 h, water *ad libitum* and fed with commercial diet for rats. To verify the estrogenic activity of EECL, four groups of ovariectomized rats were used: saline + corn oil; saline + estradiol; EECL (500 mg/kg) + corn oil; EECL (500 mg/kg) + estradiol. To study the reproductive toxicity during the fecundation and implantation phases of the embryos and also during the organogenesis phase, two groups were used for each experiment, saline and EECL (500 mg/Kg). No estrogenic or anti-estrogenic activities were observed in the EECL. The ingestion of EECL did not cause modification in the number of implantation sites, which indicates a lack of toxicity during this phase. The EECL administered orally in the dose of 500 mg/kg did not produce adverse effects on the reproductive system of the female rats.

Key words: Estrogenic activity, organogenesis, teratogens.

INTRODUCTION

Many drugs or chemical products that can cause conge-nital malformations are called teratogens. Drugs vary considerably in their teratogenic capacity. Some may cause severe morphological alterations if administered during the organogenic period; others may produce men-tal and growth retardations as well as minor malforma-tions when used in excess during development (Moore, 1990; Seip, 2008).

Toxic agents present in food and specific nutrient imba-lance in the diet can alter the maternal hormonal concen-tration and affect the composition of the secretion from the oviduct and ovary, which results in a decrease of em-bryonary survival or, in less severe circumstances, altera-tions in development and growth (McEvoy et al., 2001). With the increasing use of phytotherapics, many studies have evaluated the potential reproductive toxicity of some plants used in popular medicine. Montanari and Bevilacqua (2002) observed estrogenic activity and loss of embryos during the pre-implantation period in female murine trea-ted with an extract of *Maytenus ilicifolia* Mart. Fetal reab-sorption and an anti-implantation activity without an estro-genic activity in rats treated with extract of *Coutarea hexandra* Schum (Rao et al., 1988; Almeida et al., 1990) have been observed. *Peumus boldus* and the isolated substance boldina caused significant number of reab-sorptions and some malformations when administered to pregnant rats (Almeida et al., 2000). Phytoestrogens and estrogenic mycotoxins are widely present in nature. They

*Corresponding author. E-mail: amilfox@uol.com.br.

may cause toxic effects on the reproductive system of mammals exposed to natural conditions and, recently, their effects as endocrine disruptors had been studied (Kelce and Gray, 1997).

The estrogenic and anti-estrogenic activity of the phyto-estrogens depends upon their concentration, the endogenous sexual steroids and the specific target organ. These variations in the effects can be explained by the existence of two types of estrogen receptors (ER): α and β (Harris et al., 2005). The α-receptors (ER-α) are the main receptors found in the breast and the uterus, and the β-receptors (ER-β) are predominant in the bones and in the cardiovascular system. The 17- β-estradiol has an affinity for both receptors, while the isoflavones are more selective for the ER-β. Among the isoflavones, the Genistein has, compared to estradiol (100%), 87% of affinity with the ER- β and 4% for the ER- α, while the Daidzein has 0.5% for ER- β, 0.1% for the ER-α (Kuiper et al., 1998). Therefore, the effect level of the isoflavones over a specific tissue depends on the type of receptor that is predominant.

In this context, there are still several species that requires more information concerning their effects on the reproductive system. Among these are the species of the genus *Combretum*, which contain several confirmed pharmacological properties such as: anti-inflammatory, antiulcerogenic and antifungal (Asuzu and Adimorah, 1998; Martini and Elloff, 1998; Baba-Mossa et al., 1999). Among the species that have already been isolated are the combrestatins with potential for cancer treatment (Tozer et al., 1995; Rodrigues, 2006).

The *Combretum leprosum* species that belongs to the combretaceae family, is popularly known as Cipoaba, Mofumbo or Mufumbo (Matos, 2003). It is amply distributed in the northern region of Brazil and is used commonly as a haemostatic, sedative and for healing wounds (Silveira, 2003). The chemical study of the species revealed the presence of triterpenes and flavonoids (Facundo et al., 1993). In pharmacological trials performed with the ethanolic extract of the trunk's bark, analgesic and anti-inflammatory activities were observed (Lira et al., 2002), and also a dose-related antinociception (Pietrovski, 2006). Since the *C. leprosum* presents therapeutic potentials and there is no data related to the reproductive toxicity of this plant in the literature, the goal of this research was to evaluate the effects of an ethanolic extract of *C. leprosum* on estrogenic activity and reproductive toxicity during the fecundation, implantation and organogenesis phases of the embryo.

MATERIALS AND METHODS

Preparation of the extract

Barks from the trunk of *C. leprosum* Mart and Eicher were collected in Teresina, state of Piaui, Brazil, during the month of September 2009. The plant was identified and and deposited in the form of

excicata in the Graziela Barroso Herbarium, at the Center of Natural Sciences at the Federal University of Piaui, under number 10557.

The barks were dried in an evaporator at 40 ± 1°C and then grounded in an electrical mill. The material was submitted to a maceration process with ethanol at 70% during three successive extractions and then concentrated in a rotavap at 50°C, and later put in amber glass bottles and kept in the fridge.

Animals

In all of the protocols, immature female Wistar rats weighing 60 to 70 g or adults weighing between 180 to 250 g, were raised and kept at the experimental Vivarium at the Morphofisiological Veterinary Department –CCA, in a light/dark cycle of 12/12 h, maintained in controlled temperature conditions at 22 ± 2°C. Standard pellet food (FRI-LAB Rat – Fri-Ribe, Ribeirão Preto, SP, Brazil) and drinking water were available *ad libitum*.

All animal studies were carried out in accordance with the requirements of the Committee on Ethics in Animal Experimentation as adopted by the Federal University of Piauí (Protocol number: 33/2009). The experimental protocols were elaborated and developed based on the principle of the three R's (refine, reduce and redesign). That is, the lowest number of animals possible was used in order to determine statistical differences; the protocols developed do not overlap regarding the objective of the study. Moreover, the animals were handled by trained researchers in our laboratory only when necessary. They were not exposed to any kind of pain or stress caused by noise, lack of food, water, or variation in temperature.

Experiment 1: Estrogenic activity assay

This protocol was based on Almeida et al. (1990). Immature 32 female Wistar rats were used, being eight animals per group. The animals were separated in groups that received the following treatments:

Group I: Saline 1 mL/100 g orally and corn oil (0.1 mL/100 g) intramuscular (im)
Group II: Saline 1 mL/100g orally and estradiol (1 µg/100 g) im
Group III: Ethanolic extract of *C. leprosum* (EECL) in the dose of 500 mg/kg orally and corn oil (0,1 mL/100 g) im.
Group IV: EECL (500 mg/kg) orally + estradiol (1 µg/100 g) im.

The dose of EECL was chosen based on preliminary studies performed by the authors, where the ethanolic extract of *C. leprosum* was toxic in high doses. The LD50 (the median lethal dose) value for p.o. administration was 4722 mg/kg (Lira et al., 2002). Thus the dose of 500 mg/kg is safe for use in this protocol.

Treatment was given during three days and on the fourth day, the rats were euthanized by excess inhalation of halothane. The removal, cleaning and weighing of the uterus were done. The weights found were converted to 100 g of body weight and then analyzed statistically.

Experiment 2: Evaluation of reproductive toxicity during the fecundation and implantation phases

The protocol was carried out using 16 female adult rats, divided in two groups of eight animals each, one being the control and the other treated with EECL. The protocol executed was according to Garg et al. (1970). The rats were examined daily to identify the phase of the estrous cycle, by fresh vaginal smear. Those detected to be in proestrus were mated with a fertile male and the presence of spermatozoids in the smear the following morning of the mating

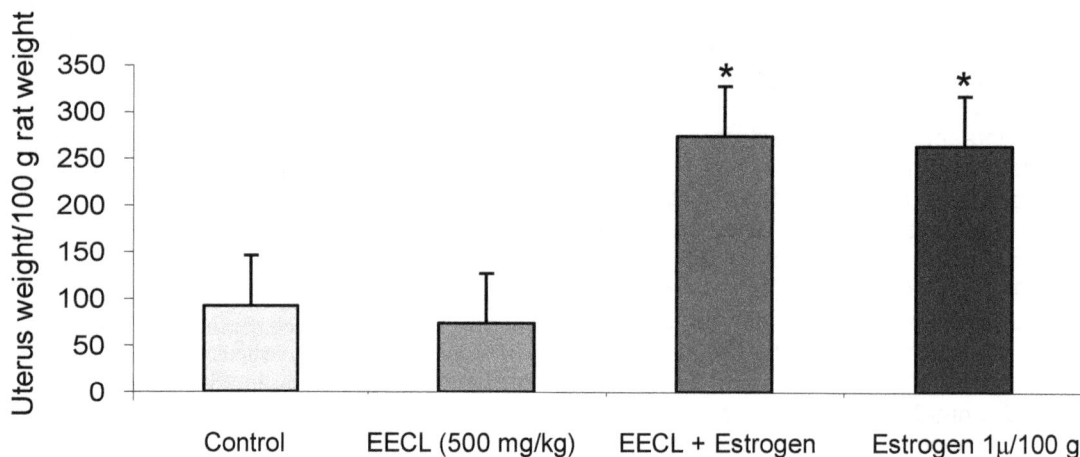

Figure 1. Effect of treatment with ethanolic extract of *C.* leprosum in the uterine weight of the prepubertal rats, with and without reposition of estradiol. The values are mean ± S.E.M; n = 8; *significant difference estatítica $p < 0.05$.

was indicative of pregnancy (day 1). The control group received 1 mL/100 g of body weight of saline orally, from the first to the seventh day of gestation. The test group received EECL 500 mg/mL, with a dose of 1 mL/100g of body weight orally, during the same period. During the eighth day, the animals were sacrificed by excessive inhalation of halothane and the uteri were removed and weighed. Then the number of corpus luteus and the sites of embryonary implantation in each uterus were verified.

Experiment 3: Evaluation of reproductive toxicity during organogenesis

The protocol was done according to Garg et al. (1970), using 16 female adult rats, divided into two groups of eight animals, one being the control group and the other the experimental group. The rats were examined and mated according to the previous protocol. They also received saline (1 mL/100 g) or EECL 500 mg/mL orally, from the sixth to the fifteenth day of gestation. After treatment, the animals were kept in observation until delivery. The number of offspring, the weight and the presence of congenital defects, macroscopically, were observed for each animal.

The parametric data was submitted to the analysis of variance (ANOVA test) and the means compared through the *Student-Newman-Keuls* test and the nonparametric data was evaluated by the Kruskal-Wallis test. Data were expressed as mean ± standard error for mean with p<0.05.

RESULTS AND DISCUSSION

EECL treated animals showed no change in uterine weight when compared to the saline group rats (Figure 1). This demonstrates an absence of estrogenic activity. Estrogenic activity of a specific substance is understood as the capacity it possesses in linking itself to estrogen receptors promoting effects similar to their own hormone. In recent years, the scientific community has turned its attention to a series of compounds known collectively as phytoestrogens. Such molecules have a structural similarity with the 17-β-estradiol and synthetic antiestrogens, such as tamoxifen and its mechanism of action are studied to

identify estrogenic and antiestrogenic activities (Mathieson and Kitts,1980; Peterson and Barnes, 1996; Morito, 2001 Müller et al., 2009), aiming at possible therapeutic applications or toxic effects.

There was no statistically significant difference between the group treated with EECL plus estradiol and the group treated with saline plus estradiol during the fecundation and implantation phases (Figure 1). This indicates the lack of antiestrogenic activity in this plant, and this goes to show that it is safe for therapeutic use in women of reproductive age. Plants that exhibit antiestrogenic activity have the capacity of interrupting the gestation of rats and mice, by the inhibition of the estrogen necessary for implantation (Ghandi et al., 1991). In mice and humans estrogen has an important role in implantation, as it participates in the estrogen/progesterone balance and in the receptivity of the uterus to the embryo (Ements, 1970).

In the experiment of reproductive toxicity, the EECL was administered from the first to the seventh day of gestation, a period that involves the stages before and after implantation. Implantation usually occurs between the fourth and fifth day of gestation in rodents and interference in this period can lead to embryonic losses (Beaudoin, 1980; Hodgen and Itskovit, 1988). The fundamental characteristic of this process is the synchronized development of the embryo to the blastocyst stage and the uterus differentiation to the receptive condition. Later interactions between the activated blastocysts and the uterine epithelium occur to begin implantation (Paria et al., 2000).

The ingestion of EECL did not cause modifications in the number of implantation sites (Figure 2). This indicates the lack of toxicity during this phase and also confirms the lack of estrogenic activity, as the presence of this activity in plant extracts can inhibit 100% of implantation in female rats, as was observed by Jagadish and Rana (2002) for *Calotropis procera*. The receptivity of the uterus

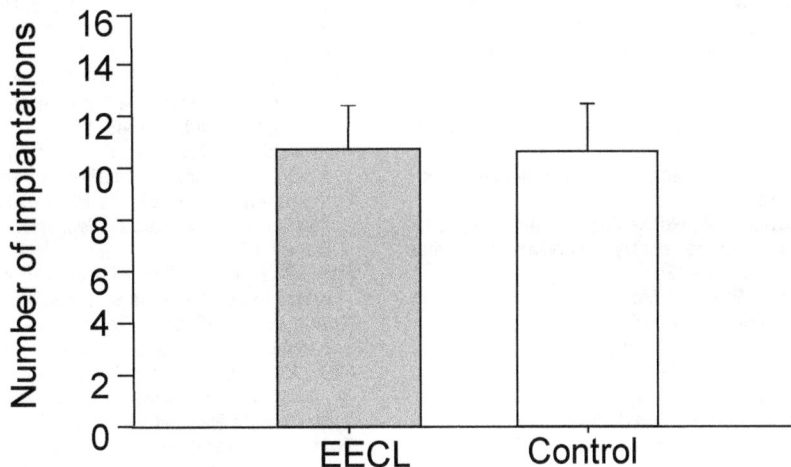

Figure 2. Number of embryonary implantation sites in rats treated with ethanolic extract of *Combretum* leprosum. The values are mean ± S.E.M; n = 8.

to the embryo depends, among other factors, on a high ratio of progesterone: estrogen (Goodman, 1994) and estrogen injections can prevent implantation (McDonald, 1989).

The number of newborn rats per litter was not altered by the treatment of the mothers with EECL (Figure 3). Since there was no fetal mortality after treatment with the extract, it can be suggested that a lack of toxicity for the normal fetus growth depends on a complex interaction between genetic, immunological, endocrinological, nutritional, vascular and environmental factors. Alterations in any of these factors can interrupt the normal growth and development of the embryo/fetus (Chahoud et al., 1999). There was also no alteration in the newborn's weight, which would indicate retardation in fetal growth (Frohberg, 1977). No congenital defects were observed, demonstrating a lack of teratogenic effects in the extract.

In summary, the data indicates that the bark extract from *C. leprosum* did not present neither estrogenic nor antiestrogenic activity. Also it did not present fetal mortality when administered during the gestational period.

ACKNOWLEDGMENTS

We are grateful to the Coordenacão de Aperfeiçoamento de Pessoal de Nível Superior (CAPES) and to the Conselho Nacional de Desenvolvimento Científico e Tecnológico (CNPq) for their financial support.

REFERENCES

Almeida FRC, Rao VSN, Gadelha MGT, Matos FJA (1990). Study on the antifertilizante activity of *Coutarea hexandra* Schum. in rats. Rev. bras. farmacogn. 71(3): 69-71.

Almeida, E.R.; Melo, A.M.; Xavier, H (2000). Toxicological evaluation of the hydro-alcohol extract of the dry leaves of *Peumus boldus* and boldine in rats. Phytother Res. 14(2): 99-102.

Asuzu JN, Adimorah RI (1998). The anti-inflammatory activity of extracts from roots of *Combretum dolichopetalum*. Phytomedicine.

5(1): 25-28.

Baba-Mossa F, Akpagana K, Bouchet P (1999). Antifugical activities of seven West African Combretaceae used in traditional medicine. J. Ethnopharmacol. 66: 335-338.

Beaudoin AR. Embryology and teratology. In: Baker HJ, Lindsey JR, Weisbroth SH (1980). The laboratory rat (Research application). New York. Academic Press. 2:75-94.

Chahoud I, Ligensa A, Dietzel L, Fagi AS (1999). Correlation between maternal toxicity and embryo/fetal effects. Reprod. Toxicol. 13: 375-381.

Ements CW (1970). Antifertility agents. Ann. Rev. Pharmacol. Toxicol. 10: 237-254.

Facundo VA, Andrade CHS, Silveira ER, Braz Filho R, Huford C (1993). Triterpenes and flavonoids from *Combretum leprosum*. Phytochemistry. 32(2): 411-15.

Frohberg H (1977). An introduction to research teratology. In: Neubert D, Merker HJ, Kwasigroch TE Methods in prenatal toxicology. Stuttgart. Georg Thieme Publishers. 1-13.

Garg SK, Saksena SK, Chaudhury RR (1970). Antifertility screening of plants. VI. Effect of five indigenous plants on early pregnancy in albino rats. Indian. J. Med. Res. 58(9):1285-9.

Ghandi M, Lal R, Sankaranarayanan A, Sharma PL (1991). Post-coital antifertility activity of *Ruta gravoleolens* in female rats and hamsters. J. Ethnopharmacol. 34: 49-59.

Goodman HM (1994). Basic medical endocrinology. 2 ed. New York; Raven Press, p.299.

Harris DM, Besselink E, Henning SM, Go VLW, Heber D (2005). Phytoestrogens induce differential estrogen receptor alpha- or beta-mediated responses in transfected breast cancer cells. Exp. Biol. Med.230: 558-568

Hodgen AD, Itskovit J (1988). Recognition and maintenance of pregnancy. In: Knobil E, Neil J The physiology of reproduction. New York. Raven Press. p.1995

Jagadish VK, Rana AC (2002). Preliminary study on fertility activity of Calotropis procera roots in female rats. Fitoterapia. 73: 111-115.

Kelce WR, Gray LE (1997). Endocrine disruptors: Effects on sex steroid hormone receptors and sex development. In: Kavlock RJ, Daston GP. Drug toxicity in embryonic development II. Germany: Springer-Verlag Berlin Heidelberg. 435-474.

Kuiper GG, Lemmen JG, Carlsson B, Corton JC, Safe SH, Van Der Saag PT, Van Der Burg B, Gustafsson JA (1998). Interaction of estrogenic chemicals and phytoestrogens with estrogen receptor beta. Endocrinology. 139: 4252–4263.

Lira SRS, Almeida RN, Almeida FRC, Oliveira FS, Duarte JC (2002). Preliminary studies on the analgesic properties of the ethanol extract of *Combretum leprosum*. Pharm. Biol. 40(30): 213-215.

Mathieson RA, Kitts WD (1980). Binding of phyto-oestrogen and oestradiol-17β by cytoplasmic receptors in the pituitary gland and hypothalamus of the ewe. J. Endocrinol. 85: 317-325.

Matos FJA (2003). O Formulário Fitoterápico do professor Dias Rocha. 2ª ed. Fortaleza; EUFC. 260p.

Martini N, Eloff JN (1998). The preliminary isolation of several antibacterial compounds from Combretum erythrophyllum. J. Ethnopharmacol. 62(3): 255-263.

McDonald LE (1989). Veterinary endocrinology and reproduction. 4º ed. Lea & Febiger, Philadelphia. p.510.

McEvoy TG, Robinson JJ, Ashoworth JA, Rooke JA, Sinclair KD (2001). Feed and forage toxicants affecting embryo survival and fetal development. Theriogenology. 55: 113-129.

Montanari T, Bevilacqua E (2002). Effect of Maytenus ilicifolia Mart. on pregnant mice. Contraception. 65(2): 171-175.

Moore KL (1990). Embriologia Clínica. 4ª edição. Rio de Janeiro, Guanabara Koogan, 355: 105-117.

Morito K, Hirose T, Kinjo J, Hirakawa T, Okawa M, Nohara T, Ogawa S, Inoue S, Muramatsu M, Masamune Y (2001). Interaction of phyto-estrogens with estrogen receptors alpha and beta. Biol. Pharm. Bull. 24: 351-356.

Müller JC, Botelho GK, Bufalo AC, Boareto AC, Rattmann YD, Martins ES, Cabrini DA, Otuki MF, Dalsenter PR (2009). Morrinda citrifolia Linn (Noni): In vivo and in vitro reproductive toxicology. J. Ethnopharmacol. 121, 229-233.

Paria BC, Lim H, Das SK, Reese J, Dey SK (2000). Molecular signaling in uterine receptivity for implantation. Semin. Cell Dev. Biol. 11(2) 67-76.

Peterson TG, Barnes S (1996). Genistein inhibits both estrogen and growth factor stimulated proliferation of human breast cancer cells Cell Growth Diff. 7: 1345-1351

Pietrovski EF, Rosa KA, Facundo VA, Rios K, Marques MCA, Santos ARS (2006). Antinociceptive properties of the ethanolic extract and of the triterpene 3β,6β,16β-trihidroxilup-20(29)-ene obtained from the flowers of Combretum leprosum in mice. Pharmacol. Biochem Behav. 83: 90-99.

Rao MV (1988). Effect of alcoholic extract of Solanum xanthocarpun seeds in adult male rats. Indian J. Exp. Biol. 26: 95-98.

Rodrigues A (2006). Perspectivas de novos tratamentos para c carcinoma tireoidiano avançado. Rev. Col. Bras. 33(3): 189-197.

Seip M (2008). Growth retardation, dysmorphic facies and mino malformations following massive exposure to phenobarbitone in utero. Acta Paediatrica. 65(4): 617-621.

Tozer GM, Kanthou C, Parkins CS, Hill SA (2001). The biology o combrestastatins as tumor vascular targeting agents. Int. J. Exp Pathol. 83: 21-38.

9

Antimicrobial activity of plant phenols from *Chlorophora excelsa* and *Virgilia oroboides*

Thiriloshani Padayachee[1]* and Bharti Odhav[2]

[1]Department of Biosciences, Vaal University of Technology, Vanderbijlpark South Africa.
[2]Department of Biotechnology and Food Technology, Durban University of Technology, South Africa.

The anti-bacterial and anti-fungal activity of four aqueous plant extracts (1 x 10^4 µg/ml) of 2,3'4,5'-tetra hydroxy-4'-geranylstilbene (chlorophorin) and 3',4, 5' - trihydroxy - 4' - geranylstilbene (Iroko) from the tree *Chlorophora excelsa* and (6aR,11aR)-3-hydroxy-8,9-methylenedioxypterocarpan (Maackiain) and 7-hydroxy-4'-methoxyisoflavone (formononetin) from *Virgilia oroboides* were evaluated by the seeded agar overlay well diffusion method. The test organisms and bioautography used included: *Bacillus coagulans, Streptococcus pneumoniae, Klebsiella pneumoniae, Escherichia coli, Mycobacteria tuberculosis, Aspergillus flavus* and *Fusarium verticilloides*. Vancomycin, the drug of choice for these organisms was used as the control at 30 µg/ml. The extracts showed that chlorophorin at 1.95 µg/ml and Iroko at 3.125 and 6.25 µg/ml respectively were active in inhibiting the growth of *S. pneumoniae* and *B. coagulans* and not active against *K. pneumoniae* and *E. coli*. Maackiain; formononetin and formononetin acetate showed little activity against *S. pneumonia, B. coagulans, K. pneumoniae* and *E. coli*. None of the extracts showed activity against *M. tuberculosis*. Maackiain, formononetin, chlorophorin and Iroko inhibited *F. vertiicilloides*, maackiain being the most active compound. Formononetin, chlorophorin and Iroko inhibited *A. flavus*. *A. flavus* was most sensitive to chlorophorin and Iroko. The bioautography method confirmed these results and was attributed to the phenolic nature of the compounds.

Key words: Plant compounds, anti-bacterial, anti-fungal, chlorophorin, Iroko.

INTRODUCTION

According to the World Health Organisation (2005), herbal medicines serve the health needs of about 80% of the world's population, especially for millions of people in the vast rural areas of developing countries. Plants have the ability to produce a wide variety of bio-active molecules through secondary metabolic pathways (Kutchan and Dixon, 2005; Cox and Balick, 1994; Mitscher et al., 1987). Current problems associated with antibiotic resistance (Lu and Collins, 2009; Recio et al., 1989), lack of availability and poverty means that one has to seek alternatives. Extracts from plants have lead to the discovery of many drugs. Africa has a large untapped resource of indigenous plants whose extracts could lead to the discovery of potentially new drugs.

These new antimicrobial agents are not restricted to products of microbial origin (Samoylenko et al., 2009; Zhu et al., 2009; McLaughlin, 2008; Hoffmann et al., 1993; Mitscher et al., 1987). Since their structures are different to those from microbial sources, they provide a new challenge to the microbe, particularly with regard to the development of resistance (Lee, 2004; Kubo and Kubo, 1995). In this study, plant compounds from two South African indigenous plants for antimicrobial activity against common bacterial pathogens and mycotoxin producing fungi were used. The compounds in this study were derived from two indigenous trees *Virgilia oroboides* and *Chlorophora excelsa*.

C. excelsa produces compounds 2,3'4,5'-tetra hydroxy-4'-geranylstilbene (chlorophorin), which was first isolated from an ether extract of the heartwood by King and Grundson (1949) and 3', 4, 5' - trihydroxy - 4' - geranylstilbene (Iroko), a stilbene classified as an aromatic phenolic compound. *V.*

*Corresponding author. E-mail: thiri@vut.ac.za.

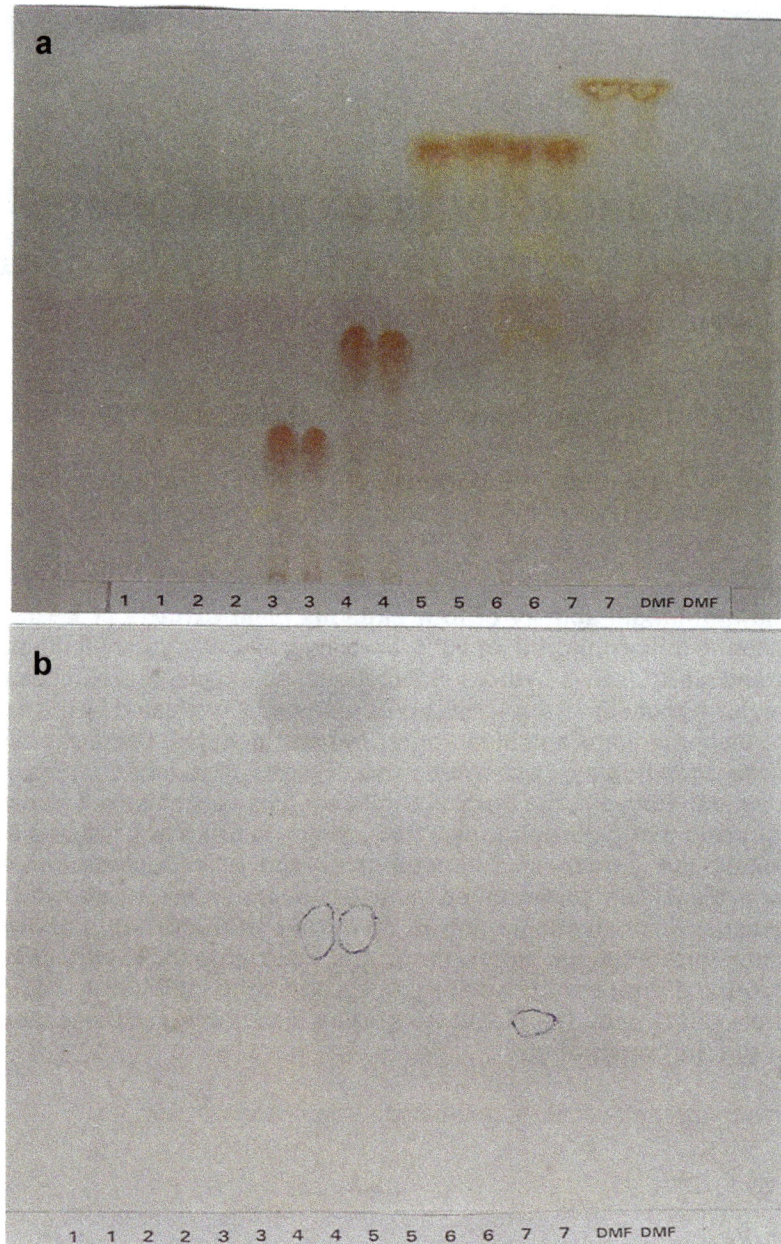

Figure 1. The effect of maackiain acetate,formononetin,chlorophorin and Iroko on *F. moniliforme*.TLC plates with separated plant compounds (a) template and (b) with *F. moniliforme* on the TLC plates showing areas of no growth (arrow) in the same region as where bands appear on the template. 1, Formononetin; 2, Formononetin acetate; 3, Iroko (stilbene); 4, Chlorophorin; 5, Maackiain; 6, Maackiain; 7, Maackiain acetate.

oroboides produces (6aR,11aR)-3-hydroxy-8,9-methylenedioxypterocarpan (Maackiain) which is classified as a pterocarpan (McMurphy et al.,1972; Pachler et al. 1967 cited by Swinny, 1989) and characterised by Swinny, (1989) and 7-hydroxy-4'-methoxyisoflavone classified as Formononetin (Bate-Smith et al.,1953; Braz et al., 1973; Letcher et al., 1976 cited by Swinny, 1989). The chemical structures of the compounds were elucidated by Swinny et

al. (1989) illustrated in and were found to be phenolic compounds as shown in (Figure 1).

Phenols have long been known to have anti-septic activity and therefore have a large potential as therapeutic agents. Since these are produced by all vascular plants and well over 4000 distinct flavonoids have now been identified some of which display therapeutic activity (Yoon et al., 2007; Ja Kim et al., 2006; Walsh, 1988). In previous studies of

Table 1. Effect of plant compounds on gram positive and gram negative bacteria.

Parameter	B. coagulans	S. pneumoniae	K. pneumoniae	E. coli
Chlorophorin	+ + +[a] MIC	+ + + MIC	-	-
Iroko	+ + + MIC	+ +MIC	-	-
Maackiain methyl ether	-	-	-	-
Maackiain acetate	+	+	-	-
Formononetin	+	-	-	-
Formononetin acetate	-	-	+	-

[a]Zone of inhibition; - = 0 - 1 mm; + = 2 - 3 mm; ++ = 4 - 5 mm; +++ = 6 - 10 mm.

Acharya and Chatterjee (1974), both Chlorophorin and Iroko have been shown to be highly active against the gram positive organisms such as *Streptococcus pneumoniae* and *Bacillus coagulans*, and less active against the gram negative organisms such as *Klebsiella pneumoniae* and *E. coli*. An isoflavone, formononetin and a pterocarpan, maackiain showed minimal activity.

Previous anti-fungal studies (Wu et al., 2008; Kordial et al., 2005; Harrison et al., 2003; Davidson, 1993) have shown that phenolic compounds do have anti-fungal properties and it was also stated by Shan et al. (2007) and McCutcheon et al. (1994) that most plants that exhibit anti-bacterial properties are also usually anti-fungal. Studies of Chipley and Uriah (1980) have also shown that the phenolic compounds, caffeic acid and chlorogenic acid inhibited the *Fusarium* species (Ja Kim et al., 2006; Harrison et al., 2003; Valle,1957) and that ferulic acid inhibits aflatoxin B_1 and G_1 production of *A. flavus* by approximately 50% and *A. parasiticus* by 75%. This study investigated the antimicrobial activity of four plant extracts: chlorophorin and Iroko from *Chlorophora excelsa* and, maackian and formononetin from *Virgrilia oroboides*, with a view to develop future biocontrol agents.

MATERIALS AND METHODS

Preparation and storage of microorganisms

B. coagulans, S. pneumoniae, K. pneumoniae, E. coli, and three strains of *M. tuberculosis, viz.,* A169 (Karachi strain- known resistance used as a control), H37 (ATCC virulent strain) and R9001 were obtained from the Department of Microbiology, University of Natal, Medical School. *A. flavus* and *F. verticilloides* were obtained as pure cultures from the Department of Physiology, University of KwaZulu Natal Medical School, Durban, RSA. The gram positive and gram negative bacteria were stored as stock cultures in Microbank vials (Prolab Diagnostics) at -70°C. *M. tuberculosis* strains were maintained on A. J. Lone Stein Jensen agar medium at 4°C. Fungal stock cultures were maintained and stored on Saboroud Dextrose agar (SDA) slants (Oxoid) at 4°C until they were required.

All experiments were conducted according to the Manual of Clinical Microbiology (Balows et al., 1991). The gram positive organisms (*B. coagulans* and *S. pneumoniae*) and gram negative organisms (*K. pneumoniae* and *E. coli*) were plated on Brain Heart Infusion agar plates and acid fast bacteria was plated on Middlebrook agar (Biolife) and supplemented with 5% glycerol and 100 ml oleic acid albumin and dextrose (OADC) (Biolab, ART No: C 70) and incubated at 37°C for

six weeks, and stored at 4°C. The inoculum was standardised using McFarlands No. 2 standard in saline. This is equivalent to a bacterial concentration of 6 x 10^8 cells/ml (Balows et al., 1991; Washington et al., 1972).

Broth cultures of the strains of *M. tuberculosis* were prepared in Dubos broth (Biolife) and were incubated at 37°C for 24 h, after which the inoculum was standardised using McFarlands No. 1 standard. This was further diluted in Dubos broth to give a final *M. tuberculosis* concentration of 3 x 10^8 cells/ml (Balows et al., 1991; Washington et al., 1972). This was used as the inoculum.

Fugal cultures (*A. flavus* and *F. verticilloides*) were prepared by plating them on SDA plates, which were incubated for at 25°C for six days. The spores were harvested in 2 ml of sterile saline and counted with a Neubauer counting chamber to give a spore concentration of 4 x 10^7 spores /ml.

Preparation and storage of plant compounds

Iroko and chlorophorin from *C. excelsa* and Maackiain and Formononetin acetate from *V. oroboides* were obtained in powder form (Swinny, 1989). They were prepared in dimethylformamide (DMF) at 1.000 µg/ml (w/v), using a non-pyrogenic sterile filter 0.22 µm, (Millipore) and stored as stock solutions at ambient temperature until needed. Their identity and purity were verified by thin layer chromatography. These plant extracts were previously tested for their toxicity to human cells by *in-vitro* toxicity test against Hep -2 cells and their mutagenic activity was ascertained using the Ames test, as outlined by Franson (1992).

Evaluation of antimicrobial activities

The antimicrobial activities of these compounds were tested using the modified well diffusion assay (Leven et al., 1979). Each test was carried out in triplicate, and the test results expressed as a mean of each triplicate. The compounds were tested against the following gram positive organisms; viz., *B. coagulans* and *S. pneumoniae*, and gram negative bacteria viz., *K. pneumoniae* and *E. coli*. Dimethylformamide was used as a negative control, vancomycin (30 µg/ml) was used as positive control for *B. coagulans* and *S. pneumoniae* and ampicillin (25 µg/ml) was used for *K. pneumoniae* and *E. coli* and Rifampicin (1.0 µg/ml) was used against *M. tuberculosis* (Table 1).

Verification against *M. tuberculosis* was also tested using a sensitivity method described by Lsenberg, (1980) using Middlebrook 7H10 (BioLife). In this method, Middlebrook agar containing 6 x 10^8 colony forming units (CFU) of the test organism was over laid onto molten sensitivity agar (Biolife). Six equidistant wells, 6 mm in dia-meter, were made in the agar plates. Four wells were filled with 25 µl of the respective positive and negative controls, each in duplicate. The MIC's were established for the most active compounds using the method described (Lsenberg, 1980). Plant compounds were diluted in

Table 2. Minimum inhibitory concentration (MIC) in µg/ml of the plant compounds and the effect of antibiotics on the test bacteria.

Compound	B. coagulans	S. pneumoniae	K. pneumoniae	E. coli	M. tuberculosis
Chlorophorin	1.953± 0.06[a]	1.953± 0.06	50± 0.06	No effect	No effect
Iroko	3.125± 0.15	6.25± 0.015	No effect	No effect	No effect
Maackiain acetate	25± 0.26	12.5± 0.015	No effect	No effect	No effect
Formononetin	50± 0.025	No effect	25± 0.15	No effect	No effect
Ampicillin	No effect	No effect	No effect	25 ± 0.15	No effect
Rifampicin	-	-	-	-	1.0
Vancomycin (control)	30	30	30	No effect	No effect

[a] ± = Standard deviation of the mean.

dimethylformamide (DMF) to give concentrations of each compound ranging from 3.125 100 µg/ml (w/v). 5 ml of molten sensitivity agar and 1 ml of the respective plant compounds were poured into sterile quadrant Petri dishes. 30 µl of the standardised inoculum was spread over the surface of each quadrant. The plates were then incubated for three weeks at 37°C in a CO_2 incubator. The last two wells were inoculated with 0.025 ml of each test compound, at concentrations ranging from of 3.125 to100 µg/ml (w/v). The plates were incubated for 24 h at 37°C. The results were analysed by measuring the zones of inhibition (clearing around each well in mm), of the test compounds relative to the clearing around DMF, using the following formula. The results were assessed according to the ability of test compounds to inhibit the growth of the microorganisms.

Activity of tested compounds (zone of inhibition of test compounds - zone of inhibition of DMF)

Preliminary screening tests were carried out to establish the activity of the plant compounds on the test fungi, using the seeded agar overlay method. 10 ml SDA plates were overlaid with 5 ml of molten SDA that was seeded with 1 ml of previously prepared spore suspension (4,07 x 10^7 spores/ml) of A. flavus and F. verticilloides respectively. Six wells (6 mm diameter) were punched into each plate (0.25 µl; 100 µg/ml) and the plant test compounds was placed in each well and incubated at 25°C for four and six days for A. flavus and F. verticilloides respectively. DMF was used as the negative control and amphotericin B (5 mg/ml) was used as the positive control. The results were analysed by measuring the zones of inhibition of the test compounds. The antifungal effect was confirmed using bioautography (Betina, 1973). Two silica gel 60 TLC aluminium sheets (Merck) were spotted with 5 µl of 10 µg/ml (w/v) of each compound, 5 µl of saline (0.9 %w/v), and amphotericin B (Oxoid) (5 µg/ml). The compounds were separated using benzene: acetone (8:4 v/v) for 1 h. The plates were then air dried. One plate, which was used as the template, was developed with anisaldehyde solution and dried in a hot air oven at 80°C for 2 min. The second plate which was used for evaluating the effect of the plant compounds on fungal growth was sprayed with a fungal spore suspension which was prepared using 2 ml of the inoculum (4 x 10^7 spores /ml) suspended in 70 ml of fungal culture media that is glucose-mineral salts medium. This plate was placed in a sterile humidified container and incubated at 25°C for 4 days. The results were obtained by comparing the zone on the template plate to the zones of inhibition on the test plate. These were compared to the zones of inhibition obtained for DMF.

Determination of minimum inhibitory concentrations (MIC)

The MIC of the active compounds was determined using the procedure described by Hailu et al. (2005) and Atlas et al. (1984). Two fold serial

dilutions of each compound were made in DMF at concentrations ranging from 3.125 µg/ml to100 µg/ml (w/v). The lowest concentration of the test compound in which no growth occurred was defined as the MIC.

The method using the seeded agar well diffusion by Hailu et al (2005) and Atlas et al. (1984) were used to measure MIC of the active compounds against A. flavus and F. verticilloide. A single well 6 mm in diameter was cut out from SDA plate, to which 0.25 µl of test compound or control was added. Two fold serial dilutions of each active compound were made in DMF at concentrations ranging from1.5625 to100 µg/ml (w/v). DMF was used as the negative control at a volume of 0.25 µl. Amphotericin B (80% purity) was used as a positive control at a concentration of 5 mg/ml. The agar was seeded with 4 x 10^7 spores/ml. The lowest concentration of the test compound in which no inhibition occurred was defined as the MIC.

RESULTS

Evaluation of antimicrobial activities

Chlorophorin and Iroko (1 x10^4 µg/ml) demonstrated high anti-bacterial activity against S. pneumoniae and B. coagulans. These compounds did not have the same effect against K. pneumoniae and E. coli. Maackiain; Formononetin and Formononetin acetate showed no activity against S. pneumonia, B. coagulan, K. pneumoniae and E. coli. The MIC's for each of the plant compounds showed that chlorophorin was the statistically more active and was more effective than Vancomycin (the drug of choice for these organisms); (1.953 µg/ml) was required to inhibit the growth of B. coagulans and S. pneumoniae as compared to that required for Vancomycin to have the same effect at 30 µg/ml. Iroko was active against B. coagulans and S. pneumoniae and their MIC were 3.125 and 6.25 µg/ml respectively.

Maackiain was active against S. pneumoniae and had a MIC of 12.5 µg/ml. Formononetin was the only compound that was active against K. pneumoniae and it had a MIC of 25 µg/ml. The MIC's are shown in Table 2. None of the compounds showed activity against M. tuberculosis or E. coli. The plant extracts demonstrated significantly higher activity as compared to the control, Vancomycin. Chlorophorin against B. coagulans and S. pneumoniae demonstrated 93.5% more activity, while Iroko against B. coagulans demonstrated 89.58% and against S.

Table 3. Activity and the MIC for maackiain acetate, formononetin acetate, chlorophorin and Iroko against *F. verticilloides and A. flavus.*

Organism	Plant compound	Zones of inhibition (1 X 10^4)	MIC (µG/ML)	Total inhibition expressed as a % of control
F. verticilloides	Maackiain acetate	13 mm	1.56	99.97
	Formononetin acetate	2.5 mm	12.5	99.75
	Chlorophorin	9 mm	6.25	99.87
	Iroko	4.5 mm	12.5	99.75
	Amphotericin B	Control	5000	100
A. flavus	Maackiain acetate	negative	No effect	0
	Formononetin acetate	1 mm	25	99.5
	Chlorophorin	3 mm	6.25	99.87
	Iroko	2.7 mm	12.5	99.75
	Amphotericin B	CONTROL	5000	100

pneumoniae 79.17% more activity.

The seeded agar overlay method showed that maackiain acetate, formononetin acetate, chlorophorin and Iroko at 1 x 10^4 µg/ml inhibited the growth of *F. verticilloide,* with maackiain being the significantly more active compound. In contrast, formononetin acetate, chlorophorin and Iroko inhibited *A. flavus* while maackiain had no effect. *A. flavus* was most sensitive to chlorophorin and Iroko. Amphotericin B at a standardized concentration of 5 mg/ml was used as the control. Chlorophorin was effective at a concentration of 6.25 µg/ml against both *F. verticilloides* and *A. flavus* and that of Iroko demonstrated a twofold enhanced activity on the same strains as compared to chlorophorin. These results and the MIC's are summarised in Table 3. The bioautography method confirmed the results of the well diffusion seeded agar overlay test and showed that *F. verticilloides* was sensitive to maackiain acetate, formononetin, chlorophorin and Iroko (Figure 1) and *A. flavus* was sensitive to chlorophorin, Iroko and formononetin acetate and resistant to maackiain (Figure 2).

DISCUSSION

Medicinal plants possess many potentially valuable therapeutic agents that provide an impetus for further research (Samoylenko et al., 2009; Zhu et al., 2009; Kutchan and Dixon, 2005; Rios et al., 2005; Recio et al., 1989). Testing micro-organisms for their susceptibility to antimicrobials serves as an important aid to chemotherapeutic intervention. The methods commonly used are agar diffusion and broth dilution method (Jawez, 1989). Pellecuer et al. (1976) and Kordali et al. (2005) showed that different results can be obtained for different samples and to demonstrate this he used phenols and essential oils. In his study, it was shown that the agar diffusion method was more effective for phenolic compounds because of its polarity and less effectiveness for non-polar extracts like essential oils. The compounds tested in this study were previously defined as phenols (Oliver-

Bever, 1986; Swinny, 1989). Maackiain was defined as a pterocarpan, formononetin an isoflavone and Iroko and chlorophorin aromatic phenolic compounds.

Hence, in this study, the agar diffusion method was used. The advantage of using this method was that small sample sizes could be used in the screening and the possibility of testing five or six compounds against a single micro-organism at any given time (Rios et al., 1988).

Maackiain and formononetin exhibited minimal activity against all the test bacteria, which could be attributed to their complex structures. Prindle (1983) reported several generalisations that can be made about the structural activity relation of simple and complexed phenols. The position of the alkyl chain may or may not influence activity; separation of the alkyl group from the phenol nucleus by an oxygen (methoxy) decreases activity (Veldhuizen et al., 2006; Ja Kim et al., 2006; Prindle, 1983; Suter, 1941; Klarmann and Shternov, 1936).

Maackiain and formononetin both have alkyl groups that are separated from the phenol nucleus by a methoxy group, which could possibly be the reason why these compounds showed minimal activity against the test bacteria.

The results in this study show that the gram negative bacteria were more resistant than the gram positive bacteria. Previous studies have shown similar results where the gram positive bacteria were more susceptible to plant compounds. In a study carried out by Taniguchi et al. (1978), it was found that of 79 plant extracts tested from 72 species of plants belonging to 35 families, 44 extracts gave positive results against gram positive bacteria, while some of the extracts showed minimal activity against the gram negative bacteria and no activity against *E. coli*. A study by Meyer and Afolayan (1995), on the anti-bacterial activity of extracts from *Helichrysum aureoniteus*, showed that the extracts were active against five gram positive bacteria tested and none inhibited the growth of five gram negative bacteria tested. None of the compounds in this study or the above studies were active against *E. coli*.

The anti-bacterial activity of chlorophorin and Iroko confirmed the results reported by Oliver-Bever (1986) on

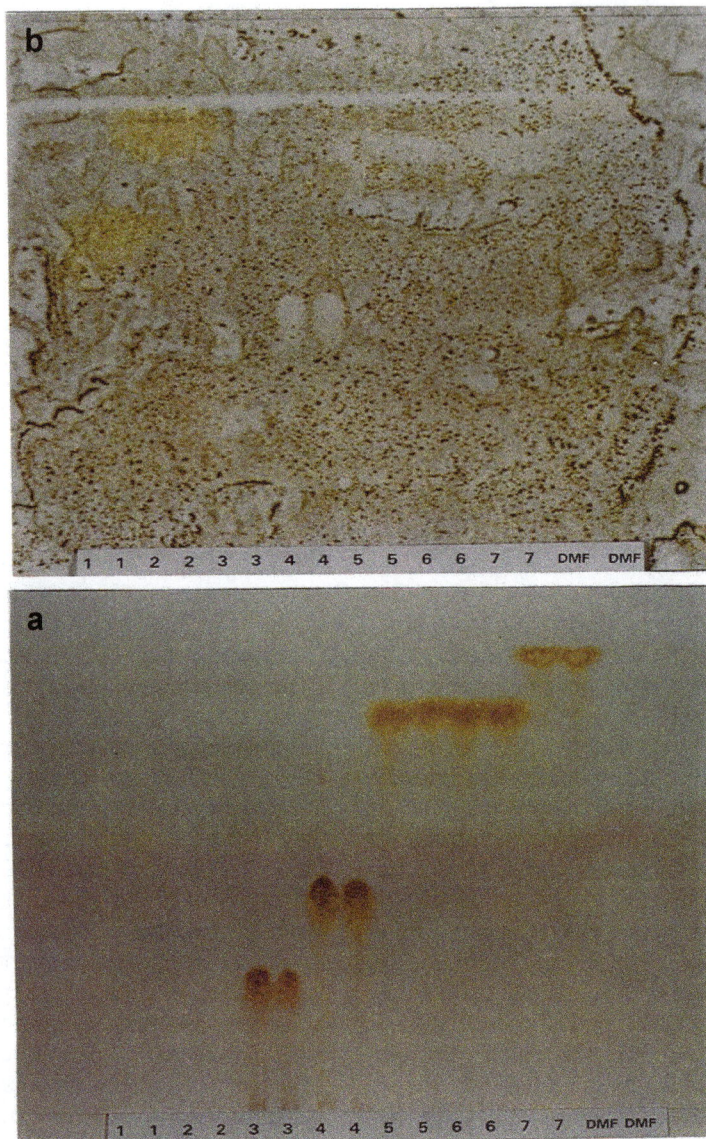

Figure 2. The effect of maackiain acetate, formononetin, chorophorin, Iroko on *A. flavus*. TLC plates with separated plant compounds (a) template and (b) with *A. flavus* grown on the TLC plates showing areas of no growth (arrow) in the same region as where the bands appear on the template.

other phenolic compounds. However, chlorophorin was more active than Iroko although both were phenolic compounds extracted from the same tree. To evaluate these compounds, the direct bioautographic method was used. According to Betina (1973), bioautography is the most important detection method for new or unidentified antimicrobial compounds. Fungi toxic products are measured qualitatively and quantitatively using this technique (Peterson and Edgington, 1969). In this study, two mycotoxin producing species *F. verticilloides* and *A. flavus* were tested against the four plant extracts maackiain, formononetin, chlorophorin and Iroko using the bioautographic method. In this study, all four compounds inhibited *A. flavus* with chlorophorin and Iroko being the

most active against *F. verticilloides* was inhibited by maackiain.

The increased activity of chlorophorin could be attributed to the fact that chlorophorin has four hydroxyl groups whereas Iroko has three hydroxyl groups. A study by Nick et al. (1995) showed that hydroxylated flavones have increased activity as opposed to those extracts that have fewer hydroxyl groups. Cushman et al. (1991), showed that hydroxylated flavones are known as inhibitors of protein kinases. This confirms the results reported by Yan et al. (2008) and Haslam (1996), that polyphenols in addition to giving the usual phenolic reaction they have the ability to precipitate proteins. In addition, Hagerman and Butler (1981) showed that phenols form irreversible complexes with proline rich proteins, which could result in the inhibition of cell wall protein synthesis. These properties of phenols, may explain the mechanism of action of the plant extracts. These complexation reactions are of intrinsic scientific interest as studies in molecular recognition as the basis of possible functions.

ACKNOWLEDGEMENTS

We thank the National Research Foundation for the financial assistance and Dr. E. E. Swinny who kindly donated the plant extracts used in this study.

REFERENCES

Acharya TK, Chatterjee IB (1974). Science Culture, via Chermical Abstracts. 82(1975) 2861X, *Cassia tora.* 40:316.

Atlas RM, Brown AE, Dobra KW, Miller L (1984). Experimental Microbiology: Fundamentals and Applications. Exercise 48: Minimum Inhibitory Concentration (MIC) Method. pp. 267-269.

Balows A, Hausler WJ, Herrmann KL, Lsenberg HD, Shadomy HJ (1991). Manual of Clinical Microbiology. Fifth edition. American Society for microbiology. Washington, pp. 326-334.

Betina V (1973). Bioautography in paper and thin-layer chromatography and its scope in the antibiotic field. J. Chromatogr. 78:41-51.

Chipley JR, Uraih N (1980). Inhibition of *Aspergillus* growth and aflatoxin release by derivatives of benzoic acid. Appl. Environ. Microbiol. 40:352.

Cox PA, Balick MJ (1994). The ethnobotanical approach to drug discovery. Scientific J. America June. pp. 60-65.

Cushman M, Nagarathnam D, Geahlen RL (1991). Synthesis and evaluation of hydroxylated flavons and related compounds potential inhibitors of the protein-tyrosine kinase P56^{lck}. J. Nat. Prod. 54:1345-1352.

Davidson PM (1993). Antimicrobials in Foods. 2nd edition, In: P. M. Davidson and A. L. Branen (Eds.), Marcel Dekker, Inc. New York, pp. 263-305.

Franson RI (1992). Culture of animal cells: A manual of basic techniques. New York, Alan R. Liss Inc. pp. 124-136.

Hagerman AE, Butler LG (1981). The specificity of proanthocyanidin - protein interactions. J. Biol. Chem. 256:4494-4497.

Hailu T, Endris M, Kaleab A, Tsige GM (2005). Antimicrobial activities of some selected traditional Ethiopian medicinal plants used in the treatment of skin disorders. J. Ethnopharm. 100:168-175.

Harrison HF, Peterson JK, Snook ME, Bohac JR, Jackson DM (2003). Quantity and Potential Biological Activity of Caeffeic Acid in Sweet Potato [*Ipomoea batatas* (L.) Lam.] Storage Root Periderm. J. Agric. Food Chem. 51:2943-2948.

Haslam E (1996). Natural Poyphenols (Vegetable Tannins) as Drugs: Possible modes of action. J. Nat. Prod. 59:205-215.

Hoffmann JJ, Timmermann BN, McLaughlin SP, Punnapayak H (1993). Potential anti microbial activity of plants from the Southwestern United States. Int. J. Pharm. 31:101-115.

Ja Kim H, Jung Kim E, Seo SH, Shin CG, Jin C, Lee YS (2006). Vanillic Acid Glycoside and Quinic Acid Derivatives from Gardeniae Fructus. J. Nat. Prod. 69(4):600-603.

Jawez E (1989). Principles of antimicrobial drug action. In: B. G. Katzung(Ed.), Basic and Clinical Pharmacology, 4th Edition. Appleton and Lange, California. pp. 542-552.

Kim HJ, Chen F, Wang X, Rajapakse N, (2006). Effect of Methyl Jasmonate on Secondary Metabolites of Sweet Basil (*Ocimum basilicum* L). J. Agric. Food Chem. 54:2327-2332.

Klarmann EG, Shternov VA (1936). Bactericidal value of coal-tar disinfectants. Limitation of the *B. typhous* phenol coefficient as a measure. Ind. Eng. Chem. 8:369-372.

Kordali S, Cakir A, Mavi A, Kilic H, Yildirim A (2005). Screening of Chemical Composition and Antifungal and Antioxidant Activities of the Essential oils from three Turkish *Artemisia* Species. 53:1408-1416.

Kubo A, Kubo I (1995). Anti-microbial agents from *Tanacetum balsamita*. J. Nat. Prod. 58:1565-1569.

Kutchan T, Dixon RA (2005). Physiology and metabolisim Secondary metabolism: nature's chemical reservoir under deconvolution. Curr. Opin. Plant Biol. 8:227-229.

Lee KH (2004). Current Developments in the Discovery and Design of New Drug Candidates from Plant Natural Product Leads. J. Nat. Prod. 67:273-283.

Leven M, Van den Berghe DA, Mertens F, Vlietinck A, Lammens E (1979). Screening of higher plants for biological activitites. In: Antimicrobial activity. Planta Medica. J. Med. Plant Res. 36:311-321.

Lsenberg HD (1980). Chemical Microbial Procedures Handbook. ASM, Publishers. Vol.15:13.1-13.13.

Lu TK, Collins JJ (2009). Team combats antibiotic resistance with engineered viruses. Proceedings of the national Academy of Sciences. March 2nd in Biology/Microbiology. Massachusetts Institute of Technology.

McCutcheon AR, Ellis SM, Hanock REW, Towers GHN (1994). Anti-fungal screening of medicinal plants of British Columbian native peoples. J. Ethnopharm. 44:157-169.

McLaughlin JL (2008). Paw Paw and Cancer: Annonaceous Acetogenins from Discovery to Commercial Products. J. Nat. Prod. 71:1311-1321.

Meyer JJM, Afolayan AJ (1995). Antibacterial activity of *Helichrysum aureonitens* (*Asteraceae*). J. Ethnopharm. 47:109-111.

Mitscher LA, Drake S, Gollapudi SR, Okwute SK (1987). A modern look at Folkloric use of anti-infective agents. J. Nat. Prod. 50:1025-1040.

Nick A, Rali T, Sticher O (1995). Biological screening of traditional medicinal plants from Papua New Guinea. J. Ethnopharm. 49:147-156.

Oliver Bever BEP (1986). Medicinal plants in tropical West Africa. University Press. Cambridge. pp. 1-8.

Pellecuer S, Allergrini J, Simeon de Buochberg M (1976). Huiles essentielles bactericides and fungicides. Review from the Pasteur Institute de Lyon. 9:135-159.

Peterson CA, Edgington, LU (1969). Quantitative estimation of the fungicide benomyl using a bioautograph technique. J. Agric. Food Chem. 17:898-899.

Prindle RF (1983). Phenolic compounds. Disinfection, Sterilization and Preservation, 3rd ed., S. S. Block and Febiger (Editors). (Marcel Dekker, Inc) Philadelphia. pp. 197-205.

Recio MC, Rios JL, Villar A (1989). A review of some anti-microbial compounds isolated from medicinal plants reported in the literature 1978-1988. J. Phytother. Res. 3:117-125.

Rios JL, Recio MC (2005). Medicinal plants and antimicrobial activity. J. Ethnopharm. 100:80-84

Rios JL, Recio MC, Villar A (1988). Screening methods for natural products with antimicrobial activity. A review of the Literature. J. Ethnopharm. 23:127-149.

Samoylenko V, Ashfaq MK, Jacob MR, Tekwani BL, Khan SI, Manly SP, Joshi VC, Walker LA, Muhammad I (2009). Indolizidine, Antiinfective and Antiparasitic Compounds from *Prosopis glandulosa* var. *glandulosa*. J. Nat. Prod. 72:92-98.

Shan B, Cai YZ, Brooks JD, Corke H (2007). Antibacterial Properties and Major Bioactive Components of Cinnamon Stick (*Cinnamomum burmannii*): Activity against Foodborne Pathogenic Bacteria. J. Agric. Food Chem. 55:5484-5490.

Suter CM (1941). Relationships between the structures and bactericidal properties of phenols. Chem. Rev. 28:269-281.

Swinny EE (1989).The structure and synthesis of Metabolites from *Virgilia oroboides* and *Chlorophora excelsa* (Iroko).M Sc. Dissertation, University of Durban-Westville, Durban, South Africa.

Taniguchi M, Chapya A, Kubo I, Nakanishi K (1978). Screening of East African Plants for Antimicrobial activity. Chem. Pharm. Bull. 26:2910-2913.

Veldhuizen EJA, Tjeerdsma-van Bokhoven JLM, Zweijtzer C, Burt SA, Haagsman HP (2006). Structural requirements for the Antimicrobial Activity of Carvacrol. J. Agric. Food Chem. 54:1874-1879.

Valle E (1957). On anti-fungal factors in potato leaf. Acta Chem. Scandinavica. 11:395-399.

Walsh G (1988). Bio-pharmaceuticals : Biochemistry and Biotechnology. John Wiley and sons, Chichester. pp. 24 -56.

Washington JA, Warren E, Karlson AG (1972). Stability of Barium sulfate Turbidity standards. J. Appl. Microbiol. 24:1013-1017.

World Health Organisation (2005). General Guidelines for Methodologies on Research and Evaluation of Traditional Medicine. WHO, Geneva, Switzerland.

Wu HS, Raza W, Fan JQ, Sun YG, Bao W, Shen QR (2008). Cinnamic Acid Inhibits Growth but Stimulates Production of Pathogenesis Factors by in Vitro Cultures of *Fusarium oxysporum* f.sp. niveum. J. Agric. Food Chem. 56:1316-1321.

Yan Y, Hu J, Yao (2009). Effects of Casien, Ovalbumin, and Dextran on the Astringency of Tea Polyphenols Determined by Quartz Crystal Microbalance with Dissipation. Langmuir American Chemical Society. 25:397-402.

Yoon KD, Jeong DG, Hwang YH, Ryu JM, Kim J (2007). Inhibitors o Osteoclast Differentiation from *Cephalotaxus koreana*. J. Nat. Prod 70:2029-2032.

Zhu Y, Zhang P, Yu H, Li J, Wang MW, Zhao W (2009). Anti-*Helicobacte pylori* and Thrombin Inhibitory Components from Chinese Dragon' Blood, Dracaena cochinchinensis. J. Nat. Prod. 70:1570-1577.

10

Efficiency evaluation of three fluidised aerobic bioreactor based sewage treatment plants in Kashmir Valley

Dilafroza Jan, Ashok K. Pandit and Azra N. Kamili

Centre of Research for Development, University of Kashmir, Srinagar-190006, Jammu and Kashmir, India.

The present investigation was conducted to monitor the physico-chemical characteristics and microbial load of sewage treatment plants (STPs) around the Dal Lake. The results show highly significant (P<0.001) reduction in some physico-chemical features and in microbial load at outlet of each STP. Order of reduction in all the STPs was found to be biochemical oxygen demand (BOD) > chemical oxygen demand (COD) > conductivity and fecal coliform (FC) > total coliform (TCC) > fecal streptococcus (FS) in the case of physico-chemical parameters and microbial characteristics, respectively. The overall performance of the wastewater treatment plant effectively removed TCC, FC and FS as follows, 52, 65 and 45%, respectively. Raw sewage showed insignificant (p>0.05) variation in some of the physico-chemical features and microbial load between the three STPs. Similarly, effluent also showed insignificant (p>0.05) variation in some of the physico-chemical features and microbial load, except conductivity which showed significant difference between the three STPs. Efficiency rates showed significant (p<0.05) differences in COD between the three STPs. The removal efficiency rate was not dependent on the type of STP and the year. It can be concluded from the study that the majority of physico-chemical features and microbial load exceeded the permissible limit as per Indian national standards. Therefore as per the results, it is suggested that the effluent should be pretreated before disposing into the environment. In addition, there is an urgent need to improve their efficiency rate by including advanced tertiary treatment processes such as rapid sand filtration, UV disinfection, chlorination, effluent polishing, construction of artificial wetlands, etc.

Key words: Microbial load, sewage treatment plants, contamination load, efficiency rate.

INTRODUCTION

Urbanization, industrialization, modernization as well as agricultural activities have put tremendous pressure on the limited freshwater resources, causing eutrophication and pollution of freshwater bodies all over the world. In recent times, abrupt increase in production and domestic use of organic chemicals has obliged sewage treatment plants (STPs) to improve their efficiency. Freshwater systems of Kashmir have not remained immune to the anthropogenic pressures and many of these, especially those located close to the human habitations, have deteriorated during the last 50 years. In addition to other anthropogenic activities, sewage discharges from STPs are considered a major contributor of contamination in this urban lake of Kashmir.

Wastewater is a major burden for water bodies and improper disposal of sewage leads to oxygen demand, increased nutrient concentration and promotion of toxic algal blooms leading to a destabilized aquatic ecosystem

(Morrison et al., 2001). It has been seen by various agencies that India wastewater is comprised of high levels of organic, inorganic and microbial contaminants (Bohdziewicz and Sroca, 2006). So far, extensive work has been carried out to study the physico-chemical removal efficiencies of STPs, whereas there is less literature regarding the microbial load in wastewater treatment plants and their removal efficiency. Sewage from households is collected via a sewer system and flows to STP for treatment of chemicals and microbial load. The high level of fecal contamination and enteric viruses present in raw sewage is a major concern for public health and the environment; and therefore assessment of sewage is essential to safeguard the public health (Okoh et al., 2005, 2007).

In most cases, STPs consist of two types of treatment systems: a physical and a biological purification steps. In physical purification, removal of the chemical is mostly due to sorption of chemicals to organic carbon. The effectiveness of the removal is directly related to the size and density of the particles. In the biological purification, removal is achieved by bacterial biodegradation, which mainly occurs via oxidation. At the end of the treatment processes, sludge and final effluent are released to the environment. Every chemical that enters the STP that is neither sorbed nor degraded will enter the environment via the effluent or evaporation from the STP. The priority objectives of wastewater treatment are to degrade organic wastes so that they do not cause oxygen demand in the receiving water body, remove nutrients to prevent eutrophication and protection of public health by destroying the pathogenic microorganisms (Gerardi, 2006; Akpor and Muchie, 2011).

Studies have shown that sewage treatment processes might also affect physico-chemical parameters of the final effluent such as biochemical oxygen demand (BOD), chemical oxygen demand (COD), electrical conductivity, total hardness, alkalinity, dissolved oxygen, some metals and non-metal ions (Rawat et al., 1998; Adami et al., 2007). Although, various microorganisms in water are considered to be critical factors in contributing to numerous waterborne outbreaks, they play many beneficial roles in wastewater influents (Kris, 2007). In addition, purification processes remove pathogenic microorganisms (Reasoner, 1982 Wang et al., 1966). Furthermore, microbiological indicators have been used for decades to monitor fecal pollution of water (Standard Methods, 1998).

Different studies have evaluated the efficiency of STPs and have compared the concentration of the chemical in the influent and that in the effluent. In most studies, significant reduction has been observed at outlet sites of STPs (Saha et al., 2012; Kumar et al., 2010; Desai and Kore, 2011). However, some studies have shown little or no reduction of pollutant concentration (Igbinosa and Okoh, 2009; Antunes, 2007; Momba et al., 2006; Akpor and Munche, 2011) which is a major concern for water

bodies as well as public health. The comparative studies between STPs have shown both significant and insignificant variation (Jamwal et al., 2009; Kumar, 2010) in efficiency rates. In the past, some studies have also shown that STPs deviate from normal permissible limit which have been given by WHO and the United States Environmental Protection Agency (EPA) (Igbinosa, 2009; Antunes, 2007; Momba et al., 2006; Akpor, 2011).

Dal Lake receives effluent from the three STPs namely Habak STP, Hazratbal STP and Lam STP as well as domestic wastewater from the surrounding settlements. These STPs were constructed by a private firm under the guidelines of Lakes and Waterways Development Authority (LAWDA) in 2004 at the cost of Rs 8.90 crore. In recent past, there has been debate on the working capability of these STPs. Although LAWDA seems to be satisfied with the working condition of these STPs, and claims that Dal lake's health would improve after all the STPs have started working, some analytical reports have raised questions about the working condition of these STPs. The research and monitoring division of LAWDA in August 2006, for example, reported increased nutrient concentration at the outflow stage, thus negating the claims made by LAWDA. So, knowing the above mentioned facts was necessary to perform a current monitoring survey on these STPs in order to know the present status of these STPs.

No past extensive study has been carried out to assess the efficiency and quality of these STPs. Because of the associated dangers of sewage, the present study was carried out to investigate the impact of the wastewater effluent discharged and to estimate the pollutant removal efficiency of STPs around the Dal Lake. We predicted that removal efficiency will depend on the characteristic features (working capability) of the individual STP, extent of aeration, hydraulic retention time, contact time and type of treatment used. In addition, microbial load will be greater in the influent than in the effluent (Figure 1).

MATERIALS AND METHODS

This study was conducted at three sewage treatment plants viz. Habak STP (34008'50"N - 74050'36"E), Hazratbal STP (34°08'06"N - 74°50'29"E) and Lam STP (34°07'42"N - 74°523'36"E) in the vicinity of Dal Lake with the design capacity of 3.2, 7.5 and 4.5 MLD, respectively (Table 1). All these plants receive domestic sewage and are treated using the fluidised aerobic bioreactor (FAB) biological treatment systems; a combined, dispersed and attached bacterial growth on fluidized media that is a modified version used in Germany, Netherlands, Europe and Canada successfully. This technology comprises of components like screening, grit removal, fluidised aerobic bioreactor followed by clarification and addition and precipitation to remove phosphates and chlorination (Figure 2). The final treated effluent is discharged into the lake. The water samples were collected on monthly basis for a period of 24 months between June 2010 and May 2012 for analysis of physico chemical features, in white plastic containers, which were previously sterilized with 70% alcohol and rinsed with distilled water. For microbial analysis, samples were collected seasonally at the three

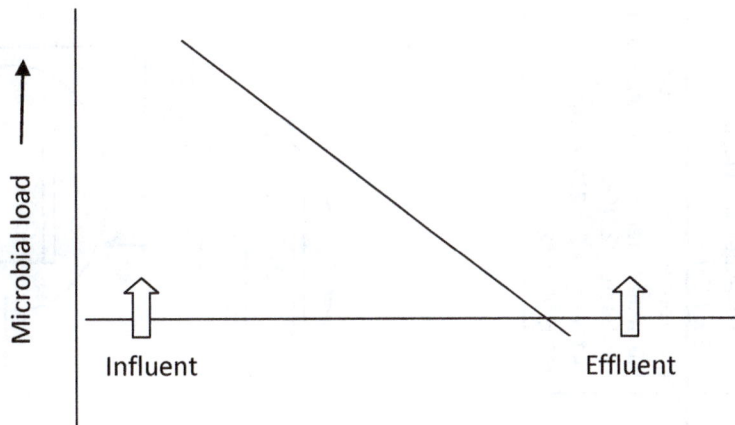

Figure 1. Hypothesis showing that microbial load decreases as the wastewater proceeds through the treatment processes.

STPs. At the sites, the containers were rinsed thrice with the wastewater before being used to collect the samples.

Physico-chemical parameters of water samples

The influent and effluent water samples were collected between 10.00 and 15.00 h from all the sampling stations in 1 L polyethylene bottles. The parameters pH and conductivity were recorded on the spot. For the estimation of dissolved oxygen, separate samples were collected in separate glass bottles and fixed at the sampling sites in accordance with the Winklers method (APHA, 1998). BOD was determined by the 5 day test method, while COD determination was carried out using the open reflux method as per Standard Methods in APHA (1998). For removal efficiency of physicochemical parameters, inlet concentrations were subtracted from outlet concentrations for each parameter.

$$\text{Removal efficiency} = \frac{\text{concentration in influent} - \text{concentration in effluent}}{\text{concentration in influent}} \times 100$$

Microbial examination of water samples

Microbiological examination of samples was conducted promptly as possible (within 24 h) after collection or were stored at 4°C in a refrigerator until use. Serial dilutions were prepared immediately after sample collection. The proper dilutions for various bacterial groups were selected so that number of colonies on plate was between 30 and 300 using spread plate method. A multiple tube fermentation technique or most probable number (MPN) technique was used to determine the bacterial indicators as faecal coliforms (FC) and faecal streptococci (FS) according to standard methods described in APHA (1998). Multiple tube fermentation method used in the present work included measurement of total plate count and MPN of coliform. After incubation for 24 h at 35°C, results were recorded when acid and gas liberated in Durham tubes had changed in color to yellow. The spread-plate method was used for all counts. FC agar and FS agar were used for enumeration of faecal coliform and faecal streptococci. Each test was done in triplicate and the geometric means were recorded. The removal efficiency of bacterial indicators was calculated using the following formula:

$$\text{Removal efficiency} = \frac{\log \text{CFU in influent} - \log \text{CFU in effluent}}{\log \text{CFU in influent}} \times 100$$

Statistical analysis

Students t test was used to assess the significant variation between the raw influent and the effluent in the different STPs. One way ANOVA test was used to analyze the significant differences in influent and effluent between the three STPs. Similarly, efficiency rate between the different STPs was tested by using one way ANOVA test.

All statistics were carried out with the SPSS 11.5 statistical software package with significance levels set at P<0.05.

RESULTS

The data shows highly significant differences (P<0.001) in physico chemical features and microbial data between the inlet and outlet samples (Tables 2 and 3; Figures 3 and 8). Statistically, insignificant differences were observed in raw sewage and effluent between the three STPs (P>0.05) (Tables 4 and 5). Similarly, insignificant differences were observed in the efficiency rate of

Table 1. Details of sewage treatment plant around Dal Lake (LAWDA).

STP name and location	STP at HABAK	STP at Hazratbal	STP at Lam Nishat
Design capacity/day (MLD)	3.2	7.5	4.5
Land required (sqm)	600	1123	850
Average flow rate at inlet	Average flow rate 133.33 (m³/h)	Average flow rate 312.5 (m³/h)	Average flow rate 187.5 (m³/h)
Peak flow rate	333 MLD	781.25 (m³ /h)	468.75 (m³ /h)
Grit chamber specifications	Long channel with Hopper bottom 6.0 x 1.2 m	Long channel with Hopper bottom 9.5 x 1.9 m	Long channel with Hopper bottom 7.5 x 1.5 m
Aeration tank i) Aeration tank volume (m³) ii) Rated aeration capacity kg/KW hr or kg/hour	Fluidized aerobic bio-reactors 5.0 m diameter, 5.0 m depth + 1 meter free board x 2 NOs 98.19 m³ x 2 = 196.38 m³ 260 m³/hour 130 x 2 + 1 Standby	Fluidized aerobic bio-reactors 7.75 m diameter, 5.0 m depth 1 meter free board x 2 NOs 236 m³ x 2 = 472 m³ 650 m³/hour 325 x 2 + 1 Standby	Fluidized aerobic bio-reactors 6.0 m diameter, 5.0 m depth + 01 meter free board x 2 NOs 141.39 m³ x 2 = 282.78 m³ 400 m³/hour 200 x 2 + 1 Standby

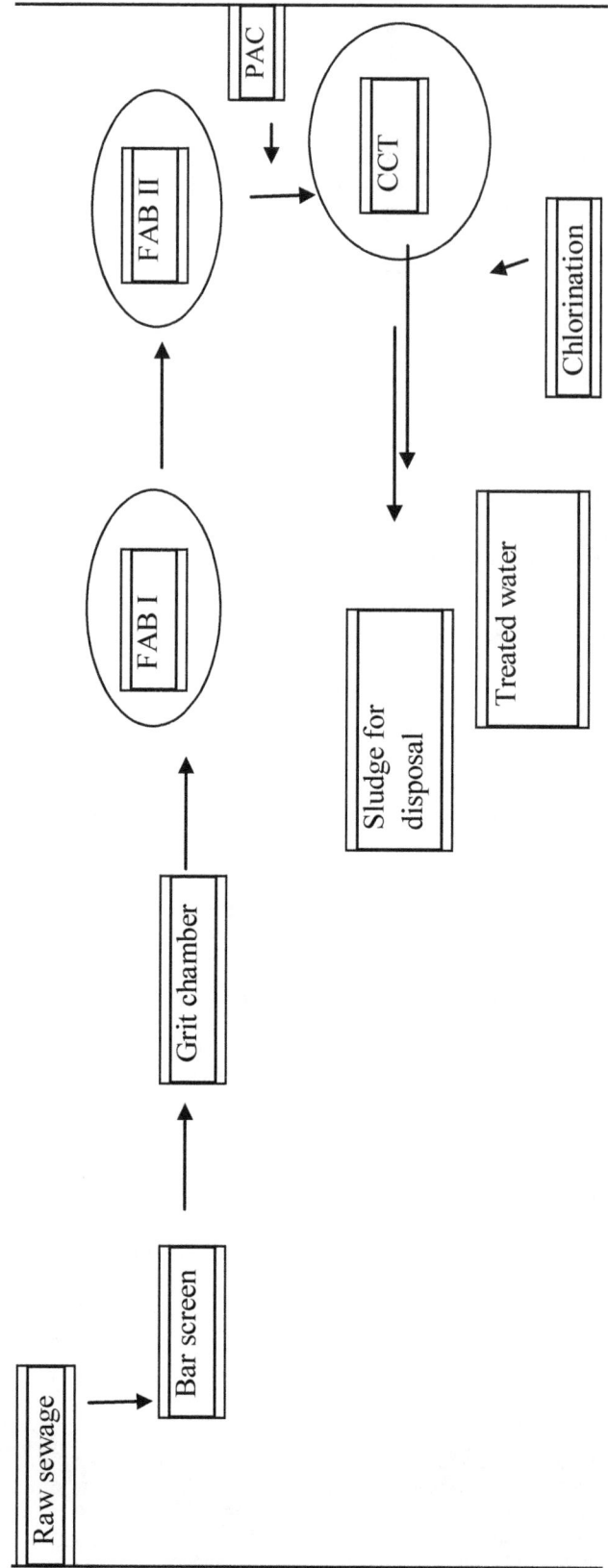

Figure 2. Process flow diagram for FAB based sewage treatment plants. FAB, Fluid aerobic bioreactor; PAC, poly aluminum chloride; CCT, calori tube settler.

Table 2. Physico chemical characteristics of raw sewage and effluent.

STPs	Raw conductivity (µScm⁻¹)	Effluent conductivity (µScm⁻¹)	P-value	Raw DO (mgl⁻¹)	Effluent DO (mgl⁻¹)	P-value	Raw BOD (mgl⁻¹)	Effluent BOD (mgl⁻¹)	P-value	Raw COD (mgl⁻¹)	Effluent COD (mgl⁻¹)	P-value
Habak												
2010-2011	787.81±136.9	533.27±86.86	P<0.001	1.5±.84	3.92±0.58	P<0.001	200.72±35.35	70.63±11.70	P<0.001	496.30±103.15	233.45±42.31	P<0.001
2011-2012	658.0±46.72	443.4±28.5	P<0.001	1.70±0.48	3.94±0.49	P<0.001	185.72±44.32	65±10.49	P<0.001	486.52±152.13	228.9±72.46	P<0.001
Hazratbal												
2010-2011	866.18±94.05	571.27±63.75	P<0.00	1.44±0.88	3.87±0.50	P<0.00	200±42.46	70.45±11.80	P<0.00	471.74±161.77	216.4±68.35	P<0.00
2011-2012	783.09±63.65	520.7±45.6	P<0.001	1.49±0.72	3.94±0.53	P<0.001	205.27±42.68	70±12.09	P<0.001	520.45±110.32	234.3±43.56	P<0.001
Lam												
2010-2011	880.54±59.9	590.90±37.86	P<0.001	1.31±.70	3.50±0.47	P<0.001	206±46.37	69.81±13.71	P<0.001	519.39±78.87	225.43±34.1	P<0.00
2011-2012	791.8±117.2	521.63±76.64	P<0.001	1.65±0.91	4.07±0.45	P<0.001	243±48.78	80.18±15.03	P<0.001	566.32±190.9	252.94±87.43	P<0.00

Table 3. Microbial load in sewage and effluent.

STPs	Raw Log TCC (×10⁷)	Effluent Log TCC (×10⁶)	P-value	Raw Log FC (×10⁵)	Effluent Log FC (×10⁴)	P-value	Raw Log FS (×10⁷⁴)	Effluent Log FS (×10³)	P-value
Habak									
2010-2011	1.89±0.14	1.96±0.20	P=0.001	1.8±0.04	0.58±0.58	P=0.002	1.0175±1.01	0.55±0.12	P<<0.001
2011-2012	0.87±0.09	0.94±0.12	P<<0.001	1.88±0.04	1.05±0.11	P<<0.001	1.92±1.27	0.7148±0.03	P<<0.001
Hazratbal									
2010-2011	2.07±0.11	2.04±0.19	P<<0.001	1.7±0.05	0.58±0.65	P<<0.001	0.99±1.25	0.66±0.05	P<<0.001
2011-2012	0.99±0.10	0.99±013	P<<0.001	1.91±0.07	1.05±0.11	P<<0.001	1.87±0.12	0.75±0.10	P<<0.001
Lam									
2010-2011	1.99±0.16	2.03±0.18	P=0.001	1.82±0.07	0.58±0.65	P<<0.001	1.02±1.26	0.71±0.12	P<<0.001
2011-2012	0.96±0.10	0.97±0.12	P<<0.001	1.87±0.07	1.05±0.11	P<<0.001	1.92±0.13	0.70±0.1	P<<0.001

different physico chemical features and microbial loads between the three STPs (P>0.05), except COD, which showed significant variation between the three STPs (F = 4.6; P = 0.013) (Table 6).

The results obtained showed the pH on alkaline sideand ranged between 7.48 ± 0.08 and 7.53 ± 0.11 in raw sewage. Conductivity was found to be vary from 1.31 ± 0.70 to 1.70 ± 0.48 mg/L, 185.72 ± 44.32 to 243 ± 48.78 mg/l and 486.52 ± 152.13 to

880.54 ± 59.9 (Figure 3). In the raw sewage, DO, BOD (Figure 4) and COD (Figure 5) were found to fluctuating in sewage from 658.0 ± 46.72 to

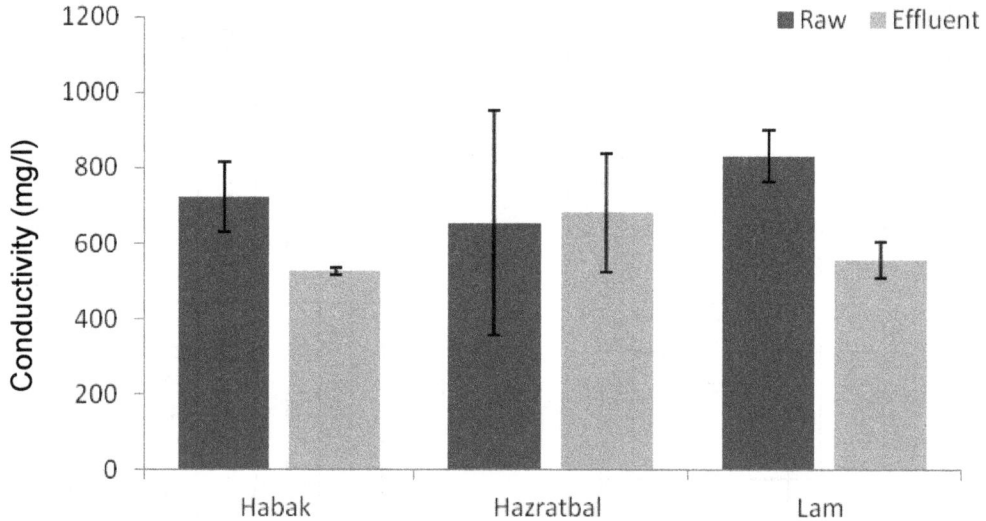

Figure 3. Conductivity of influent and effluent sewage from different STPs.

Table 4. Removal efficiencies (%) of different STPs.

STP location	Conductivity (μScm^{-1})	BOD (mgl^{-1})	COD (mgl^{-1})	TCC (cfu/100 ml)	FC (MPN/100 ml)	FS (MPN/100 ml)
Habak						
2010-2011	32.28±1.4	64.62±1.07	52.62±3.3	53.59±5.38	69.99±6.3	47.49±3.54
2011-2012	32.47±1.58	64.47±3.08	52.64±4.33	50.82±2.3	62.06±1.13	45.35±3.31
Hazratbal						
2010-2011	34.05±1.00	64.32±2.28	53.52±4.08	52.28±2.66	62.98±2.11	47.52±3.55
2011-2012	33.48±1.36	65.54±2.39	54.65±3.5	50.53±3.33	60.70±4.05	44.16±3.28
Lam						
2010-2011	33.03±1.56	65.99±2.82	56.55±1.8	51.67±4.78	60.82±5.32	46.10±3.53
2011-2012	34.01±0.95	66.61±1.79	55.34±4.2	52.30±4.38	62.29±3.8	45.52±2.77

Table 5a. Results of ANOVA test in raw sewage between three STPs.

Test parameter	Conductivity (μScm^{-1})	BOD (mgl^{-1})	COD (mgl^{-1})	TCC (cfu/ ml)	FC (MPN/100 ml)	FS (MPN/100 ml)
F	2.5	1.22	0.95	0.44	0.13	0.09
P-value	0.08ns	0.30ns	0.39ns	0.64ns	0.87ns	0.91ns

*Indicates significant at 0.05; ns indicates not significant.

Table 5b. Results of ANOVA test in effluent between three STPs.

Test parameter	Conductivity (μScm^{-1})	BOD (mgl^{-1})	COD (mgl^{-1})	TCC (cfu/ml)	FC (MPN/100 ml)	FS (MPN/100 ml)
F	6.12	2.29	0.28	0.37	1.01	0.00
P-value	0.004*	0.10ns	0.75ns	0.69ns	0.37ns	1.0ns

*Indicates significant at 0.05; ns indicates not significant.

Table 6. Results of ANOVA test in efficiency rate between three STPs.

Test Parameter	Conductivity (μScm^{-1})	BOD (mgl^{-1})	COD (mgl^{-1})	TCC (cfu/ ml)	FC (MPN/100 ml)	FS (MPN/100 ml)
F	0.13	0.61	4.6	0.47	1.57	0.10
P-value	0.87^{ns}	0.54^{ns}	0.013^{*}	0.63^{ns}	0.22^{ns}	0.90^{ns}

*Indicates significant at 0.05; ns indicates not significant.

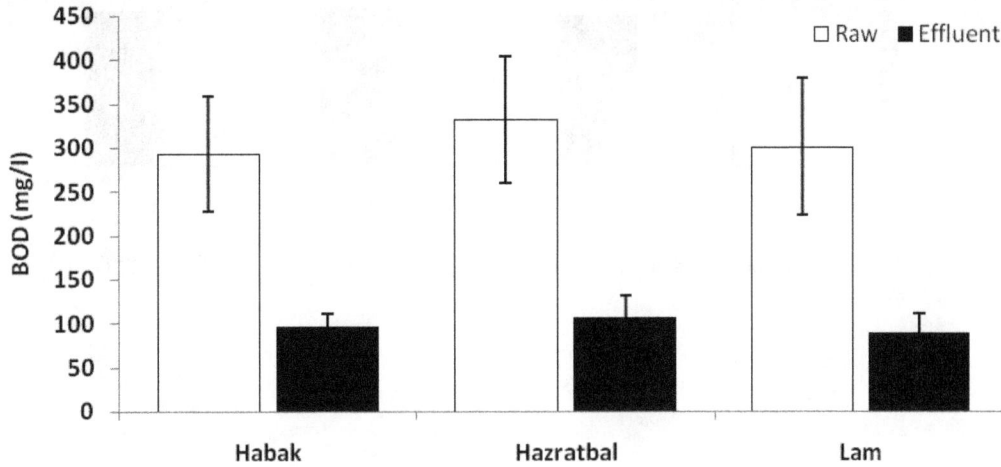

Figure 4. BOD of influent and effluent sewage from different STPs.

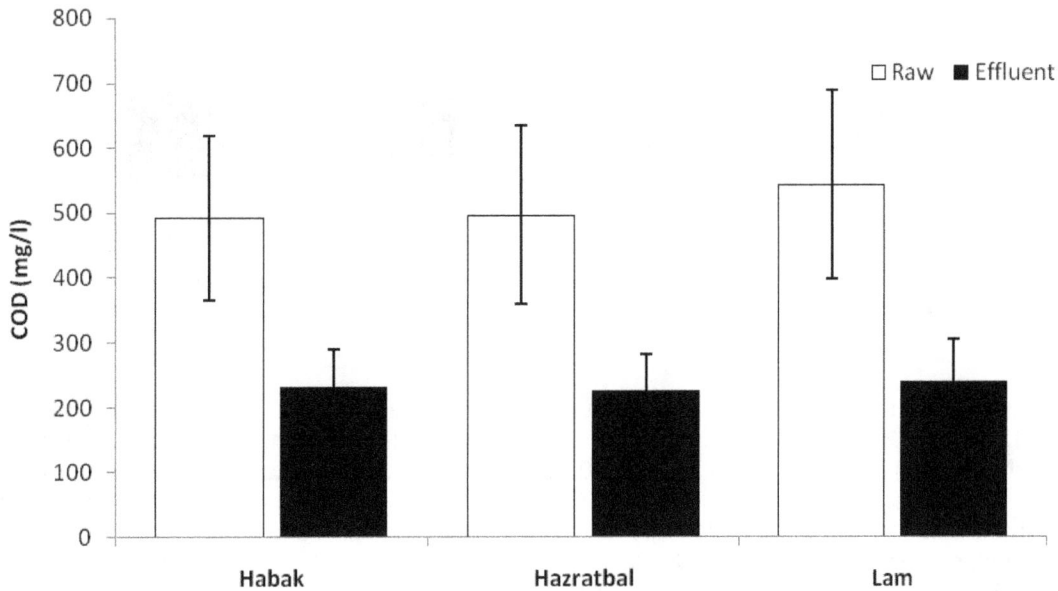

Figure 5. COD of influent and effluent sewage from different STPs.

to 566.32 ± 190.9 mg/L, respectively. The pH in the effluent was found to vary in the range of 7.64 ± 0.12 to 7.73 ± 0.09. The effluent concentrations of conductivity ranged from 443.4 ± 28.5 to 590.90 ± 37.86 μScm^{-1}. The DO, BOD and COD concentration in effluent ranged from 3.50 ± 0.47 to 4.07 ± 0.45 mg/L, 65 ± 10.49 to 80.18 ± 15.03 mg/L and 216.4 ± 68.35 to 234.3 ± 43.56 mg/L, respectively at all the sites. The overall removal efficiency of conductivity, BOD and COD was found to be 32.28 ± 1.4 to 34.05 ± 1.00, 64.32 ± 2.28 to 66.61 ± 1.79 and 52.62 ± 3.3 to 56.55 ± 1.8, respectively.

In the case of microbial indicators, TCC, FC and FS

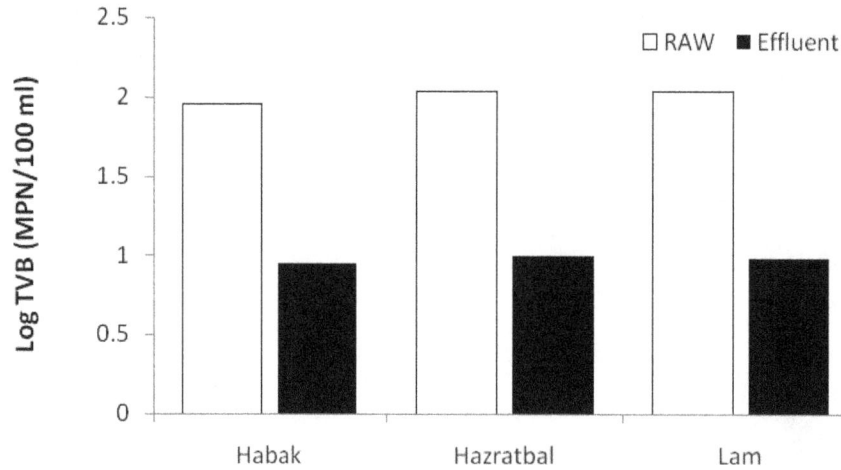

Figure 6. Microbial quality of raw and effluent sewage from different STPs (TCC in 2010-2011).

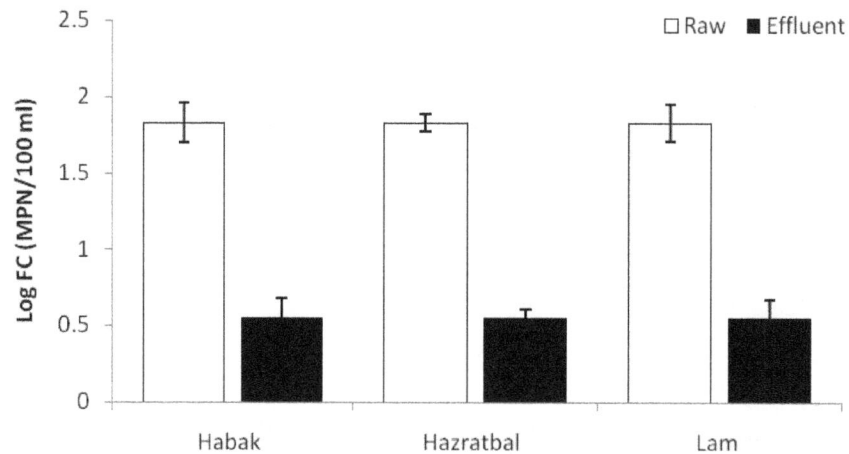

Figure 7. Microbial quality of influent and effluent sewage from different STPs (FC in 2010-2011).

concentrations in the inlet sample were found to fluctuate from 0.94 ± 0.12 to 2.07 ± 0.11 (cfu/ml), 1.7 ± 0.05 to 1.91 ± 0.07 (MPN/100 mL) and 0.99 ± 1.25 to 1.92 ± 0.13 (MPN/100 mL), respectively (Figures 6, 7 and 8). TCC, FC and FS concentrations were found to vary in the outlet water samples from 0.94 ± 0.12 to 2.04 ± 0.19 (cfu/ml), 0.58 ± 0.58 to 1.05 ± 0.11 (MPN/100 mL) and 0.55 ± 0.12 to 0.75 ± 0.10 (MPN /100 mL), respectively. Observations also revealed that the percent removal efficiency of TCC, FC and FS in all the STPs ranged from 50.53 ± 3.33 to 53.59 ± 5.38, 60.70 ± 4.05 to 69.99 ± 6.3 and 44.16 ± 3.28 to 47.52 ± 3.55, respectively.

DISCUSSION

Although data shows significant differences in physico-chemical features and microbial load between the inlet and outlet, nevertheless, these variations do not meet the Indian national standards. As per Indian standards, to discharge effluents into water bodies, BOD, COD and faecal coliform should be less than 30 and 250 mg/L and 2500 MPN/100ml, respectively.

The insignificant differences observed in the efficiency rate between the three STPs may be due to the fact that the same treatment technologies are employed for treatment of sewage at the three plants. Jamwal et al. (2009) while evaluating the efficiency of STPs in Delhi observed that effluents from all STPs exceeded FC standard of 10^3 MPN/100 ml for unrestricted irrigation criteria set by the National River Conservation Directorate (NRCD). However, they investigated STPs which were using different types of technologies including activated sludge process (ASP), extended aeration (EA),

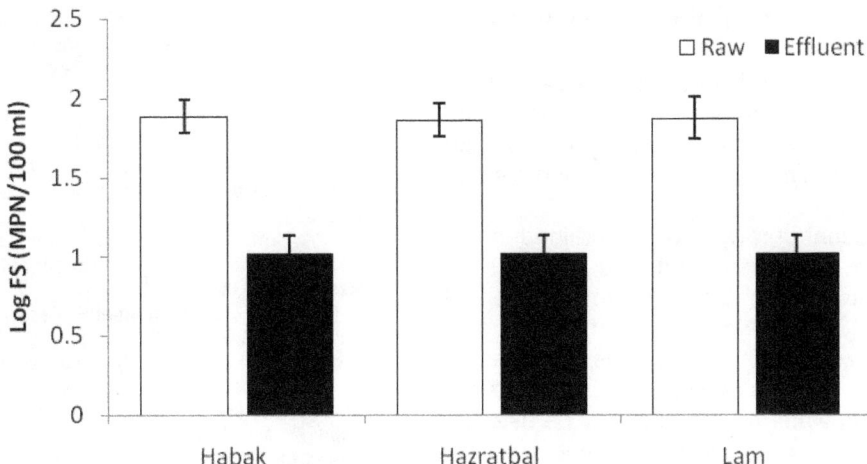

Figure 8. Microbial quality of influent and effluent sewage from different STPs (FS in 2011-2012).

BIOFORE, trickling filter and oxidation pond for the treatment of sewage. In our case, all STPs used the FAB based technology, which is a recent technique for removal of pollutants from municipal sewage. The comparatively low efficiency rate in the three STPs may be due to the fact that primary sedimentation is not carried out in these STPs and as such, the efficiency rate of pollutant removal is affected. Furthermore, the high bacterial load in the effluent may be attributed to the fact that an appropriate dose of chlorine for disinfection is not given to the effluent. The results show that faecal streptococci are more resistant and persistent even after chlorination as compared to faecal coliforms. These results coincide very well with the findings of Cohen and Shuval (1972). The reduction of microbes depends on protozoan predation, settlement of suspended solids, inactivation due to sunlight, activity of filamentous bacteria, turbidity and sedimentation.

It seems from the results that pH increased slightly from raw to final effluent which may be due to the reduction of free CO_2 level in final effluent and addition of polyaluminium chloride. The pH range recorded for all the sampling sites lie within the WHO pH tolerance limit, that is, 6 to 9 for wastewater to be discharged into water body. Our results coincide with the findings of Morrison et al. (2001) who also found increase in pH level during treatment process. Occasional absence of DO concentration in sewage was recorded in some STPs due to heavy organic loading and septic conditions. Kumar et al. (2010) found similar sewage characteristics while working on two STPs in Karnataka. The DO concentrations of the effluents in our study were less than 5 mg/l. Consequently, these water sources would not be suitable for use of aquatic ecosystems (Rao, 2005). However, there was a slight increase in the effluent DO content due to the treatment processes, which may reduce the amount of impurities present in sewage

through oxidation of organic matter (Prescott et al., 2002).

The high values of conductivity in raw sewage in the present study indicate increased salt concentration and pollution level of STPs. Reduction in BOD in all STPs during the treatment process may be due to the oxidation of organic matter by microorganisms that are used in FAB treatment as well as coagulation and flocculation brought about in the Claritube settler, which is a clarification cum flocculation chamber. Similar results were observed by Jamwal et al. (2009). The percentage removal of BOD in all the STPs was 67.86 ± 2.6 to $70.00 \pm 3.6\%$, which is below the expected value of 85 to 90%, thus showing that BOD reduction is less than the expected. The same rate of reduction in BOD in all STP's could be due to similar technology which is currently used in all three STPs. COD shows similar trend as BOD. The significant variation in efficiency rate in COD between three STPs could be due to different types of sewage coming from different catchment areas.

Similarly, results show decrease in COD level at outlet during the treatment process due to the above mentioned facts. High COD and BOD concentration observed in the wastewater might be due to the use of chemicals, which are organic or inorganic that are oxygen demanding in nature (Akan et al., 2008). The values for most of the parameters in the discharged effluent were almost higher than the acceptable limits. This shows the inefficiency of the treatment plant in removing the pollutants in the sewage.

Tertiary treatment methods are essential to remove nitrogen, phosphorus, suspended solids, dissolved solids, refractory organics and heavy metals. Tertiary treatment process involves the addition of certain chemicals such as alum and polyelectrolytes which convert the dissolved substances into a solid settle able form. These chemical coagulates, mix up with the sewage water to form an insoluble gelatinous floc of suspended solids which settles down quickly.

In our case, it was observed that tertiary treatment is not capable of removing pollutants to a large extent. The reason may be inflow of sewage beyond designed capacity and improper maintenance and lack of expertise in the field.

The high coliform count in raw sewage obtained in our results may be an indication that the sewage is comprised of faecal matter coming from household latrines (APHA, 1998). The presence of pathogenic bacteria in treated wastewater effluent is a potential public health hazard, as this water source is directly discharged in receiving water bodies and may be used by communities for multiple purposes. At the inlet, physico-chemical parameter values are relatively high which causes microbial biomass development, in particular, increases in faecal coliform and faecal streptococci (Rajib et al., 2011). It is clear from our results that some amount of microbial load is retained even after the purification treatment process. So, it is essential to include a tertiary treatment step in STPs so that the purification process results in bacterial concentrations that are in compliance with discharge (Koivunen et al., 2003).

CONCLUSION AND SUGGESTIONS

All the municipal wastes that may have otherwise gone untreated into the environment are restricted from entering. Thus, these treatment plants play an important role in the control of pollution level. The results show highly significant reduction in some of the physic chemical features and microbial load. However, performance of these STPs do not meet the permit standards set by the Indian standard and WHO. Therefore, it is suggested that authorities should improve their efficiency capability by including tertiary treatment processes such as rapid sand filtration, UV disinfection, artificial lagoons, wetlands, adequate contact time, etc. In addition, there is need of trained and technical staff for proper monitoring and operation of these STPs in a standardized manner. Furthermore, regular monitoring of these STPs by national experts is necessary in order to improve the operational capability of these STPs.

REFERENCES

Adami G, Cabras I, Predonzani S, Barbieri P, Reisenhofer E (2007). Metal pollution assessment of surface sediments along a new gas pipeline in the Niger Delta (Nigeria). Environ. Monit. Assess. 125:291-299.

Akpor OB, Muchie M (2011). Environmental and public health implications of wastewater quality Afr. J. Biotechnol. 10:2379-2387.

American Public Health Association (APHA) (1998). Standard Methods for the examination of water and waste water, 19th edition. American Water works Association Water Pollution Control Federation publication, Washington, D.C.

Antunes S, Dionisio L, Silva MC, Valente MS, Borrego JJ (2007). Proc. of the 3rd IASME/WSEAS Int. Conf. on Energy, Environment Ecosystems and Sustainable Development, Agios Nikolaos, Greece pp. 24-26.

Bohdziewicz J, Sroka E (2006). Application of hybrid systems to the treatment of meat industry wastewater. Desalination 198(1-3):33-40.

Cohen J, Shuval HI (1972). Coliforms, faecal coliforms and faeca streptococci as indicators of water pollution. Water Air Soil Pollu 2:85-95.

Desai PA, Kore VS (2011). Performance evaluation of effluent treatmen plant for textile industry in Kolhapur of Maharashtra. Univ. J. Environ Res. Technol. 1:560-565.

EPA (2003). US Environmental Protection Agency Safe Drinking Wate Act. EPA 816-F-03-016.

Igbinosa EO, Okoh AI (2009). Impact of discharge wastewater effluent on the physico-chemical qualities of a receiving watershed in a typica rural community. Int. J. Environ. Sci. Technol. 6(2):175-182.

Jamwal P, Mittal AK, Mouchel JM (2009). Efficiency evaluation c sewage treatment plants with different technologies in Delhi (India) Environ. Monit. Assess. 153:293-305.

Koivunen J, Siitonen A, Heinonen-Tanski H (2003). Elimination o enteric bacteria in biological-chemical wastewater treatment ani tertiary filtration units. Water Res. 37:690-698.

Kris M (2007). Wastewater pollution in China. Available from http www.dbc.uci/wsu stain/suscoasts/krismin.html

Kumar PR, Pinto LB, Somashekar RK (2010). Assessment of the efficiency of sewage treatment plants: a comparative study betweer nagasandra and mailasandra sewage treatment plants. Kathmandi Univ. J. Sci. Eng. Technol. 6:115-125.

Momba MNB, Osode AN, Sibewu M (2006). The impact of inadequate wastewater treatment on the receiving water bodies – Case study Buffalo City and Nkokonbe Municipalities of the Eastern Cape Province. Water SA. 32:687-692.

Morrison GO, Fatoki OS, Ekberg A (2001) Assessmetn of the impact o point source pollution from the Keiskammahock sewage treatmen plant on the Keiskamma River. Water SA 27:475-480.

Okoh AI, Barkare MK, Okoh OO, Odjadjare E (2005). The cultura microbial and chemical qualities of some waters used for drinking anc domestic purpose in a typical rural setting of Southern Nigeria. J Appl. Sci. 5:1041-1048.

Okoh AI, Odjadjare EE, Igbinosa EO, Osode AN (2007). Wastewate treatment plants as a source of microbial pathogens in the receiving watershed. Afr. J. Biotechnol. 6:2932-2944.

Prescott LM, Harley JP Klein DA (2002). Microbiology, 5th edition. Mc Graw Hill, New York. pp. 651-658.

Rajib AB, Kallali H, Saidi N, Abidi S (2011). Physico-chemcial anc microbial characterization performancy in waste water treated unde aerobic reactor. Am. J. Environ. Sci. 7(3):254-262.

Rao PV (2005). Textbook of environmental engineering. Eastern Economy Ed., Prentice-Hall of India Private Limited, New Delhi Chapter 3, p. 280.

Rawat KP, Sharma A, Rao SM (1998). Microbiological anc physicochemical analysis of radiation disinfected municipal sewage. Water Res. 32:737-740.

Reasoner DJ (1982). Microbiology: detection of bacterial pathogens anc their occurrence. J. WPCF. 54:942.

Saha ML, Alam A, Khan MR, Hoque S (2012). Bacteriological, physica and chemical properties of the Pagla sewage treatment plant's water. J. Biol. Sci. 21:1-7.

Wang WLL, Dunlop SG, Munson PS (1966). Factors influencing the survival of Shigella in wastewater and irrigation water. J. Wate Pollut. Control Fed. 38:1775-1781.

11

Demulsification capabilities of a Microbacterium species for breaking water-in-crude oil emulsions

Hossein Salehizadeh[1,2] , Aida Ranjbar[2] and Kevin Kennedy[1]

[1]Department of Civil Engineering, University of Ottawa, Ottawa, ON, K1N 6N5, Canada.
[2]Chemical Engineering Group, Faculty of Engineering, University of Isfahan, Isfahan, Iran.

A bacterium strain belonging to *Microbacterium* sp., isolated from oily sludge samples of Siri Island in the south of Iran, produced a strong, thermo stable microbial demulsifier ($Y_{x/s}$=0.663, $Y_{p/s}$=0.204, productivity=0.185 g L^{-1} h^{-1}) on glucose as a sole carbon source supplemented with yeast extract. The optimum values of temperature, inoculum concentration, pH and culture age for microbial demulsifier production were 25°C, 10^8 CFU mL^{-1}, 7 and 24 h, respectively. The maximum demulsification activity and the half-life value ($t_{1/2}$) of culture broth measured for a water-in-crude oil (W/CO) emulsion were 96.4% and 36 h at 80°C in flask. The demulsifier was purified to homogeneity using cold ethanol. For 4.33 mg mL^{-1} of partially purified microbial demulsifier, the half-life value for the W/CO model emulsion was 3 h.

Key words: Biopolymer, demulsification, *Microbacterium*, demulsifier, petroleum emulsion, water-in-crude oil.

INTRODUCTION

An emulsion is a thermodynamically unstable system in which liquid drops are dispersed in another immiscible liquid phase. Petroleum emulsions are generally formed in reservoirs or during refining processes and in oil transportation through pipelines. They are commonly classified, based on the continuous phase, into two groups: i) water-in-oil emulsions (W/O), and ii) oil-in-water emulsions (O/W) (Manning and Thampson, 1995; Scharmam, 2005). The formation of oilfield emulsions at various stages of exploration, production and recovery leads to many problems, such as corrosion and scaling on pipelines and equipment used in production or recovery (Mouraille et al.,1998). Recently, microbial demulsifiers have been attracting more attention due to their excellent surface properties, low toxicity, biodegradability, low cost, high specificity at extreme temperatures, and environmental compatibility (Desai and Banat, 1997). So far, microbial demulsifiers have

been commonly produced by pure cultures such as *Alcaligenes* sp., *Corynebacterium petrophilum*, *Nocardia amarae*, *Rhodococcus aurantiacus*, *Mycobacterium* sp., *Bacillus subtilis*, *Torulopsis bombicola*, *Acinetobacter calcoaceticus*, *Arthrobacter* sp., *Micrococcus* sp. (Carins et al., 1982; Cooper et al., 1982; Stewart et al., 1983; Das et al., 2001; Huang et al., 2009; Wen et al., 2010; Li et al., 2012, Long et al., 2012), and mixed bacterial culture (Kosaric and Duvanjak, 1987; Nadarajah et al., 2002) for demulsifying, or at least destabilizing, O/W and W/O emulsions.

This work aimed to investigate the behavior of a demulsifier isolated from *Microbacterium* sp. on water and crude oil (W/CO) model emulsions, optimization of growth conditions and factors influencing microbial demulsifier production as well as the effect of the W/CO demulsification, assay temperature and inoculum concentration on the model emulsion.

Table 1. Properties of crude oil used for preparing the model emulsion.

Property	Value
API[1] gravity	30.2
Kinematic viscosity at 20°C (cP)	11.93
Density (Kg m^{-3})	875
BS&W[2] (%v)	0.5
Sulfur content (%w)	1.3
Asphaltene content (%w)	2
Wax content (%w)	11
Vanadium and nickel (%w)	1.22

[1]American petroleum institute; [2]Bottom sediment and water.

MATERIALS AND METHODS

Screening of microorganism

Samples of polluted soil and oily sludge were collected from the Siri Island and Aghajari oil field located in the south of Iran, in the area near the Esfahan Oil and the Tehran Oil Refineries. 1 mL of oily sample was added to 99 mL of mineral salt (MS) solution in a 250 mL glass flask containing nutrient broth (Merck, Germany). This mixture was shaken at 160 rpm and 30°C for 12 h. Screening was carried out by transferring serial dilutions of samples onto nutrient agar plates, incubating at 30°C for 48 h and selection of individual colonies for culturing. The capability of microbial demulsifier production was determined by measuring the demulsification activity of water-in-oil emulsions.

The isolate was identified in the National Laboratory of Industrial Microbiology (NLIM) at Alzahra University according to Bergey's Manual of Systematic Bacteriology and other literature (Holt and Williams, 1989; Evtushenko and Takeuchi, 2006).

Medium and culture conditions

The medium for microbial demulsifier production included MS solution (g L^{-1}): NH$_4$NO$_3$, 4; K$_2$HPO$_4$, 4; KH$_2$PO$_4$, 6; MgSO$_4$.7H$_2$O, 0.2; CaCl$_2$, 0.0001; FeSO$_4$, 0.0001, supplemented with yeast extract (1.0 g L^{-1}) and glucose (30 g L^{-1}) as the carbon source. Production of the microbial demulsifier was optimized using the Taguchi experimental design (Qualitek 4, Demo version) software. The effect of various carbon sources on microbial demulsifier production was examined by culturing with addition of various separate sugars (glucose, sucrose, molasses) at 3% (w/v) and hydrocarbon substrates (hexadecane, kerosene, crude oil) at final concentration of 0.4% (v/v) to the MS solution. Hydrocarbons were sterilized by filtration using Millipore membranes (0.22 μm). All chemicals used were of analytical grade obtained from Merck, Germany.

Preparation of W/CO emulsion

W/CO model emulsion was prepared according to the method described by Nadarajah et al. (2002) with some modifications. The stock solution of Tween 80 in water was prepared by dissolving 40 mL of Tween 80 (HLB=15) in 1 L of de-ionized water on a stirring plate for 1 min. To prepare the W/CO emulsion, 8 mL of crude oil was added dropwise to 2 mL Tween 80 in water solution in a test tube and vortexed at maximum speed for 3 min before each use Crude oil was obtained from the Esfahan Oil Refinery, Esfahan Characteristics of the crude oil sample are presented in Table 1.

Measurement of demulsification activity

1 mL of the culture broth was added to 9 mL W/CO emulsion in a test tube and vortexed for 30 s to form a homogeneous culture emulsion mixture, then transferred to a 10 mL graduated cylinder covered with aluminum foil and incubated at a defined temperature. For all experiments, the controls used 1 mL uninoculated production medium and 9 mL W/CO emulsion. The demulsification activity of the microbial demulsifier was determined as follows: demulsification activity (%) = ([Initial emulsion volume – final emulsion volume at interface]/Initial emulsion volume) ×100 (Nadarajah et al., 2002). All experiments were conducted in triplicate and the mean of the results was reported. W/CO demulsification assay temperature was 80°C for 48 h.

Optimization of microbial demulsifier production using the Taguchi method

Taguchi experimental design software (Qualitek4, Demo version) was used to optimize the demulsification activity of the culture. A standard array (L16) was applied to evaluate the effect of four factors and four levels (Table 2). Analysis of data was done using the analysis of variance (ANOVA) method in order to determine which factors were statistically significant.

Extraction and biochemical analysis of microbial demulsifier

Extraction of the crude microbial demulsifier was carried out according to the method of Peat et al. (1961) with some modifications. The culture broth was centrifuged at 5000 rpm for 10 min. The precipitate was washed with a phosphate buffer (5%, w/v) and then suspended in a solution containing 20% (v/v) potassium citrate (0.1 M) and potassium meta bisulfate (0.02 M). The pH of the solution was adjusted to 6, and autoclaved at 121°C for 20 min, then centrifuged at 4000 rpm for 10 min. To obtain the precipitate of the microbial demulsifier, three volumes of 95% cold ethanol containing 1% (v/v) acetic acid were added. The solution was kept at 4°C for 12 h, after which the microbial demulsifier was centrifuged and freeze dried.

The phenol/sulphuric acid method was used to determine the

Table 2. Comparison of microbial demulsifier production using glucose (3%) and molasses (3%) in the presence of yeast extract (1.0 g L^{-1}) in 5 L fermenter.

Carbon source	Culture age (h)	$Y_{x/s}$ (g g^{-1})	$Y_{P/S}$ (g g^{-1})	Productivitiy (g L^{-1}h^{-1})	Demulsification activity[1] (%)	Half-life[2] $t_{1/2}$ (h)
Glucose	24	0.663	0.204	0.185	98.6	3
Molasses	64	0.92	0.124	0.0469	64.7	18

[1]Data obtained using culture broth; [2]data obtained by adding 1 mL of the partially purified microbial demulsifier (4.33 g/L) to 9 mL 20:80 W/CO emulsion. W/CO demulsification assay time was 48 h, and W/CO demulsification assay temperature was 80°C.

Table 3. Relationship between W/CO mixing ratio and the added Tween 80 volume on stability of model emulsion.

Tween 80 volume (μl)	W/CO ratio 40:60 (v/v)	W/CO ratio 30:70 (v/v)	W/CO ratio 20:80 (v/v)
800	- - - -	- - - -	- - - -
600	- - - -	- - - -	- -
400	- - - -	- - - -	- -
200	- - - -	- - -	-
80	- - -	- - -	+
60	- - -	-	-
40	- - -	-	-
20	- - -	- - -	- -

W/CO phase separation percent: stable emulsion 0% (+); 20% (-); 30 (--); 40% (---); more than 50% phase separation (----).

total sugar content of the microbial demulsifier (Dubois et al., 1956). The protein content of the microbial demulsifier was measured by the Lowry-Folin (1951). The Lipid content of the microbial demulsifier was determined according to the method of Floch (1975).

RESULTS

Screening of microorganism

Out of the 11 strains that were isolated from polluted soil and oily sludge samples, only one strain (isolate S2) exhibited considerable demulsification activity and grew in a liquid medium containing glucose as the sole carbon source. Emulsification activity of culture broth was growth dependent achieving its highest value at the beginning of stationary phase. Isolate S2 was aerobic, rod-shaped, gram-positive, motile, catalase-positive and oxidase-negative. Based on the morphological and biochemical properties in NLIM's report, the isolate belongs to *Microbacterium* sp.

Demulsification characteristics of emulsion

Although W/CO emulsion was not the best model to test due to non-uniformity in crude oil, it was used in all experiments as the most closely approximated actual field emulsions. The stability of different volumetric ratios of the W/CO emulsions (20:80, 30:70 and 40:60) was

studied. The stability of the W/CO emulsion was found to decrease with increasing dispersed phase (water) content from a relative percentage of 20 to 40%. The W/CO ratio of 20:80 exhibited the highest emulsion stability (Table 3). The effect of the Tween 80 concentration in the range of 0.02 to 0.8 mL on the stability of the W/CO emulsions was also examined. Results show that the most stable emulsions were achieved by adding 80 μl Tween 80 to 2 mL of water and mixing it with 8 mL of crude oil. Stability increased as pH increased from 5 towards the higher pHs and the best results were obtained at pH 6.3. Concomitantly, the pH of the W/CO emulsion was adjusted to 6.3 in all experiments. The uninoculated control emulsion was stable for 30 days at 80°C and 240 days at 25°C.

Effect of demulsification assay temperature on demulsification of the W/CO model emulsion

The influence of the demulsification assay temperature in breaking the W/CO emulsions using the culture broth of *Microbacterium* sp. was evaluated at five different temperatures. According to Figure 1, the demulsification rate enhanced linearly with increasing W/CO demulsification temperature. Demulsification activity was negligible at 25°C. The highest percentages of demulsification were observed at 80 and 100°C. Since the difference in the demulsification activity at 80 and 100°C was small, the W/CO demulsification tempera-

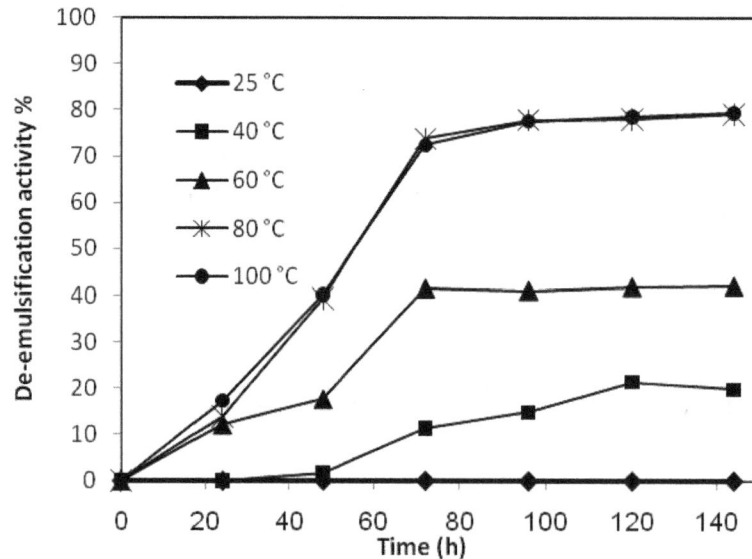

Figure 1. Effect of W/CO demulsification assay temperature on demulsifying the model emulsion by *Microbacterium* culture under unoptimized production. Inoculum concentration was 10% v/v (bacterial counts: 5×10^8 CFU mL^{-1}). Cultivation temperature was 30°C. W/CO demulsification assay time was 48 h. Control experiment carried out by adding uninoculated production medium was negligible.

ture was selected as 80°C throughout this research. For control test, medium containing yeast extract was added to the emulsion and its effect was negligible and the emulsion was stable.

Effect of inoculum concentration on demulsification of the W/CO model emulsion

Figure 2 shows that demulsification activity intensified by increasing the culture broth concentrations in the assay, but correlation was not linear. Both 10 and 15% (bacterial counts: 5×10^8 CFU mL^{-1}) of the culture broth samples indicated similar demulsification activity. After 48 h of W/CO demulsification time, the maximum demulsification activity of the model W/CO emulsion was 82 with 10% inoculum under unoptimized conditions, while only 11% of the phase separation was achieved using 1% culture broth.

Optimization of demulsifier production

Production of the microbial demulsifier by the *Microbacterium* sp. in flasks was optimized using the Taguchi experimental design statistical method by evaluating demulsification activity when varying four factors (inoculum concentration, culture age, pH, and temperature). The effect of each factor on demulsification

was determined by ANOVA. The effect of temperature was the most important factor followed by inoculum concentration, pH, and culture age and was determined as 70.4, 12.7, 7.1 and 5.8%, respectively. The temperature and inoculum concentration were the most significant interactive factors in the production of the demulsifier. The optimum values of temperature, inoculum concentration, pH, and culture age were 25°C, 10^8 CFU mL^{-1}, pH 7 and 24 h, respectively. The value error was 4% and the selected values in the experimental design were significant at more than a 95% confidence limit. To evaluate the proposed optimum conditions repeated experiments on microbial demulsifier production were conducted under optimum conditions. The demulsification activity obtained was always close to the expected value predicted by the Taguchi software. Cell free supernatant of *Microbacterium* did not exhibit demulsification activity while culture broth and washed cells displayed considerable demulsification activity.

The best demulsification activity by the culture broth of *Microbacterium* sp. growing on glucose (3% w/v) and yeast extract (1.0 g L^{-1}) was 96.4 % ($t_{1/2}$=36 h) (in flask) at W/CO demulsification assay temperature of 80°C and pH 6.3. With addition of 4.33 mg mL^{-1} of purified microbial demulsifier, half-life value ($t_{1/2}$) of the W/CO model emulsion was 3 h at 80°C. Demulsification of water-in kerosene system (30:70 v/v) using 24 old culture broth was also examined and the half-life value ($t_{1/2}$) was reduced to 1 h at 25°C.

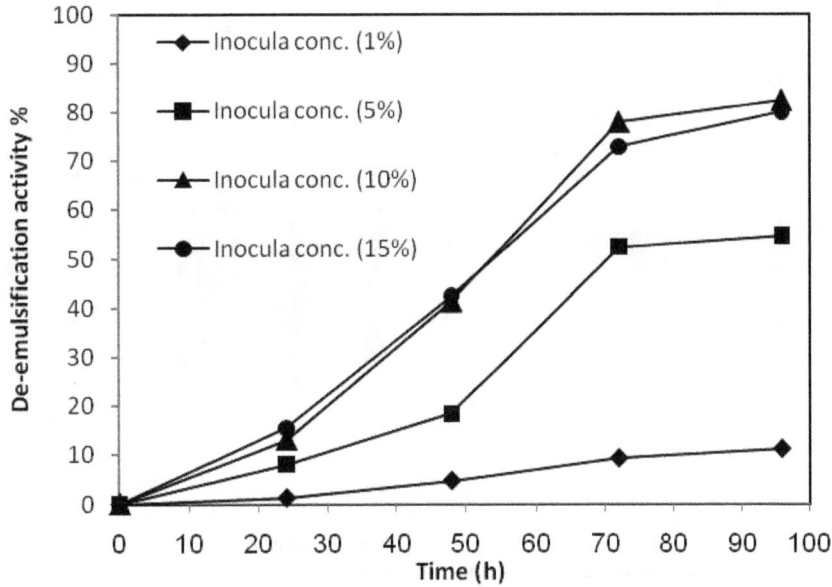

Figure 2. Effect of inoculum concentration on the demulsification of the model emulsion by *Microbacterium* culture under unoptimized production. Bacterial counts was 5×10^8 CFU mL^{-1}. W/CO demulsification assay temperature was 80 °C for 48 h. Control experiment carried out by adding uninoculated medium was negligible.

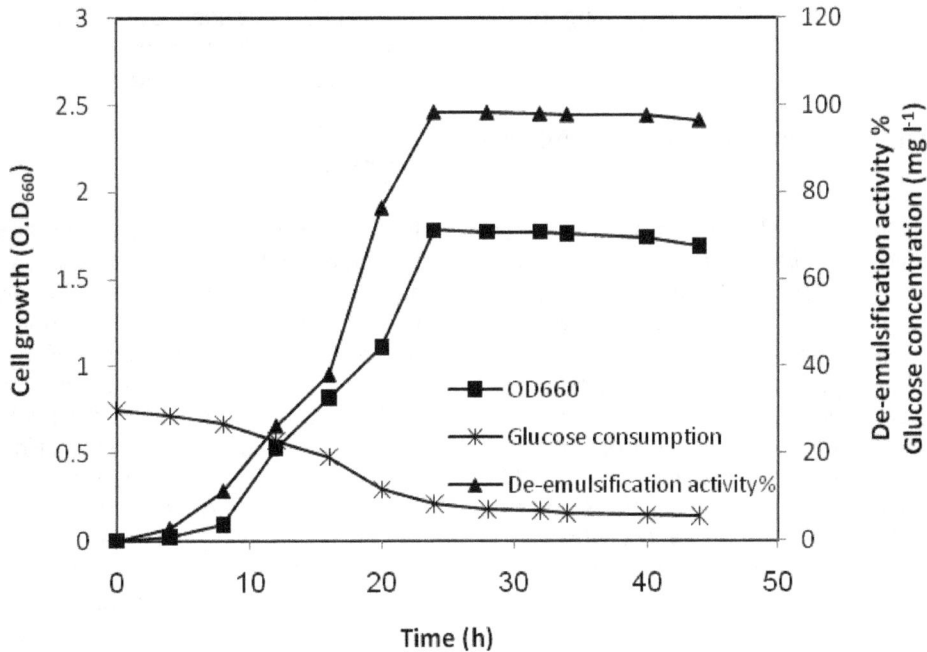

Figure 3. The cell growth, glucose consumption, demulsification activity of *Microbacterium* in a 5 L fermenter at 25°C, 200 rpm, and 0.5 vvm. Glucose: 30 g L^{-1}, yeast extract: 1.0 g L^{-1}.

Figure 3 shows bacterial growth, glucose consumption, and demulsification activity of *Microbacterium* sp. under optimum batch conditions when grown in a 5 L fermenter (Minifors, Infors, Switzerland, aeration at 0.5 vvm with turbine impellers stirring at 200 rpm) using 3% glucose in the presence of yeast extract (1.0 g L^{-1}) as an organic nitrogen source. Table 2 summaries the result of the production of the microbial demulsifier using glucose and

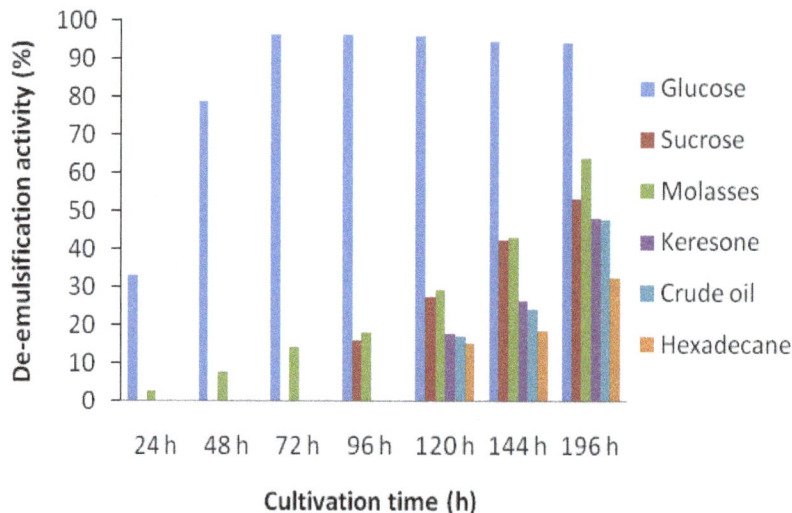

Figure 4. Effect of carbon source on demulsification activity of *Microbacterium* under optimum conditions. Initial nitrogen content in medium included NH_4NO_3 (4 g L^{-1}) and yeast extract (1.0 g L^{-1}).

molasses supplemented with yeast extract in the 5 L fermenter. Using molasses as the carbon source, demulsification activity of approximately 65% was achieved after 64 h. As with glucose, activity was growth associated and was maximum at the beginning of the stationary phase.

Effect of carbon and nitrogen sources on demulsifier production

Additional tests were run in shake flasks to evaluate demulsification activity of *Microbacterium* sp. using a variety of carbon sources such as glucose, molasses, sucrose, kerosene, crude oil, and hexadecane with and without supplementary yeast extract (Figure 4). *Microbacterium* sp. grown on sugar substrates displayed higher demulsification activities as compared with hydrocarbon substrates with best results obtained with glucose. Therefore, glucose was selected as the carbon source for further experiments. In fact, no significant demulsification activity was observed by *Microbacterium* sp. in the absence of yeast extract in the growth media.

Characterization of microbial demulsifier

Following extraction of the crude demulsifier, it was found to have a high sugar component of 73% and protein content of 15.6%. The infra-red (IR) spectrum of the microbial demulsifier (Figure 5) showed absorptions at 1539 and 1641.5 cm^{-1}, representing peptide groups. The peaks at 1641 and 1402 cm^{-1} indicate carboxylate groups. The peaks at 2914 and 2956 cm^{-1} confirm the presence

of aliphatic carbon-hydrogen bonds. The peaks at 3278 and 3396 cm^{-1} represent broad amine and hydroxyl groups. The strong adsorption at 1090 exhibits the C–O stretching and the presence of a methoxyl group. The demulsifier produced by *Microbacterium* sp. is a thermo stable, cell-surface associated component and 38% of its activity remained after heating at 100°C for 30 min.

DISCUSSION

In this study, an aerobic and gram positive strain belonging to *Microbacterium* sp. was isolated from an oily sludge sample from Siri Island in the south of Iran. Growth conditions play a key role on the production of microbial demulsifier produced by *Microbacterium* sp. *Microbacterium* was able to grow on glucose as a sole carbon source and produced microbial demulsifier only in the presence of yeast extract. Kosaric (1996) obtained similar results and showed that *Rhodococcus auranticus*, *Rhodococcus rubropertinctus* and *Nocardia amarae* had better demulsifying ability when grown on carbohydrates such as glucose than by growing on hydrocarbons such as hexadecane in the presence of yeast extract.

Demulsification activity of the culture broth obtained from the 5 L fermenter increased in parallel with the cell growth (10 to 20 h) and reached a maximum value of 98.6% at the end of the exponential growth phase (approximately 24 h) and did not change during the stationary phase which was stopped after 48 h (Figure 3). Similar results have been reported by other researchers (Das et al., 2001; Nadarajah et al., 2002). Demulsifier produced by *Alcaligenes* sp. which was a hydrophilic cell-wall associated bioproduct, achieved 96.5%

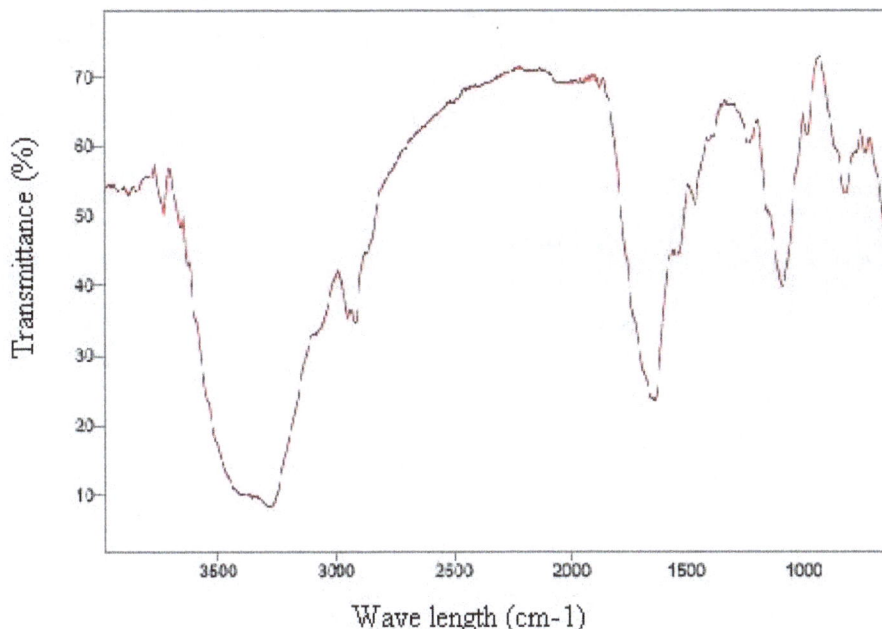

Figure 5. The IR spectrum of the partially purified microbial demulsifier produced by *Microbacterium* sp.

demulsification in W/O emulsion after 24 h ($t_{1/2}$= 2 h) (Das et al., 2001). Nadarhajah reported the culture containing *Acientobacter calcoaceticus* exhibited high demulsification activity (96%) in W/O emulsion after a 24 h incubation time which corresponded to the beginning of the stationary phase. *Acientobacter radioresistans* were capable of providing more than 90% demulsification, while *P. aeruginosa, P. carboxydohydrogena* and *Alcaligenes latus* demonstrated up to 80% demulsification (Nadarajah et al., 2002). *Alcaligenes* sp. S-XJ-1 showed demulsification activity of 81.3% for W/O emulsion within 24 h of incubation with cell concentration of 500 mg L^{-1} (Wen et al., 2010).

In summary, *Microbacterium* sp. was found to produce a highly effective, thermostable microbial demulsifier using glucose at the end of the exponential growth phase. *Microbacterium* sp. demulsifier promises potential biotechnological application for use in the oil industry and for environmental remediation.

ACKNOWLEDGEMENTS

The authors gratefully acknowledge the Petroleum Engineering and Development Company (PEDEC) and University of Isfahan for their financial supports.

REFERENCES

Carins WL, Cooper DG, Zajic JE, Wood JM, Kosaric N (1982). Characterization of *Nacordia amarae* as a potent biological coalescing agent of water-in-oil emulsions. Appl. Environ. Microbiol. 43:362-366.

Cooper DG, Akit J, Kosaric N (1982). Surface activity of the cells and extracellular lipids of *Corynobacterium fasciens* CF 15. Ferment. Technol. 60:19-24.

Das M (2001). Characterization of demulsification capabilities of a *Micrococcus* species. Biores. Technol. 79:15-22.

Desai JD, Banat IM (1997). Microbial production of surfactants and their commercial potential. Microbiol. Mol. Biol. 61:47-64.

Dubois M, Gilles KA, Hamilton JK, Smith F, Rebers PA (1956). Colorimetric method for determination of sugars and related substances. Anal. Chem. 28:350-356.

Evtushenko L, Takeuchi M (2006). The family of *Microbacteriaceae*. In: Dworkin M (Ed,) The prokaryotes: A handbook on the biology. pp.1020-1098. Springer-Verlag, Berlin.

Folch J, Lees M, Stanley GH (1975). A simple method for the isolation and purification of total lipids from animal tissues. Biol. Chem. 226:497-509.

Huang XF, Liu J, Lu LJ, Wen Y, Xu JC, Yang DH, Zhou Q (2009). Evaluation of screening methods for demulsifying bacteria and characterization of lipopeptide bio-demulsifier produced by *Alcaligenes* sp. Biores. Technol. 100:1358-1365.

Kosaric N, Duvanjak Z (1987). Demulsification of water-in-oil emulsion with sludge. Wat. Pollut. Res. 22:437-443.

Kosaric N (1996). Biosurfactants. In: Rehm, H.J. (Ed.), Biotechnology. pp. 697-717. VCH Verlag, Wienheim.

Holt JG, Williams ST (1989). Bergey's manual of systematic bacteriology. Lippincott Williams and Wilkins, Baltimore.

Li X, Li A, Liu C, Yang J, Ma F, Hou N, Xu Y, Ren N (2012) Characterization of the extracellular biodemulsifier of Bacillus mojavensis XH1 and the enhancement of demulsifying efficiency by optimization of the production medium composition. Process Biochem. 47:626-634.

Long X, Zhang G, Shen C, Sun G, Wang R, Yin L, Meng Q (2012) Application of rhamnolipid as a novel biodemulsifier for destabilizing waste crude oil. Bioresour. Technol. 131, 1-5.

Lowry OH, Rosebrough NJ, Farr AL, Randall RJ (1951). Protein measurement with the Folin-Phenol reagents. Biol. Chem. 193:265-275.

Manning FS, Thompson RE (1995). Oilfield processing. Penn Well, Oklahoma.

Mouraille O, Skodvin J, Peytavy JL (1998). Stability of water-in-oil emulsions. Dispersion Sci. Technol. 19:339-367.

Nadarajah N, Singh A, Ward OP (2002). Demulsification of petroleum oil emulsion by a mixed bacterial culture. Proc. Biochem. 37:1135-1141.

Peat S, Whelan WJ, Edwards TE (1961). Polysaccharides of baker's yeast. Part IV. Mannan, Chem. Soc. 1:28-35.

Scharmam LL (2005). Emulsions, foams suspensions: fundamentals and applications. Wiley-VCH Verlag GmbH, Wienheim.

Stewart AL, Gray NCC, Cairns WL, Kosaric N (1983). Bacteria-induced demulsification of water-in-oil petroleum emulsions. Biotechnol. Let 5:725-730.

Wen Y, Cheng H, Lu LJ, Liu J, Feng Y, Guan W, Zhou Q, Huang X (2010). Analysis of biological demulsification process of water-in-o emulsion by *Alcaligenes* sp. S-XJ-1. Biores. Technol. 101:8315-8322

12

Xanthine oxidase activity during transition period and its association with occurrence of postpartum infections in Murrah buffalo (*Bubalus bubalis*)

Bhabesh Mili[1] , Sujata Pandita[2], Bharath kumar B.S.[2], Anil Kumar Singh[2], Madhu Mohini[3] and Manju Ashutosh[2]

[1]Division of Physiology and Climatology Indian Veterinary Research Institute, Izatnagar-243122 (U.P.) India. [1]Division of [2]Dairy Cattle Physiology, National Dairy Research Institute, Karnal, Haryana, India. [3]Division of Dairy Cattle Nutrition, National Dairy Research Institute, Karnal, Haryana, India

The aim of this study was to quantify xanthine oxidase (XO) levels during the transition period in Murrah buffalo (*Bubalus bubalis*) and determine its association with certain postpartum infections. For this, six healthy buffaloes were selected from the National Dairy Research Institute (NDRI) herd and managed under standard managemental practices as followed at the institute. Blood samples were drawn weekly from each buffalo from day -21 to +21 relative to parturition by jugular vein puncture. Additional blood samples were collected from buffaloes suffering from metritis (n=5), endometritis (n=6) and mastitis (n=8) on alternate days. XO activity followed a defined pattern with values gradually declining from day -21 up to the day of calving followed by an increase to day +21, but the difference was statistically not significant between pre- and post-partum stages. The activity significantly declined on the day of calving when compared to the pre-partum mean value (p<0.05). The activity was significantly enhanced in buffaloes with bacterial infections, endometritis, and mastitis compared to healthy controls (p<0.05). However, the levels were not significantly altered among buffalo with metritis. The results indicate that increased XO activity during the postpartum period was associated with some bacterial infections in buffaloes, which could be due to increased phagocytic activity as a part of the innate defense system.

Key words: Xanthine oxidase, transition period.

INTRODUCTION

The transition period is the most interesting stage of postpartum health. During this period, host defense mechanisms can be compromised directly because of numerous genetic, physiological and environmental factors that can affect the cow's immunological defenses (Sordillo, 2005). The physiological stress associated with rapid differentiation of secretory parenchyma, intense mammary gland growth and the onset of copious milk synthesis and secretion during this period are accom-

panied by a high energy demand and an increased oxygen requirement (Gitto et al., 2002). Increased oxygen demand augments the production of oxygen-derived reactive oxygen species (ROS) and result into oxidative stress to the animal. A relationship between the physiological changes associated with parturition and a loss in overall antioxidant potential was established in both humans and dairy cows (Bernabucci et al., 2005; Sordillo et al., 2007). The oxidative stress therefore plays a pivotal role in the pathophysiology of inflammation (Kabe et al., 2005).It is well known that inflammatory events involve the generation of free radicals through NADPH oxidase and myeloperoxidase in immunoparti-cipating cells for the purpose of bacterial defense and phagocytosis (Hellsten et al., 1997). Superoxide radicals (O_2^-) are also of importance in neutrophil attraction and adherence to endothelium. The enzyme xanthine oxidase is one of the proposed sources of superoxide radicals during inflammatory, an event which in most of the tissues, is localized in the vascular walls. In diabetic rats, xanthine oxidase was reported to play an important role in the development of oxidative stress (Desco et al., 2002). Xanthine oxidase also converts nitrate to nitrite under anoxic conditions and therefore, acts as an alternate source of nitric oxide (NO) generation which are comparable to those produced from nitric oxide synthase (Li et al., 2003). In this process, nitrite and reducing substrate concentrations constitute important regulators of XO catalyzed NO generation.

There is no information available on XO activity in buffaloes and its quantitative importance in biological system or as a marker for detection of bacterial infections. Thus, the present investigation was aimed to quantify XO activity during transition period in Murrah buffaloes and its association with certain postpartum infections in buffaloes.

MATERIALS AND METHODS

Selection and feeding management of buffaloes

Six apparently healthy pregnant Murrah buffaloes were selected during the dry period from NDRI herd. All these buffaloes were maintained under general managemental practices as followed for the herd. The feed and water was available *ad libitum* to these buffaloes throughout the day. 19 symptomatic buffaloes were also selected from same herd, with symptoms indicative of metritis (n=5), endometritis (n=6) and mastitis (n=8). The diagnosis of metritis, endometritis, and mastitis was confirmed by the institute's herd veterinary officer based on symptoms of metritis and endometritis as described by Sheldon et al. (2006). Metritis was diagnosed by the presence of systemic signs of sickness, including fever, red-brown watery foul-smelling uterine discharge, dullness, elevated heart rate, and low production, whereas clinical endometritis was diagnosed by the presence of purulent (>50% pus) or mucopurulent (approximately 50% pus, 50% mucus) uterine exudates in the vagina, 21 days or more post partum. Clinical mastitis was diagnosed by an elevated somatic cell count in milk and visual signs of inflammation such as clumpy, watery, bloody or yellowish milk.

Blood sampling and processing

Blood samples (15 ml) were drawn in sterile heparinized vacutainer tubes from each buffalo at 6.00 A.M. in the morning by jugular venipuncture on days -21, -14, -7, 0, +7, +14, +21 relative to parturition. Immediately after collection, the tubes were transported to the laboratory in an ice box for further processing. Blood samples

from infected buffaloes postpartum were collected on the day o confirmation of metritis, endometritis and mastitis respectively followed by one more sample on an alternate day. The heparinizec tubes were centrifuged at 3000 rpm for 15 min, plasma aliquotec and stored at -20°C till analysis of Xanthine oxidase (XO) activity XO was determined using the xanthine oxidase colorimetric assay kit from BioVision. It involves oxidation of xanthine to hydroger peroxide (H_2O_2) by XO, which reacts with OxiRed Probe tc generate 570 nm light. The detection range of the test was 1-100 mU/ml. Kit reagents were prepared and stored in accordance with the manufacturer's instructions.

Statistical analysis

The data was presented as mean ± standard error (SE) in mU/m units. The unpaired and paired student "t" test was performed ir graph prism version 5. The student "t" test was applied to determine the significant changes in XO activity during transition period. The unpaired "t" test was used to compare the normal post partum XC activity with the activity during different bacterial infections. The normal XO activity was computed by grouping the data from day 7 to 21 postpartum.

RESULTS AND DISCUSSION

Changes in plasma XO activity during transition period ir healthy buffaloes and in those with postpartum bacteria infections are shown graphically in Figures 1 and 2 respectively. Although plasma xanthine oxidase activity constituted a definite pattern with values gradually declining from 41.01±1.47 mU/ml on day -21 relative tc parturition to 34.04±4.05 mU/ml on the day of calving followed by a rise to 43.04 mU/ml on day +21 postpartum, this difference did not achieve statistica significance during transition period in these buffaloes. The decline did not reach statistical significance on the day of calving in relation to the pre-partum day 21 value (p<0.05). The xanthine oxidase activity was significantly elevated in buffaloes suffering from endometritis (44.31±0.48 mM/ml) and mastitis (44.74±0.51 mM/ml) compared against 39.52±1.59 mM/ml in healthy ones. However, there was no significant difference in mean xanthine oxidase activity between metritis and healthy buffaloes.

The activation of local and systemic host defense mechanisms requires cross-talk between numerous types of immune cells. One component of this response is inflammation. The host of signaling molecules releasec by activated immune cells includes inflammatory media-tors such as nitric oxide, xanthine oxidase, prostaglan-dins and cytokines. While many of these molecules promote local inflammation and increased blood flow, xanthine oxidase play a key role in stimulating oxidative reactions. It is a highly versatile enzyme that is widely distributed among species (from bacteria to man) and within the various tissues of mammals. In healthy tissue, xanthine oxidase mainly exists in a dehydrogenase form not capable of producing superoxide radicals, but the enzyme may be modulated to its superoxide-generating

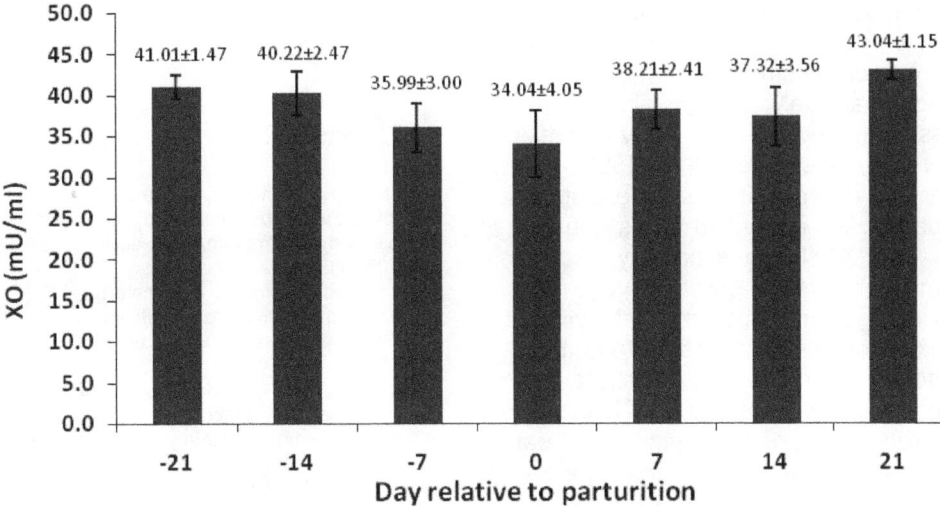

Figure 1. XO activity (mean ± SE) in healthy buffaloes during transition period.

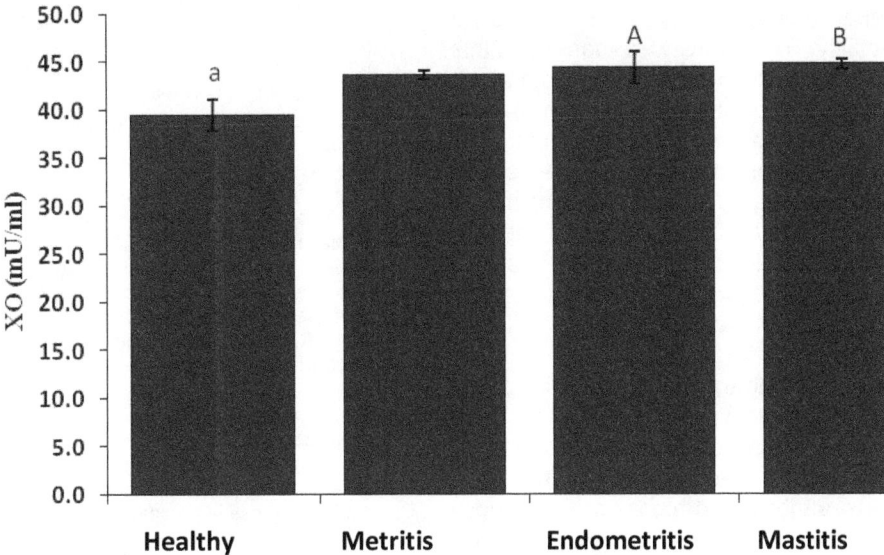

Figure 2. Plasma XO activity (mean ± SE) in buffaloes exhibiting postpartum infections (Bar with superscript aA and aB differ significantly from each other).

oxidase form via oxidation of critical sulfhydryl groups (Della and Stirpe, 1972) or through limited proteolysis (Della and Stirpe, 1968).

Xanthine oxidase has been implicated in several physiological and pathological cases. It is a form of xanthine oxidoreductase that generates reactive oxygen species (Ardan et al., 2004). To the best of our knowledge this is the first report that has quantified the levels of XO during transition period. We did not find significant difference in xanthine oxidase activity duringthe pre- or post-partum period. However, XO activity demonstrated a defined trend with lowest value on the day of parturition. After calving, the level remained low for first 14 days followed by a rise on day 21. Earlier reports have indicated a relationship between the physiological changes associated with parturition and a loss in overall antioxidant potential in both humans and dairy cows (Bernabucci et al., 2005; Sordillo et al., 2007). The possibility that oxidative stress during the transition period might be a major underlying cause of inflammatory and immune dysfunction in dairy cattle has been supported by various studies conducted either *in vivo* or *in vitro* (Sordillo and Aitken, 2009). Since Xanthine oxidase is involved in the generation of reactive oxygen species (Ardan et al., 2004) through NADPH oxidase and myeloperoxidase in immunoparticipating cells for the purpose of bacterial defense and phagocytosis during inflammation (Hellsten et al., 1997), therefore it could be used as a marker of oxidative stress.

We recorded comparatively higher activity during transition period as compared to earlier study in buffaloes (24.5 and 1.4 U/L) and cattle (Chen et al., 1996). Due to versatile nature of this enzyme and its wide distribution among various species, the enhanced levels could have other benefits as well. XO being a primary enzyme in purine metabolism, higher levels might have a special role to play in nitrogen metabolism also which in ruminants has exceptional importance. Alternatively, it could also enhance substantially the substrate source (nitrite) for NO generation (Li et al., 2003; Haitao et al., 2004) under normoxic or hypoxic conditions. In an earlier study in our laboratory, the nitrite level remained elevated throughout the pregnancy in buffaloes, reaching maximum during the last trimester (Huozha et al., 2010) which further confirmed its association with NO generation.

The present study shows for the first time that buffaloes suffering from bacterial infections express an increased xanthine oxidase activity. The activity response in this investigation was significantly higher ($p<0.02$) in buffaloes suffering from endometritis and mastitis but not in metritic case. These results are in agreement with the findings of Kataria et al. (2010) who reported low activity in healthy cows (51 ± 2.0 mU/L) as compared to high (79 ± 6.0 mU/L) levels in brucellosis infected cows. Elevated activity levels of xanthine oxidase following various disease conditions were attributed to a protease-induced conversion of xanthine dehydrogenase to xanthine oxidase (Lindsay et al., 1990; Smith, et al., 1991) as demonstrated in whole-tissue homogenates and could occur during increased leukocyte activity as a part of body innate defense mechanism or through enhanced phagocytic activity during chronic bacterial infections. The higher activity was also attributed to higher rate of oxidetive reactions (Kataria et al., 2010) which might contribute to the generation of reactive oxygen species thus leading to oxidative stress. In these studies, the actual sites for the increase in xanthine oxidase were not assessed. However, in muscle injury, the elevated expression occurred mainly in the endothelial cells of microvessels and also in leucocytes present in the muscle (Hellsten et al., 1997). Thus, the elevated levels could be used as a marker for detection of some bacterial infections in buffaloes during post partum period.

Conclusion

In conclusion, the changes in XO could be used as a marker for oxidative stress during transition period and was related to total leukocytic activity. It could also be used as an inflammatory marker for postpartum bacterial

infections in buffaloes. Further studies are required to confirm it as a risk factor for uterine infections.

REFERENCES

Ardan T, Kovaceva J, Cejková J (2004). Comparative histochemical and immunohistochemical study on xanthine oxidoreductase/xanthine oxidase in mammalian corneal epithelium. Acta. Histochem 106(1):69-75.

Bernabucci U, Ronchi B, Lacetera N, Nardone A (2005). Influence of body condition score on relationships between metabolic status and oxidative stress in periparturient dairy cows. J. Dairy Sci. 88:2017-2026.

Chen P F, Tsai A L, Berka V, Wu K K (1996). Endothelial nitric-oxide synthase. Evidence for bidomain structure and successful reconstitution of catalytic activity from two separate domains generated by a baculovirus expression system. J. Biol. Chem 271:14631-14635.

Della Corte E, Stripe F (1968). The regulation of rat-liver xanthine oxidase: activation by proteolytic enzymes. FEBS Letters. 2(2):83-84

Della Corte E, Stripe F (1972). The regulation of rat liver xanthine oxidase. Involvement of thiol groups in the conversion of the enzyme from dehydrogenase (type D) into oxidase (type 0) and purification of the enzyme. J. Biochem. 126:739-745.

Desco M C, Asensi M, Márquez R, Martínez-Valls J, Vento M, Pallardó F V, Sastre J, Viña J (2002). Xanthine oxidase is involved in free radical production in type 1 diabetes: protection by allopurinol Diabetes. 51:1118-1124.

Gitto E, Reiter RJ, Karbownik M, Tan DX, Gitto P, Barberi S, Barberi I (2002). Causes of oxidative stress in the pre- and perinatal period Biol. Neonate. 81:146-157.

Haitao Li, Alexandre S, Xiaoping L, Jay LZ (2004). Characterization of the effects of oxygen on xanthine oxidase mediated nitric oxide formation. J. Biol. Chem. 279:16939-16946.

Hellsten Y, Frandsen U, Orthenblad N, Sjodint B, Richter E A (1997) Xanthine oxidase in human skeletal muscle following eccentric exercise: a role in inflammation. J. Physiol. 498:239-248.

Huozha R, Pandita S, Manju A (2010). Production of nitric oxide by Murrah buffalo lymphocytes during gestation. Res. Vet. 21(1):895-899.

Kabe Y, Ando K, Hirao S, Yoshida M, Handa H (2005). Redox regulation of NF-kappa B activation: distinct redox regulation between the cytoplasm and the nucleus. Antioxid Redox Signal. 7:395-403.

Kataria N, Kataria A K, Maan R, Gahlot A K (2010). Evaluation of oxidative stress in brucella infected cows. J. Stress Physiol. Biochem 6:209-216.

Li H, Samouilov A, Lium X, Zweier J L. (2003). Characterization of the magnititude and kinetics of xanthine oxidase catalyzed nitrate reduction: evaluation of its role in nitrite and nitric oxide generation in anoxic tissues. Biochemistry. 42:1150-1159.

Lindsay TF, Liauw S, Romaschin AD, Walker PM (1990). The effect of ischemia/reperfusion on adenine nucleotide metabolism and xanthine oxidase production in skeletal muscle. J. Vasc. Surg.12:8-15.

Sheldon I M, Lewis G S, LeBlanc S, Gilbert R O (2006). Defining postpartum uterine disease in cattle. Theriogenology. 65:1516-1530.

Smith J K, Carden D L, Korthuis R J (1991). Activated neutrophils increase microvascular permeability in skeletal muscle: role of xanthine oxidase. J. Appl. Physiol. 70:2003-2009.

Sordillo L M, Aitken S L (2009). Impact of oxidative stress on the health and immune function of dairy cattle. Vet. Immunol. Immuno. 128:104-109.

Sordillo LM, O'Boyle N, Gandy JC, Corl CM, Hamilton E. (2007). Shifts in thioredoxin reductase activity and oxidant status in mononuclear cells obtained from transition dairy cattle. J. Dairy Sci. 90:1186-1192.

Molecular study on extended spectrum β-lactamase-producing Gram negative bacteria isolated from Ahmadi Hospital in Kuwait

Eshaq A. Mohmid[1] , El-Sayed A. El-Sayed[2] and Mahmoud F. Abdel El-Haliem[2]

[1]Department of Microbiology, Medical Laboratory, Ahmadi Hospital, Kuwait Oil Company, Ahmadi, Kuwait.
[2]Department of Botany, Faculty of Science, Zagazig University, Zagzig, Egypt.

During the period from November 2009 to April 2010, 84 out of 560 extended spectrum β-lactamase (ESBL) producing negative bacteria were isolated from patients in different departments of the Ahmadi Hospital in Kuwait. The isolates were collected from urine catheter, wound, sputum, blood and other different samples. The ESBL infection rate in the in-patients was 62% and part of them (19%) were in the intensive care unit. All the isolated bacteria were identified and tested for antimicrobial susceptibility using an automated system (VITEK 2) and different antibiotic discs (15) by standard disc diffusion. The number of the recorded isolated multi-resistant Gram's negative bacteria was 54 isolates of *Escherichia coli*, 18 of *Klebsiella pneumoniae*, 11 of *Pseudomonas aeruginosa*, six of *Proteus mirabilis*, five of *Enterobacter cloacae*, four of *Acinetobacter baumanii* and one of *Enterobacter aerogenes*. They were resistant to the third generation of cephalosporins; Ceftazidime, Cefotaxime and Ceftriaxone. Meropenam (MEM) was the highest effective antibiotic against all the isolated bacteria (86%). The production of the ESBL was detected by phenotypic methods using E-test (96.4%), double disk synergy test (95%) and VITEK 2 (84.5%) in all multi-resistant isolates except *A. baumanii* and *P. aeruginosa*. All ESBL producing isolates were extracted and subjected to PCR using blaSHV, blaCTX-M and blaTEM primers. The bla-CTX-M (63.1%) was the most predominant ESBL gene that was produced in abundance by 42 isolates of *E. coli*. The most predominant ESBL isolates producing bla-TEM, bla-CTX-M and bla-SHV genes were successfully identified by 16 S rDNA. The conjugation assay between *E. coli* HB101 and the most predominant ESBL producing *E. coli* showed that the bla-CTX-M gene was able to be transferrable suggesting that they were plasmid mediated.

Key words: Extended spectrum β-lactamase (ESBL), VITEK 2, E-Test, DDST, polymeric chain reaction (PCR), 16S rDNA.

INTRODUCTION

Extended spectrum of β -lactamases (ESBLs) are enzymes produced by a variety of Gram negative bacteria which confer an increased resistance to commonly used antibiotics. They are a worrying global public health issue as infections caused by such enzyme producing organisms are associated with a higher morbidity and mortality with greater fiscal burden. Coupled with increasing prevalence rates worldwide and an ever diminishing supply in the antibiotic armamentarium, these enzymes represent a clear and present danger to public health (Dhillon and Clark, 2011). The introduction of the third-generation of cephalosporins into clinical practice in the early 1980s is heralded as a major breakthrough in the fight against β -lactamase-mediated bacterial resistance to antibiotics.

These cephalosporins have been developed in response to the increased prevalence of ß-lactamases in certain organisms (for example, ampicillin hydrolyzing TEM-1 and SHV-1 β -lactamases in *E. coli* and *K. pneumoniae*). The third generation of cepha-losporins against most β-lactamase-producing organisms has major advantages of lessened nephrotoxic effects compared to aminoglycosides and polymyxins. The first report of plasmid-encoded β -lactamases capable of hydrolyzing the extended-spectrum cephalosporins, SHV-2, was published from a strain of *K. ozaenae* (Knothe et al., 1983). Among *Enterobacteriaceae*, extended spectrum beta lactamases (ESBLs) have been found mainly in *Klebsiella* spp. and *E. coli*, but have been also reported in another genera, such as *Citrobacter*, *Enterobacter*, *Morganella*, *Proteus*, *Providencia*, *Salmonella*, *Serratia* and *Pseudomonas* spp. (Arlet and Philippon, 1991; Ivanova et al., 2008). Thereafter, the number of ESBL variants occurring through amino acid mutations has progressively increased while demonstrating geographic variations (Winokur et al., 2001).

SHV-types of ESBLs are mostly derivatives of a non-ESBL SHV-1 and quickly invaded several continents (Bradford, 2001; Paterson et al., 2003). The majority of plasmid-mediated beta-lactamases, namely, TEM-1 or less frequently, TEM-2 are broad-spectrum beta-lactamases which do not hydrolyze oxyimino-cephalosporins or aztreonam. The (CTX-M) family, first described in 1992, is known to be the most dominant non-TEM, non-SHV ESBL among *Enterbacteriaceae* and it was recognized as a rapidly growing family of ESBLs that prefer to hydrolyze cefotaxime rather than ceftazidime (Bauernfeind et al., 1992; Bonnet, 2004). Several researchers used genotypic methods for the identification of the specific gene responsible for the production of the ESBLs, which have the additional ability to detect low-level resistance; that is, it can be missed by phenotypic methods (Woodford and Sundsfjord, 2005). Furthermore, molecular assays also have the potential to be done directly on clinical specimens without culturing the bacteria, with subsequent reduction of detection time (Tenover, 2007). Olsen and Woese (1993) showed that the 16S ribosomal ribonucleic acid (16S rRNA) gene, since the discovery of polymeric chain reaction (PCR) and deoxyribonucleic acid (DNA), was highly conserved within and among species of the same genus in compar-

ring the gene sequences of the bacterial species. Hence it can be used as the new gold standard for identification of bacteria to the species level of bacterial strains that have posed problems for the accurate identification of such isolates.

The genes that encode ESBLs were frequently found on the same plasmids as genes that encode resistance to aminoglycosides and sulfonamides. Many bacteria species possess changes that confer high-level resistance to quinolones. As a result, ESBL-producing bacterial species in hospitals and intensive care units (ICU) are commonly multidrug resistant, which possess a particular challenge for the treatment of nosocomial infections, especially in critically ill patients. Inappropriate empiric antimicrobial treatment for nosocomial- or community- acquired infections has been reported to contribute to significantly greater mortality rates in the ICU, and inadequate antimicrobial treatment of infection was the most important independent determinant of hospital mortality (Kollef et al., 1999). ESBLs are increasingly spreading among *Enterobacteriaceae* (clinical isolates) throughout the world due mostly to their presence on highly conjugative plasmid. Surveys that are done in Canada, Greece, United Kingdom and Italy showed an association between the CTX-M type of ESBL and resistance to other antimicrobial agents (Bonnet, 2004). This is explained by a number of findings showing that bla-CTX-M genes are commonly found on large plasmids that often carry other genes conferring resistance to other antimicrobial agents including aminoglycosides, fluoroquinolones, chloramphenicols, tetracyclins and others; particularly, bla-OXA-1, bla-TEM-1 (Leflon-Guibout et al., 2004; Bratu et al., 2005).

The aim of this work was to determine the prevalence of ESBL producing members of the Gram negative isolates which are isolated from samples collected from different departments at the Ahmadi Hospital in Kuwait by means of phenotypic and genotypic methods and also, characterize the genetic basis of ESBL producing isolates and compare them with the universal isolates of the gene bank.

MATERIALS AND METHODS

Bacterial isolates and sample collection

Totally, 560 clinical significant bacterial strains belonging to the family *Enterobacteriaceae* were isolated from patients seen and treated at the Ahmadi Hospital in Kuwait from November 2009 to April 2010. Isolates found to be resistant or with decreased susceptibility to any of the third generation of cephalosporins such as ceftazidime, cefotaxime or ceftriaxone were selected for ESBL testing (Duttaroy and Mehta, 2005; Paterson and Bonomo, 2005). The samples were collected from different departments and sent to the microbiology laboratory. They were (465) urine, (17) sputum, (23) catheter, (11) blood, (20) wound and (24) different samples.

The selected isolates were reviewed and recorded for the work of the study; including patient file number, diagnosis, age, sex, type of sample, location and type of infected organism.

Identification of gram negative bacteria

The samples were cultured on blood and MacConkey agar media (OXOID). The plates were incubated at 37°C for 24 h, and the Gram negative bacterial colonies were selected for identification and antimicrobial susceptibility testing by VITEK 2 (BioMerieux, Marcy L'Etoile, France) using ID-GNI and AST-GN27 cards according to the manufacturer's instruction (NCCLS, 1999). Then, the results were interpreted by using software version VTK2-R 4.01, an advanced expert system (AES) (Livermore, 1995; Canton et al., 2001; Sanders et al., 2001). *E. coli* (ATCC 25922) and *P. aeruginosa* (ATCC 27853) were used for negative quality control processing in all experimental tests as recommended by clinical and laboratory standards institute (CLSI, 2005).

Criterion for selection of ESBL producing strains

Routine disc diffusion susceptibility test of the isolates was performed by standard disk diffusion; Bauer-Kirby method (Bauer et al., 1966). All the isolates with the control strains were inoculated according to McFarland standards (McFarland, 1907; Farmer et al., 2007) and streaked on Mueller Hinton agar medium (Oxoid Ltd., Basingstoke, UK) using sterile cotton swabs. Then, 15 different antibiotic discs (MAST GROUP Ltd, UK), belonging to 8 groups of β- and non β -lactam agents, -were added onto the plates and incubated at 37°C for 24 h. The activity of each antibiotic disc was determined by measuring the antibiotic zone diameter (NCCLS, 2003). Following the CLSI criteria; any resistant with one of the third generation cephalosporins is selected for ESBL detection and confirmation (CLSI, 2005).

ESBL phenotypic tests

All the isolated resistant gram negative bacteria that are resistant to the third generation cephalosporins were tested by VITEK 2, E-Test (AB Biodisk, Sweden) and double disc synergy test (DDST) methods. VITEK2 is based on the detection of the inhibitory effect of clavulanic acid on ESBLs in the presence of either cefotaxime or ceftazidime. E-test strips of cefotaxime/cefotaxime + clavulanic acid (CT/ CTL) and ceftazidime/ceftazidime + clavulanic acid (TZ/TZL) are designed to confirm the presence of clavulanic acid that is able to inhibit ESBL enzymes in *E. coli*, *K. pneumoniae*, *K. oxytoca* and other relevant species (Cormican et al., 1996).

DDST is used to reconfirm the strains that are ESBL positive tested by E- test and VITEK 2 and employed three discs, ceftazidime (30 µg), cefotaxime (30 µg) and amoxicillin with clavulanic acid (30 µg) from Oxoid. The discs were placed with their centers at the recommended distance (25 to 30 mm) apart from each other on the plat. When the inhibition zone of ceftazidime or cefotaxime in combination singly with clavulanic acid is enhanced (zone diameter ≥ 5), it confirms an ESBL producing organism as recommended by the NCCLS (2000).

ESBL molecular tests

The ESBL isolates were tested *in vitro* as follows:

Extraction of ESBLs

One loopful of overnight growth ESBL isolates with the positive control strains was inoculated into 1.0 ml sterile distilled water. The bacterial suspension was emulsified and boiled for 10 min at 100°C and then kept for 5 min on ice. The emulsified suspension was centrifuged for 10 min at 12000 rpm and 300 to 400 ul from the supernatant was stored at -20°C (Lou et al., 1993; Merk et al., 2006).

Amplification and sequencing of ESBL genes

PCR analysis *in vitro* was performed on all isolates with the positive control strains containing specific primers to confirm the presence of ESBL. The ESBL coding regions were amplified using the primers listed in Table 1 and a cell suspension as template containing 2.5 ul buffer, 2.0 ul MgCl^2 (Promega, Ltd, UK), 0.5 ul dNTP, 0.25 ul Taq DNA polymerase (Promega, Ltd, UK), 1.0 ul forward and retained primers, 2.0 ul DMSO (only for blaSHV primer), 2.0 ul extracted DNA and sterile water for the remaining 25.0 ul (Dashti et al., 2009). The PCR products were analyzed by agarose gel electrophoresis using 1.5% (wt/vol) in a programmable PCR master cycler gradient (Eppendorf, Germany) following the technique described by Tenover et al. (1995) and photographed using a gel documentation system (UVP Company, Upland, CA, USA). *E. coli* strain ATCC 25922 was used as a negative control in all PCR assays. The positive control for SHV was *K. pneumoniae* 6064, whilst the positive control for TEM and CTX-M was *E. coli* 971 (Dashti et al., 2010). The blaTEM and blaCTX-M PCR products were purified using the NucleoSpin Extract II (Macherey-Nagel, Duren, Germany) and sequenced using the AB13100 DNA sequencing system (Applied Biosystem, Foster City, CA, USA). The resulting DNA sequences were compared with ESBL producing gram negative bacteria in the gene bank using BLAST at the website (http:// www.ncbi.nih.gov/blas).

Identification by 16SrDNA

Total DNA was extracted from the most predominant ESBL producing isolates and amplified using 16SrRNA primers listed in Table 1. Presence and yield of specific PCR products (16S rRNA gene) were monitored by running 1% agarose gels and cleaned up by using GeneJET™ PCR purification kit (Fermentas). The amplified DNA fragments were sequenced at GATC Biotech AG (Konstanz, Germany) using ABI 3730xl DNA sequencing system and the 16S rDNA sequences were analyzed by using the basic local alignment search tool (BLAST) program (http://www.ncbi.nlm.nih.gov/blast).

Conjugation assay

The recipient cell in this experiment was *E. coli* HB101 (Maniatis et al., 1982) and the donor cell was *E. coli* No. 22 (The predominant isolated blaCTX-M producing Gram's negative bacteria in this study). The strains that exhibited resistance to any of the tested antibiotics were examined for the ability to transfer resistance by conjugation. The recipient and donor strains *in vitro* were grown to mid-log phase. Equal volumes of donor and recipient cells were spread on a Mueller-Hinton plate and incubated for 24 h at 37°C. Positive and negative controls were prepared and examined. The resulting biomass was harvested, plated on Luria-Bertani agar

Table 1. The sequences and sizes of the used primers.

Primer	Forward	Reverse	PCR condition
blaTEM (858 bp) (Arlet et al., 1995)	5'- ATG AGT ATT CAA CAT TTC CG -3'	5'- CCA ATG CTT ATT CAG TGA GG-3'	95°C for 5 min, followed by 30 cycles. Denaturation: 94°C for 30 s. Annealing: 55°C for 1 min. Extension: 70°C for 1 min. Final extension: 75°C for 10 min.
bla CTX-M (499 bp) (Johann et al., 2004)	5' GAC GAT GTC ACT GGC TGA GC 3'	5'- AGC CGC CGA CGC TAA TAC A- 3'	
bla SHV (827 bp) (Al Naiemi et al., 2005)	5'- CTG GGA AAC GGA ACT GAA TG-3'	5' -GGG GTA TCC CGC AGA TAA AT-3'	95°C for 5 min, followed by 32 cycles. Denaturation: 94°C for 1 min. Annealing: 57°C for 1 min. Extension: 70°C for 1 min. Final extension: 72°C for 10 min.
16S rRNA (1500 bp) (Weisburg et al., 1991)	5'-AGA GTT TGA TCC TGG CTC AG-3'	5'-GGT TAC CTT GTT ACG ACT T-3'	94°C for 10 min, followed by 35 cycles. Denaturation: 95°C for 30 s. Annealing: 56°C for 1 min. Extension: 72°C for 1 min. Final extension: 72°C for 10 min.

plates (LB) containing streptomycin (16 mg/ml) and ceftazidime (8 mg/ml), and incubated for 24 h at 37°C (Sigma, USA). *E. coli* HB101 showed resistance to Streptomycin (recipient cell) while *E. coli* No. 22 was sensitive to Streptomycin (donor cell). However, *E. coli* No. 22 was resistant to ceftazidime while *E. coli* HB101was sensitive to ceftazidime.

RESULTS

Bacterial isolates

During the period from November 2009 to April 2010, 99 out of 560 clinically significant bacterial strains belonging to the family *Enterobacteriaceae* were resistant to most of the 15 different antibiotic discs including the third gene-ration cephalosporins. The multi-resistant isolates were collected from urine catheter, wound, sputum, blood and different samples. The ESBL infection rate in the hospi-talized patients (62%) was higher than out-patients and most of them were in the intensive care unit (data are not shown).

Identification of the bacterial isolates

The red (lactose) and colorless (non-lactose) fermenting colonies on MacConkey agar or creamy (mucoid) colonies on blood agar were selected for gram stain, oxidase test and analyzed in VITEK 2 for bacterial identification. Gram's stained films showed gram's negative bacilli and oxidase test was negative for the

lactose fermenting colonies and positive for non-lactose fermenting colonies. The number of the multi-resistant isolates was identified as *E. coli* (54 out of 358), *K pneumoniae* (18 out of 96), *P. aeruginosa* (11 out 58), *P mirabilis* (6 out of 17), *E. cloacae* (5 out of 18), *A baumanii* (4 out of 9) and *E. aerogenes* (one out of 4) Most of the multi-resistant gram negative bacteria were isolated from urine samples (465 out of 560), whereas blood samples were the lowest one (11 out of 560).

Antimicrobial susceptibility test

The isolated gram negative bacteria (Table 2) were resistant to the first, second and third generation o cephalosporins; CZL, CXM, CAZ, CTX and CRO. The antibiotic sensitivity of tigecycline (TGC), amikacin (AK) nitrofurantion (NI), gentamicin (GM), tazocin (PTZ) Amoxicillin / clavulanic acid (AUG) and ciprofloxacin (CIP) reached 84.8, 77.7, 42.4, 43.4, 71.7, 22.2 and 21.2%, respectively. All the multiresistant gram negative bacteria were sensitive to MEM (86.8%) except the isolates of *P. aeruginosa* and two isolates of *A. baumanii* The antibiotic susceptibility of *E. coli* to TGC, AK, PTZ NI, GM, AUG and CIP reached 98, 96, 89, 72, 63, 39 and 18.5%, respectively. The antibiotic susceptibility rate of *K pneumoniae* showed 100% against TGC, 83% against PTZ, 66.6% against AK, 28% against CIP, 22% against GM, 16.5% against NI and 5.5% against AUG. MEM and AUG were the effective antibiotics against *E. coli* and *K. pneumoniae*, but not active against other gram negative

Table 2. The antibiotic activity (%) on the multi-resistant isolates.

Bacteria	Antibiotic								
	MEM	TGC	AK	NI	GM	PTZ	AUG	CIP	TS
E. coli (n = 54)	100	98	96	72	63	89	39	18.5	20.3
K. pneumoniae (n = 18)	100	100	66.6	16.5	22	83	5.5	28	11.1
P. aeruginosa (n = 11)	0	0	9	0	9	9	0	9	0
Pr. mirabilis (n = 6)	100	100	100	0	0	66.6	0	0	0
A. baumanii (n = 4)	50	25	50	0	50	50	0	25	50
E. cloacae (n = 5)	100	100	80	0	80	40	0	80	60
E. aerogenes (n = 1)	100	100	0	0	0	0	0	0	0
Total (n = 99)	86.8	84.8	77.7	42.4	43.4	71.7	22.2	21.2	18.2

AUG, Amoxicillin / clavulanic acid; CIP, ciprofloxacin; TS, Trimethoprime / sulfamethoxazole; MEM, meropenam; PTZ, tazocin; NI, nitrofurantion; AK, amikacin; GM, gentamicin; TGC, tigecycline; n, number of isolates.

Table 3. The phenotypic and molecular characterizations of the isolated ESBL producing gram negative bacteria.

Organism	VITEK	D.D.S.T. AMC/CAZ/CTX	E-test		PCR		
			TZ/TZL	CT/CTL	TEM	CTX-M	SHV
E. coli (n = 54)	53	53	53	53	20	42	37
K. pneumoniae (n = 18)	18	18	18	18	15	9	12
E. cloacae (n = 5)	0	2	1	5	1	0	0
E. aerogenes (n = 1)	0	1	1	1	1	1	1
P. mirabilis (n = 6)	0	6	6	6	3	1	1
Total (n = 84)	71	80	79	83	40	53	51
Total (%)	84.5	95	94	98.8	47.6	63.1	60.7

isolates. The isolates of *P. aeruginosa* were resistant to the tested antibiotics except CIP, PTZ, AK and GM, but MEM and TGC have antimicrobial activities against *E. aerogenes*.

Phenotypic detection of ESBLs

The VITEK2 (Table 3) detected the presence of ESBL activity of the tested isolates (71%) represented by *K. pneumoniae* (100%) and *E. coli* (98%). Using DDST, the enhanced inhibition zone of the ceftazidime or cefotaxime in combination with clavulanic acid detected the presence of ESBL activity by 80 isolates (81%). The rate of ESBL detection reached 100% by *K. pneumoniae*, *P. mirabilis* and *E. aerogenes*, whist 98% by *E. coli* and *E. cloacae*. The production of ESBL by DDST was not detected in one isolate of *E. coli* and three isolates of *E. cloacae*. On the other hand, the synergic activity of clavulanic acid combined with ceftazidime (TZ) or cefotaxime (CT) was confirmed by two different E-test strips containing ceftazidime and cefotaxime with or without clavulanic

acid. The ceftazidime ESBL strips (TZ/TZL) detect the presence of ESBL activity by 80% of the tested isolates. Furthermore, the rate of the ESBL detection was 100% by *K. pneumoniae*, *P. mirabilis* and *E. aerogenes*; 98% by *E. coli*; and 20% by *E. cloacae*. The production of ESBL was not detected by one isolate of *E. coli* and four isolates of *E. cloacae*. With cefotaxime ESBL strips (CT/CTL), ESBL activity was detected by all the tested isolates (84%) represented by 100% in *K. pneumoniae*, *P. mirabilis*, *E. aerogenes*, *E. cloacae* and 98% in *E. coli*. The production of ESBL by ceftazidime ESBL strips was not detected in one isolate of *E. coli*. The production of ESBL was not recorded by all the tested *P. aeruginosae* and *A. baumanii* isolates in this study.

The analysis of the phenotypic ESBLs showed that there are differences between the VITEK 2 and both of the E-test and DDST. Out of 99 tested multiresistant isolates, 71 isolates recorded ESBL production by VITEK 2 according to two species. On the other hand, 79 isolates were ESBL producers by TZ\TZL, 83 isolates were ESBL producers by CT\CTL and 80 isolates were ESBL producers by DDST according to five species.

Genotypic detection of ESBLs

The PCR products that were amplified by specific primers of ESBL were detected in 75 isolates out of 84 tested ESBL producing gram negative bacteria with 89.3%. The blaTEM genotypes (858 bp) were detected in 40 isolates with 47.6% (Table 3). They were recorded in 20 isolates of *E. coli* with 37%, 15 isolates of *K. pneumoniae* with 83.3%, 3 isolates of *P. mirabilis* with 50%, one isolate of *E. cloacae* with 20% and one isolate of *E. aerogenes* with 100%. The blaCTX-M primers manifested 499 bp amplicon in 53 isolates with 63.1% of the 84 tested isolates. The blaCTX-M genotypes were reported in 42 isolates of *E. coli* with 78%, 9 isolates of *K. pneumoniae* with 50%, one isolate of *P. mirabilis* with 17% and one isolate of *E. aerogenes* with 100%. All the isolates of *E. cloacae* did not detect the blaCTX-M gene. The blaSHV primers detected the characteristic of 827 bp amplicon in 51 isolates out of the 84 tested bacteria by 60.7%. The blaSHV genotypes were detected in 37 isolates of *E. coli* with 68.5%, 12 isolates of *K. pneumoniae* with 66.6%, one isolate of *P. mirabilis* with 17% and one isolate of *E. aerogenes* with 100%. They were not detected in all isolates of *E. cloacae*. Overall, the blaCTX-M was the commonest genotype (63.1%) followed by blaSHV (60.7%) and blaTEM (47.6%). The most predominant ESBL producer was *E. coli* (62%) followed by *K. pneumoniae* (20.2%). The blaSHV, blaTEM and blaCTX-M genotypes in combination were present in 18 isolates with 21.4%. They were detected in *K. pneumoniae* (6 isolates), *E. coli* (11 isolates) and one isolate of *E. aerogenes*. On the other hand, eight isolates out of the 84 bacteria confirmed the presence of blaSHV and blaTEM combination in *K. pneumoniae* (4 isolates), *E. coli* (3 isolates) and one isolate of *P. mirabilis*. In addition, the blaSHV and blaCTX-M enzymes in combination were produced by 18 isolates of *E. coli*. Four isolates of *E. coli* and 3 isolates of *K. pneumoniae* produced ESBL by both blaTEM and blaCTX-M enzymes. The amplified PCR product for each blaTEM, SHV and blaCTX-M genes of the four ESBL producing gram negative bacteria were selected for sequencing according to the resulting antibiotic susceptibility test, phenotypic and genotypic characterization (Figures 1, 2, 3 and 4). Agarose gel electrophoresis results show that all the four selected ESBL producing isolates had identical banding patterns in accordance to the tested primers (Figures 5 and 6).

From Table 4, the forward and reverse DNA nucleotide sequences of the blaTEM producing *P. mirabilis* (No. 54) are identified as blaTEM producing *P. mirabilis* from gene bank No. JN043376 (78 to 99%). The nucleotide sequences of DNA forward and reverse revealed that the blaCTX-M producing *E. coli* (No. 22) is identified as blaCTX-M of *E. coli* gene bank No. FJ668785 (99%). The forward and the reverse DNA nucleotide sequences of

the blaSHV producing *K. pneumoniae* (No. 93) are identical to blaSHV producing *K. pneumoniae* gene bank No. FJ815288 and HM002660, respectively (98%). The forward and reverse DNA nucleotide sequences of the bla-TEM producing *E. cloacae* (No. 21) are identified as blaTEM producing *Enterobacter* sp. gene bank No FJ349257 (91%) and bla-TEM producing *E. cloacae* gene bank No. AY302260 (90%), respectively. The identical band patterns of the selected ESBL isolates were identified by 16SrDNA at a molecular size of 1500 bp The resulting DNA sequences of the selected ESBL isolates including F27 and R1492 primers were deposited by the gene bank nucleotide sequence database (Table 5).

The gene bank accession numbers of the forward and reverse ESBL producing *P. mirabilis* (No. 54) are BankIt1534618 JX17256 and BankIt1534627 JX17257 respectively. BankIt1534363 JX017250 and BankIt1534367 JX017251 were the gene bank accession numbers for the forward and reverse ESBL producing *E. coli* (No. 22), respectively. BankIt1534372 JX17252 and BankIt1534380 JX17253 were the gene bank accession numbers for the forward and reverse ESBL producing *K. pneumoniae* (No. 93), respectively. BankIt1534608 JX17254 and BankIt1534614 JX17255 were the gene bank accession numbers of the forward and reverse ESBL producing *E. cloacae* (No. 21), respectively.

Conjugation assay

Based on the conjugation assay in this experiment between the recipient cell (*E. coli* HB101) and the donor cell (*E. coli* No. 22), the blaCTX-M gene was able to be transferrable to transconjugant successfully suggesting that they were plasmid mediated. The gene bank numbers of the resulting DNA sequences of the forward and reverse CTX primers for the transconjugant were identified as bla-CTX-M producing *E. coli* genes bank No. FJ668785 (99%) similar to the forward and reverse DNA nucleotide sequences of the bla-CTX-M producing *E. coli* (No. 22).

DISCUSSION

Eighty four (84) out of 560 isolates are ESBL producing gram negative bacteria including *E. coli*, *K. pneumniae*, *P. mirabilis*, *E. cloacae* and *E. aerogenes*. The specimens were collected from different departments while urine samples were from elderly females patients above 50 years old having urinary tract infections. The ESBL infection rate in the hospitalized patients (62%) is higher than the out-patients mainly in the intensive care unit. These results are correlated with that obtained by

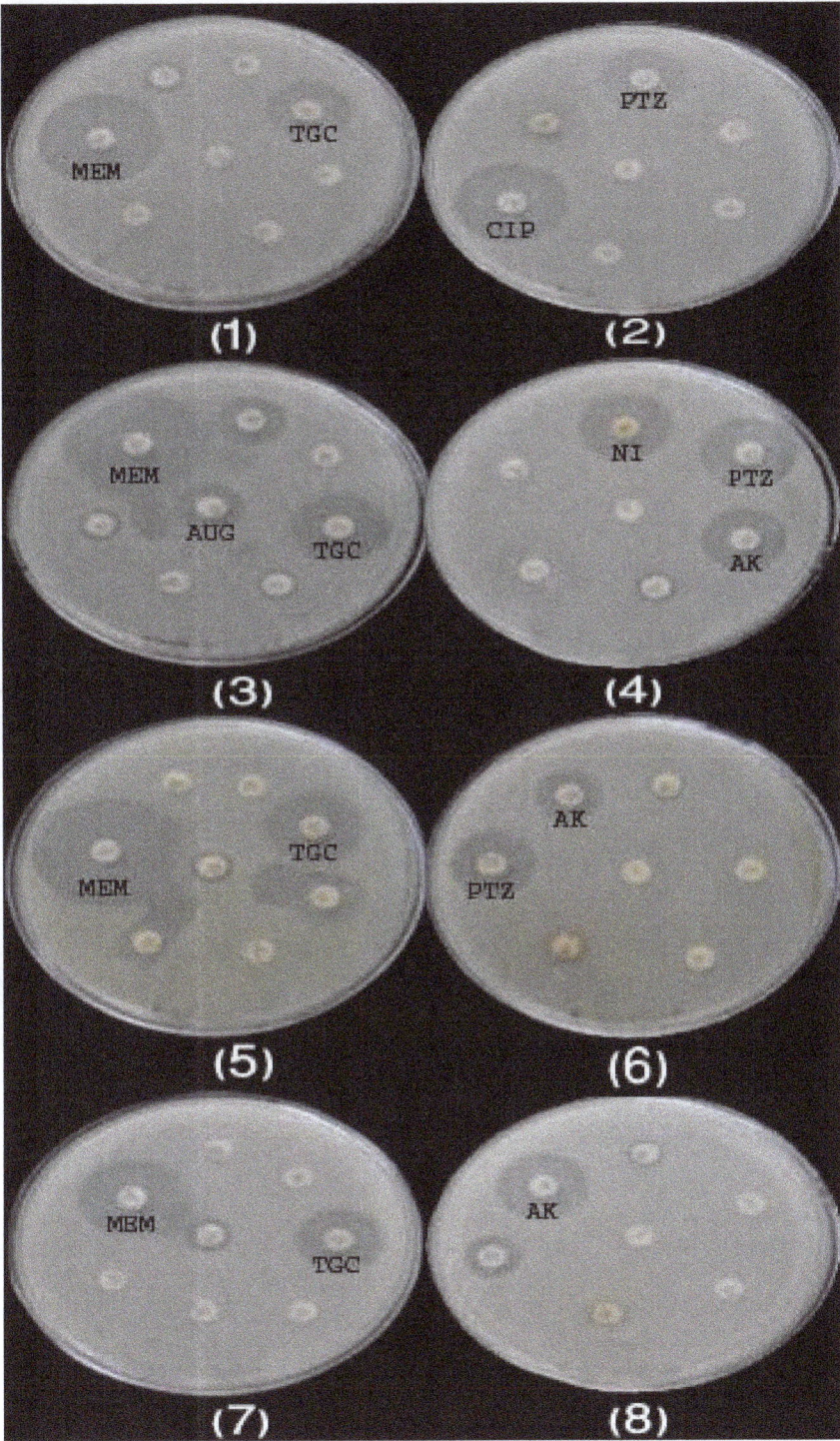

Figure 1. The antibiotic activity of the selected multi-resistant isolates. Plates 1 and 2, *E. cloacae* (No. 21); 3 and 4, *E. coli* (No. 22); 5 and 6, *P. mirablis* (No. 54); 7 and 8, *K. pneumoniae* (No. 93).

Kiratisin et al. (2008) where a total of 2,777 patients were identified as having infections due to *E. coli* or *K. pneumoniae* at both major tertiary-care centers. The majority of patients were over 60 years old and most of

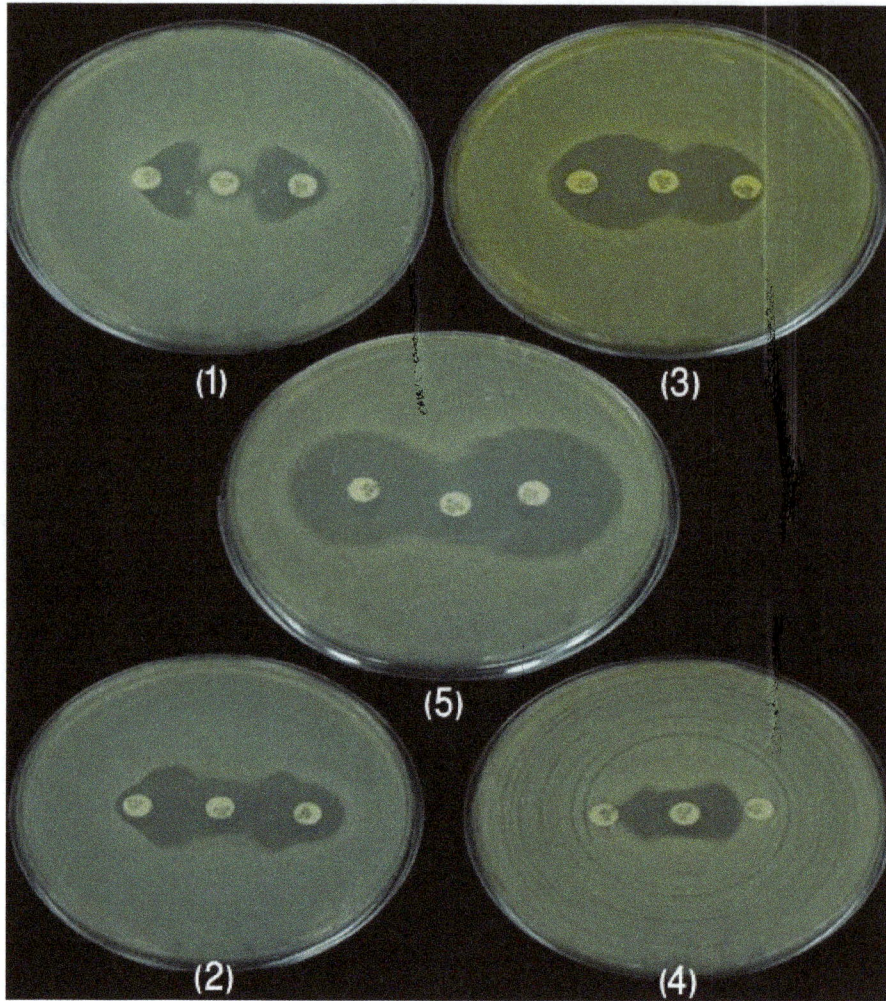

Figure 2. The enhancement bacterial zone of the selected ESBL isolates by DDST. Plate 1, *E. cloacae* (No. 21); 2, *E. coli* (No. 22); 3, *P. mirablis* (No. 54); 4, *K. pneumoniae* (No. 93); 5, *E. coli* ATCC No. 25922. Left disc: CAZ; middle disc: AUG; right disc: CTX.

them had been hospitalized in the intensive care units. All the ESBL producers were recovered from urine specimens and female patients were the predominant for particularly ESBL producing *E. coli* infection. Our results are in agreement with those of Dechen et al. (2009) who identified 81 ESBL producing isolates out of 238 Gram's negative bacilli by DDST and phenotypic confirmatory test. The isolates were *E. coli* (n = 34), *K. pneumoniae* (n = 20), *P. aeruginosa* (n = 15), *P. mirabilis* (n = 3), *Morganella morganii* (n = 5) and *Citrobacter frundii* (n = 4). The isolates were collected from 152 urine, 70 wound, 12 blood, 22 sputum and 2 cerebrospinal fluid samples. It was concluded that ESBL producers can be detected by DDST and phenotypic confirmatory test with equal efficacy. Data has shown that VITEK 2 GNI-GNB system reported all the members of the tested gram negative

bacilli successfully. *E. coli* were the predominant gram's negative bacilli among the others and MEM was the only susceptible antibiotic against all the multi-resistant gram negative isolates.

Caroline and Michael (2003) demonstrated that the VITEK 2 instrument has an accuracy of 93.0% for the identification of gram's negative bacilli. Khalid et al (2009) reported that out of a total of 11,886 isolated gram negative bacilli, 2695 were ESBL producers. *E. coli* and *K. pneumoniae* were the predominant comparatively in the hospital wards while *Proteus* spp. was predominant in medical wards. Urine was the major source with low occurrence in blood cultures. Shah and Mulla (2012) found that MEM and ETP were effective against ESBL producers especially *E. coli* and *K. pneumoniae* and remain good choices for the treatment of suspected

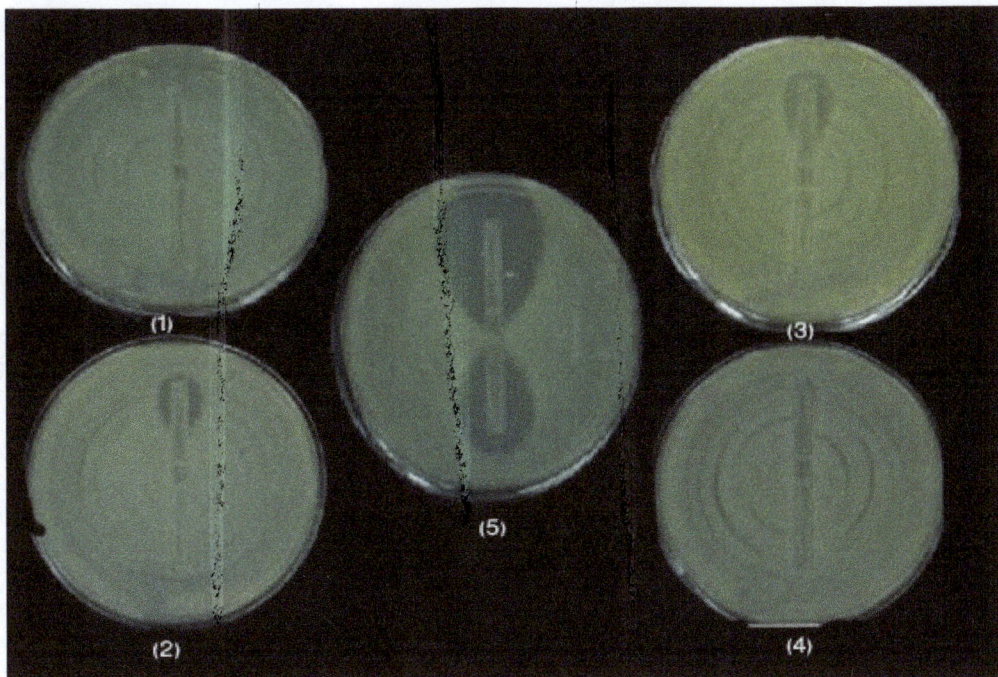

Figure 3. The phenotypic determination of the selected ESBL isolates by E-test strips (CT/CTL Plate 1, *E. cloacae* (No. 21); 2, *E. coli* (No. 22); 3, *P. mirablis* (No. 54); 4, *K. pneumoniae* (No. 93); 5, *E. coli* ATCC No. 25922.

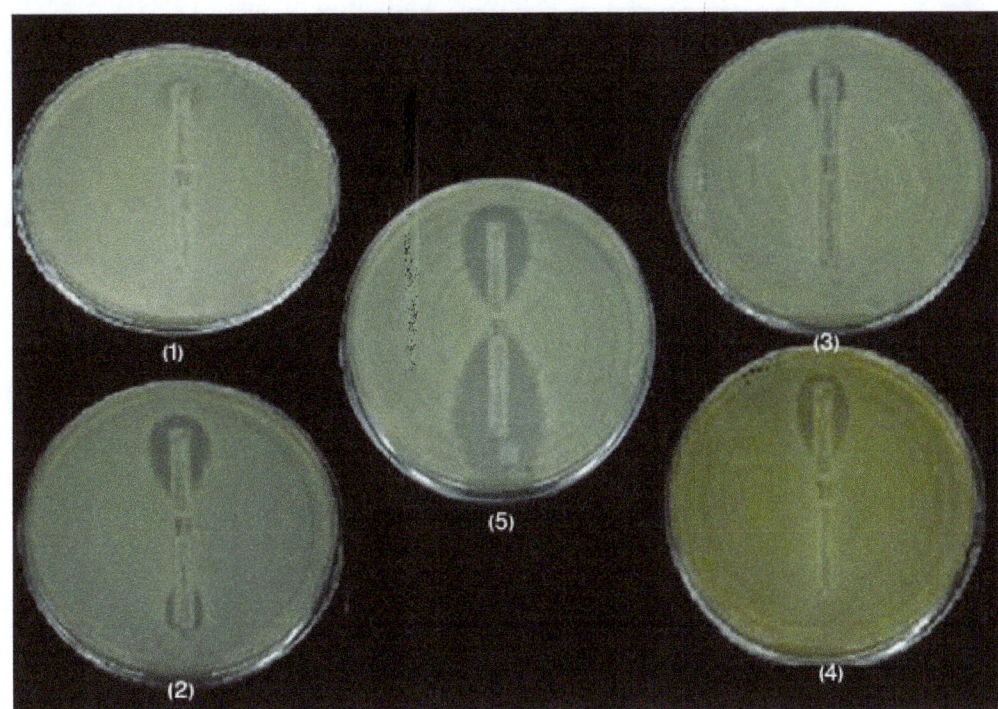

Figure 4. The phenotypic determination of the selected ESBL isolates by E-test strips (TZ/TZL). Plate 1, *E. cloacae* (No. 21); 2, *E. coli* (No. 22); 3, *P. mirablis* (No. 54); 4, *K. pneumoniae* (No. 93); 5, *E. coli* ATCC No. 25922.

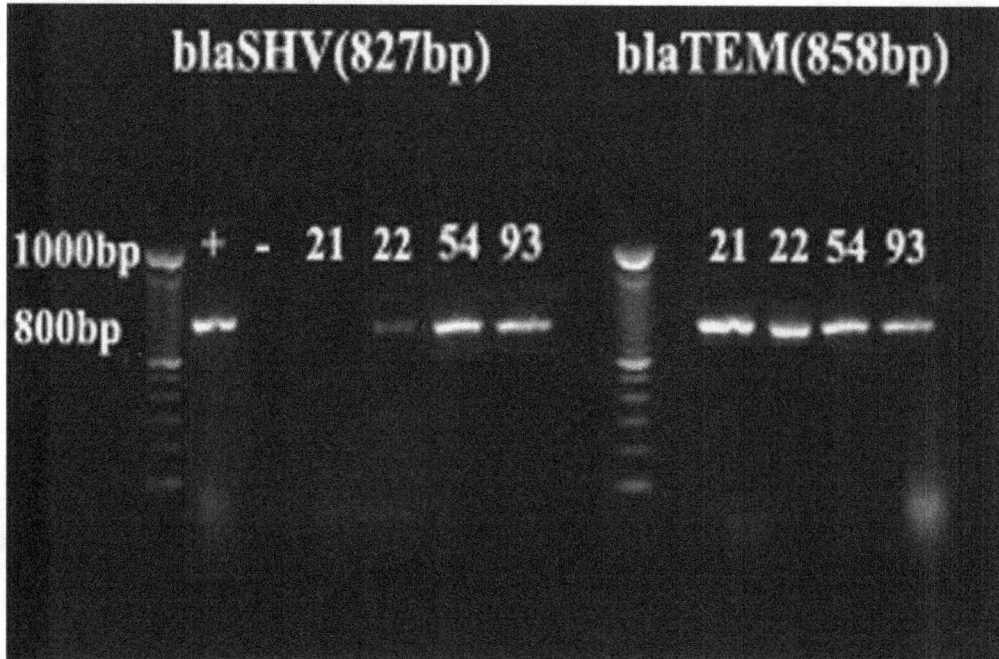

Figure 5. The agarose gel electrophoresis showing the identical band patterns of the blaTEM (858 bp) and blaSHV (827 bp) produced by *E. cloacae* (No. 21), *E. coli* (No. 22), *P. mirablis* (No. 54) and *K. pneumonia* (No. 93). Negative band of blaSHV is seen by *E. cloacae* (No. 21).

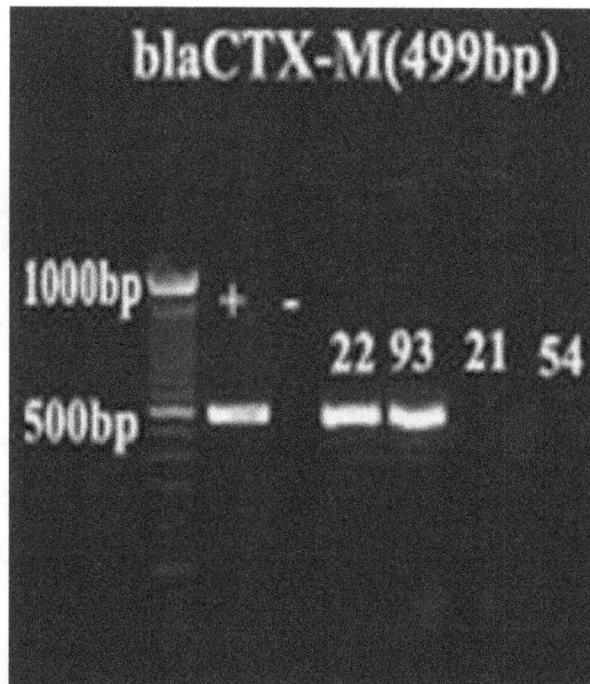

Figure 6. The agarose gel electrophoresis showing the identical band patterns of the blaCTX-M (499 bp) produced by *E. coli* (No. 22) and *K. pneumoniae* (No. 93). Negative bands of blaCTX-M are detected by *E. cloacae* (No. 21) and *P. mirablis* (No. 54).

Table 4. The analysis of the DNA sequences of the ESBL primers using NCBI/BLAST.

Organism	ESBL primer	Forward ESBL gene bank no.	Reverse ESBL gene bank number
E. coli (No. 22)	CTX-M	**FJ668785** Escherichia coli strain 61 plasmid bla-CTX-M15 gene. Identity 99%.	
K. pneumoniae (No. 93)	SHV	**FJ815288** Klebsiella pneumonia strain HB100 plasmid bla-CTX-M15 gene. Identity 98%.	**HM002660** Klebsiella pneumonia 379T ESBL CTX-M15 gene. Identity 98%.
P. mirabilis (No. 54)	TEM	**JN043376** Proteus mirabilis Strain M-0928-11 ESBL TEM-1 gene Identify by 78-99%.	
E. cloacae (No. 21)	TEM	**FJ349257** Enterobacter species ESBL TEM-150 gene. Identity 91%.	**AY30226** Enterobacter cloacae strain 212 TEM-1 gene. Identity 90%.

ESBL infections. In the present study, the detection of ESBLs varies between VITEK 2 (84.5%), TZ/TZL (94%), CT/CTL (98.8%) and DDST (95%) on 560 isolates of E. coli, K. pneumoniae, P. aeruginosa, P. mirabilis, E. cloacae, E. aerogenes and A. baumanii. The performance characteristics of the conventional methods for the detection of ESBLs (E-test and DDST) were better than the automated system (VITEK2). This finding is in agreement with those of Irith et al. (2007) who compared commercially microbiological identification testing system (VITEK 2) with the conventional phenotypic confirmatory tests (E-test and DDST) to detect ESBL production in gram's negative bacteria. The E-test showed the highest specificity for the detection of ESBLs followed by DDST and VITEK2. They recommended the use of a manual test for confirmation once an organism is reported positive for ESBL production by any of the semi-automated system. Our results also agree with those of Maria et al. (2011) who evaluated the accuracy of positive ESBL results by VITEK 2 regarding clinical isolates of E. coli and K. pneumoniae using DDST and E-test displaying distinct results by which the VITEK 2 system was in disagreement in 23.9% of cases with DDST and in 15.3% with E-test.

Molecular detection in the present work showed that the tested isolates are positive ESBL produced by blaSHV, blaCTX-M and blaTEM genotypes. The bla-CTX-M is the most predominant ESBL gene (63.1%) followed by blaSHV (60.7%) detected in E. coli (62%) and K. pneumoniae (20.2%). Two or more genotypes of ESBL were present in 34 isolates of E. coli, blaCTX-M and blaSHV being the most common combination genes (33.3%) followed by bla-SHV with bla-TEM and bla-CTX-M (11%). ESBL were present in 17 isolates of K. pneumoniae, bla-SHV and bla-TEM being the most

combination genes (39%) followed by blaSHV with blaTEM and blaCTX-M (28%). This finding is in agreement with those of Ankur et al. (2009) who examined ESBL production phenotypically for a total of 200 consecutive clinical isolates of E. coli (n = 143) and K. pneumoniae (n = 57) collected at a tertiary care hospital followed by further typed for the blaTEM / SHV / CTX-M genes by PCR using specific primers. ESBLs were found in 63.6% of E. coli isolates and 66.7% of K. pneumoniae isolates and majority of them harboured two or more ESBL genes (57.3%). Overall, blaCTX-M was the commonest genotype (85.4%) followed by blaTEM (54.9%) and blaSHV (32.9%) either alone or in combination. Two or more genes for ESBL were present in 47 out of 82 ESBL isolates including the blaTEM with blaCTX-M being the most common combination (28.1%). They concluded high ESBL occurrence with CTX-M as the emerging type in the selected hospital. The results in the present study identified successfully the selected ESBL producing gram negative bacteria by 16SrDNA. Identification of bacteria by 16S rDNA gene sequences analysis in the present work discriminate more finely among the selected strains than is possible with phenotypic methods. Thus, it provides an accurate identification at the species level and can clarify the clinical importance of the isolated bacteria of the infectious diseases (Fredricks and Relman, 1996; Clarridge et al., 2001).

Tang et al. (2000) compared a variety of identification systems including cellular fatty acid profiles, carbon source utilization and conventional biochemical identification with the 16SrRNA gene sequence to evaluate both unusual aerobic gram negative bacilli isolated from clinical specimens. They found that 16SrRNA gene sequence provided more rapid, unambi-

Table 5. The characteristics and identifications of the selected ESBL gram negative bacteria by 16S rDNA.

Organism	Phenotypic test			PCR			Gene bank accession number (16S rDNA)	
	VITEK	E-Test	DDST	TEM	CTX-M	SHV	F27	R1492
E. coli (No. 22)	+	+	+	+	+	+	BankIt1534363 JX017250	BankIt1534367 JX017251
K. pneumoniae (No. 93)	+	+	+	+	+	+	BankIt1534372 JX017252	BankIt1534380 JX017253
P. mirabilis (No. 54)	-	+	+	+	-	+	BankIt1534618 JX017256	BankIt1534627 JX017257
E. cloacae (No.21)	-	+	+	+	-	-	BankIt1534608 JX017254	BankIt1534614 JX017255

guous identification of the difficult bacterial isolates than did conventional methods and that this identification could translate to improve clinical outcomes.

Bosshard et al. (2003) found that only a minority of the clinical laboratory isolates of aerobic gram negative rods could be correctly identified by phenotypic methods whereas 16SrRNA gene sequencing is an excellent method for identifying these organisms which are difficult to identify by conventional methods. Our results show that the blaCTX-M genotype of E. coli (No. 22) is able to be transferred to transconjugant successfully when conjugated with E. coli (HB101); this suggests that they are plasmid mediated.

Iroha et al. (2010) reported that ESBLs are carried on bacterial chromosomes or plasmids and plasmid-mediated ESBLs can carry genes on them that have the ability to transfer a replica of themselves to other bacteria. They also can carry genes conferring resistance to other classes of antibiotics that make the recipient bacteria resistant to multiple antibiotics. Furthermore, these plasmids can emerge on strains that do not cause human diseases and then the non-pathogenic strains could transfer their plasmids to strains that can cause human diseases. Plasmid conjugation is an important mechanism

of disseminating drug resistance among bacteria population. In conclusion, the prevalence rate of ESBL producing organisms is high globally. The ESBL producing organisms are known to cause serious nosocomial infections, long term carriage in the community, community-acquired infections such as urinary tract infections and intra-abdominal abscess. The findings from the present study reveal high prevalence of ESBL from Ahmadi hospital in Kuwait where E. coli is the highest producer followed by K. pneumoniae, P. mirabilis, E. cloacae and E. aerogenes, respectively. ESBL producers can be detected by E-test and DDST with equal efficacy. The blaCTX-M is the most predominant ESBL genotypes among the multiresistant gram negative bacteria including the third generation of cephalosporins. Thus, this study emphasizes the inclusion of ESBL detection in routine laboratory tests in hospitals and clinics especially in the developing countries.

ACKNOWLEDGEMENT

The authors wish to thank Professor Dr. Ali A. Dashti, Professor of Microbiology, Medical Laboratory Science Department, Faculty of Allied Health Science, Health Science Center, Kuwait University, Kuwait, for his excellent technical assistance.

REFERENCES

Al Naiemi N, Duim B, Savelkoul P, Spanjaard L, Jongr E, Bart, A, Vandenbroucke C, Grauls D, Menno D (2005). Widespread Transfer of Resistance genes between Bacterial species in an Intensive Care Unit: Implication for Hospital Epidemiology. J. Clin. Microbiol. 34(9):4862-4864.

Ankur G, Prasad KN, Amit P, Sapna G, Ujjala G, Archana A (2009). Extended spectrum β-lactamase in Escherichia coli and Klebsiella pneumoniae, and associated risk factors. Indian J. Med. Res. 129:695-700.

Arlet G, Brami G, Decre D, Flippo A, Gaillot O, Lagrange PH, Philippon A (1995). Molecular characterization by PCR-restriction fragment length polymorphism of TEM beta-lactamases. FEMS Microbiol. Lett. 134: 203-208.

Arlet G, Philippon A (1991). Construction by polymerase chain reaction and intragenic DNA probs for three main types of transferable β-lactamases (TEM, SHV, GARB). FEMS. Microbiol. lett. 82:19-26.

Bauer AW, Kirby WM, Sherris JC, Turck M (1966). Antibiotic susceptibility testing by a standardized single disk method. Am. J. Clin. Pathol. 45:493-496.

Bauernfeind A, Casellas JM, Goldberg M, Holley M, Junwirth R, Mongold P, Rohnush T, Schweighart S, Wilhelm R (1992). A new plasmidic cefotaximase from patients infected with Salmonella typhimurium. Infect. 20:158-163.

Bonnet R (2004). Growing group of extended-spectrum β-lactamases: the CTX-M enzymes. Antimicrob. Agents Chemother. 48(1):1-14.

Bosshard PP, Abels S, Zbinden R, Bottgar EC, Altwegg M (2003). Ribosomal DNA sequencing for identification of aerobic Gram-positive rods in the clinical laboratory (an 18-month evaluation). J. Clin. Microbiol. 4:4134-4140.

Bradford PA (2001). Extended spectrum β-lactamases in the 21st century: characterization, epidemiology and the detection of this important resistance threat. Clin. Microbiol.Rev. 14:933-951.

Bratu S, Landman D, Haag R, Recco R, Eramo A, Alam M, Quale J (2005). Rapid spread of carbapenem-resistant Klebsiella pneumoniae in New York City: a new threat to our antibiotic armamentarium. Arch. Intern. Med. 165(12):1430-1443.

Canton R, Perez-Zazquez M, Oliver A, Coque TM, Loza E, Ponz F, Baquero F (2001). Validation of VITEK-2 and the advanced Expert System with a collection of Enterobacteriaceae harbouring extended spectrum or inhibition resistant β-lactamases. Diagn. Micr. Infec. Dis. 41:65-70.

Caroline M, Michael J (2003). Evaluation of the Vitek 2 ID-GNB Assay for identification of members of the family Enterobactericaea and other nonenteric Gram negative bacilli and comparison with the Vitek GNI cards. J. Clin. Microbiol. 41(5):2096-2101.

Clarridge JE, Attorri S, Musher DM, Hebert J, Dunbar S (2001). Streptococcus constellatus and Streptococcus anginosus ("Streptococcus milleri group") are of different clinical importance and are not equally associated with abscess. Clin. Infect. Dis. 32:1511-1515.

Clinical and Laboratory Standards Institute (CLSI) (2005). Performance standards for antimicrobial susceptibility testing: 15th informational supplememnt. CLSI document M100-S15. CLSI. Wayne, PA.

Cormican MG, Marchall SA, Jones RN (1996). Detection of extended spectrum β -lactamases (ESBL) producing strains by the E-test ESBL screen. J. Clin. Microbiol. 34(8): 1880-1884.

Dashti A, Jadan M, Abdulsamad A, Dashti H (2009). Heat treatment of bacteria: DNA extraction for molecular techniques. KMJ. 41(2):117-122.

Dashti A, Jadaon M, Gomaa H, Noronha B, Udo E (2010). Transmission of Klebsiella pneumoniae clone harbouring genes for CTX-15-LIKE and SGAV-112 enzymes in a neonatal intensive care unit of a Kuwaiti hospital. J. Med. Microbiol. 59:687-692.

Dechen CT, Shyamasree D, Luna A, Ranabir P, Takhellambam SK (2009). Extended spectrum β -lactamase detection in Gram-negative bacilli of nosocomial origin. J. Global Infect. Dis. 1(2): 87-92.

Dhillon R, Clark J (2011). ESBL: A clear and present danger? Crit. Care Res. Pract. 2012 (2012):1-11.

Duttaroy B, Mehta S (2005). Extended spectrum β -lactamases (ESBL) in clinical isolates of Klebsiella pneumoniae and Escherichia coli. Indian J. Pathol. Micr. 48(1):45-48.

Fredricks DN, Relman DA (1996). Sequence-based identification of microbial pathogens: a reconsideration of Koch's postulates. Clin. MicrobioL. Rev. 9:18-33.

Irith W, Heinrich KG, Dietrich M, Enno S, Harald S (2007). Detection of extended-spectrum β -lactamase among Enterobacteriaceae by use of semiautomated microbiology system and manual detection procedures. J. Clin. Microbiol. 45(4):1167-1174.

Iroha IR, Amadi ES, Oji AE, Nwuzo AN, Ejike-Ugwa PC (2010). Detection of plasmid borne extended spectrum β -lactamase enzymes from blood and urine isolates of Gram negative bacteria from a University teaching hospital in Nigeria. Curr. Res. Bacteriol. 3(2):77-83.

Ivanova D, Markovska R, Hadjiera N, Mitov I, Bauernfeind A (2008). Extended spectrum β-lactamase-producing Serratia marcescens outbreak in a Bulgarian hospital. J. Hosp. Infect. 70:60-65.

Johann D, Pitout D, Ashfaque H, Nancy D (2004). Phenotypic and Molecular Detection of CTX-M- β -Lactamases produced by Escherichia coli and Klebsiella spp. J. Clin. Microbiol. 42(12):5715-5721.

Khalid MB, Abiola CS, Afaf EJ (2009). Prevelence of extended-spectrum β-lactamase-producing Enterobacteriacaeae in Bahrain. J. Infect. Public Health. 2:129-135.

Kiratisin P, Apisarnthanark A, Laesripa C, Saifon P (2008). Molecular Characterization and Epidemiology of extended-spectrum β -lactamase-producing Escherichia coli and Klebsiella pneumoniae isolates causing health care-associated infection in Thailand, where the CTX-M family is endemic. Antimicrob. Agents Ch. 52(8):2818-2824.

Knothe H, Shah P, Krcmery V, Antal M, Mitsuhashi S (1983).Transferable resistance to cefotaxime, cefoxitin, cefamandole and cefuroxime in clinical isolates of Klebsiella pneumoniae and Serratia marcescens. Infect.11:315- 317.

Kollef MH, Sherman G, Ward S, Fraser VJ (1999). Inadequate antimicrobial treatment of infections: a risk factor for hospital mortality among critically ill patients. Chest. 115:462-474.

Leflon-Guibout V, Jurand C, Bonacorsi S, Espinasse F, Guelfi MC, Duportail F, Heym B, Bingen E, Nicolas-Chanoine MH (2004). Emergence and spread of three clonally related virulent isolates of CTX-M-15-producing Escherichia coli with variable resistance to aminoglycosides and tetracycline in a French geriatric hospital. Antimicrob. Agents Chemother. 48(10):3736-3742.

Livermore DM (1995). β-lactamases in laboratory and clinical resistance. Clin. Microbiol. Rev. 8:557-584.

Lou YK, Qin H, Molodysky E, Morris BJ (1993). Simple microwave and thermal cycler boiling methods for preparation of cervicovaginal lavage cell samples prior to PCR for human papillomavirus detecteion. J. Virol Methods. 44:77-81.

Maniatis T, Fritsch EF, Sambrook J (1982). Molecular cloning: a laboratory manual. Cold Spring Harbor Laboratory, Cold Spring Harbor, N.Y.

Maria JE, Rita R, Manuela R, Acacio GR, Cidalia P (2011). Extended-spectrum β -lactamases of Escherichia coli and Klebsiella pneumoniae screened by the VITEK 2 system. J. Med. Microbiol. 60(6):756-760.

McFarland J (1907). Nephelometer: an instrument for estimating the number of bacteria in suspensions used for calculating the opsonic index and for vaccines. J. Am. Med. Ass. 14:1176-1178.

Merk S, Meyer H, Greiser-Wilke I, Spraque LD, Neubauer H (2006). Detection of Burkholderia cepacia DNA from artificially EDTA-blood and lung tissue comparing different DNA isolation methods. J. Vet. Med. 53:281-285.

(NCCLS) National Committee for Clinical Laboratory Standards (1999). Performance Standards for Antimicrobial Susceptibility; Ninth Informational Supplement M100-S9. Villanova, PA, USA: NCCLS.

(NCCLS) National Committee for Clinical Laboratory Standards (2003). Approved standard M2-A8. Performance Standards for Antimicrobial Susceptibility. 8th ed. Wayne, PA.

(NCCLS) National Committee for Clinical Laboratory Standards (2000). Performance Standards for Antimicrobial Disk Susceptibility Tests. 4th Edn. Approved Standard M2-A7. NCCLS. Villanova Pa.

Olsen G, Woese C (1993). Ribosomal RNA: a key to phylogeny. FASEB. 7:113-123.

Paterson DL, Bonomo RA (2005). Extended-spectrum β-lactamases: a clinical update. Clin. Microbiol. Rev.18:657-686.

Paterson DL, Hujer KM, Hujer AM, Yeiser B, Bonomo MD, Rice LB, Bonomo RA (2003). Extended-spectrum ß -lactamases in Klebsiella pneumoniae bloodstream isolates from seven countries: Dominance and widespread prevalence of SHV- and CTX-M-type β -lactamases. Antimicrob. Agents Chemother. 47:3554-3560.

Sanders CC, Peyret M, Moland ES, Cavalieri SJ, Shubert C, Thomson KS, Boeufgras JM, Sanders WE (2001). Potential impact on VITEK 2 system and Advanced Expert System on the clinical laboratory of a university-based hospital. J. Clin. Microbiol. 39:2379-2385.

Shah K, Mulla SA (2012). Susceptibility of ESBL-producing Enterobacteriaceae to ertapenem, meropenem and piperacillin-tazobactam. Nati.J. of Med. Res. 2(2):223-225.

Tang YW, Von Graevenitz A, Waddington MG, Hopkins MK, Smith DH, Li H, Kolbert CP, Montgomery SO, Persing DH (2000). Identification of coryneform bacterial isolates by ribosomal DNA sequence analysis. J. Clin. Microbiol. 38:1676-1678.

Tenover FC (2007).   Rapid detection and identification of bacterial pathogens using novel molecular technologies: infection control and beyond. Clin. Infect. Dis. 44: 418-423.

Tenover FC, Areibt RD, Goering RV, Mickelsen PA, Murray BE, Persing DH, Swaminathan B (1995). Interpreting chromosomal DNA restriction patterns produced by pulsed-field gel electrophoresis. J. Clin. Microbiol. 33: 2233-2239.

Weisburg WG, Barns SM, Pelletier DA, Lane DJ (1991). 16S ribosomal DNA amplification for phylogenetic study. J. Bacteriol. 173(2):697-703.

Winokur PL, Canton R, Casellas JM, Legakis N (2001). Variations in the prevalence of strains expressing an extended-spectrum beta lactamase phenotype and characterization of isolates from Europe the Americas, and the Western Pacific region. Clin. Infect. Dis. 32 S94-S103.

Woodford N, Sundsfjord A (2005).   Molecular detection of antibiotic resistance: when and where? J. Antimicrob. Chemother. 56:259-261.

14

Marine bacterial prodigiosin as dye for rubber latex, polymethyl methacrylate sheets and paper

Jissa G. Krishna[1,2], Ansu Jacob[3], Philip Kurian[3], Elyas KK[1,4] and M. Chandrasekaran[1]*

[1]Microbial Technology Laboratory, Department of Biotechnology, Cochin University of Science and Technology, Cochin 682 022, Kerala, India.
[2]National Centre for Biological Sciences, Bangalore 560 065, India.
[3]Department of Polymer Science and Rubber Technology, Cochin University of Science and Technology, Cochin-682 022, Kerala, India.
[4]Department of Biotechnology, Calicut University, Kerala, India.

Prodigiosin is known for its immunomodulatory, antibacterial, antimycotic, antimalarial, algicidal and anticancer activities. Here, we reported the evaluation of prodigiosin pigment as a dyeing agent in rubber latex, paper and polymethyl methacrylate (PMMA) so that it can be considered as an alternative to synthetic pigments. Maximum color shade was obtained in rubber sheet prepared with 0.5 parts per hundred gram of rubber (phr) pigment and PMMA sheet incorporated with 0.08 µg pigment. Results indicate scope for utilization of prodigiosin as dye for PMMA and rubber and also prodigiosin dyed paper as a pH indicator. Further, being a natural and water insoluble pigment, it is ecofriendly.

Key words: Prodigiosin, *Serratia* sp., dye, rubber, polymethyl methacrylate.

INTRODUCTION

Natural pigments and synthetic dyes are extensively used in various industries including food, textile, paper, rubber, in agricultural practice, and water science and technology. However, the effluents released from the dyeing units of these industries contain synthetic dyes that are toxic and cause extensive environmental pollution besides polluting the ground water resources of drinking water and agriculture practices. Consequently, synthetic dyes have a significant negative impact on the environment (Tibor, 2007; Balakrishnan et al., 2008). To alleviate the problems caused by the synthetic dyes and chemicals, alternative eco-friendly technologies and use of natural pigments that are biodegradable are preferred. In this context, natural pigments have drawn the attention of industries as safe alternative. As per the available data, Europe imports US $ 53 million worth of natural

dyes and the major importing countries include Germany (32%), France (17%), Italy (14%), USA (12%) and U.K. (10%). The largest dye suppliers include Mexico, Peru, China and India, each exporting dyes worth US $ 15 million to Europe.

Natural biocolorants obtained from plants, animals and microorganisms are possible alternatives to synthetic dyes and pigments currently employed in various industries (Mapari et al., 2005). Among the natural sources, microorganisms offer great scope and hope as compared to other resources. The genetic diversity in microbes, ease of their cultivation, extraction and sophistication in technologies has made their choice more feasible (Juailova et al., 1997). In fact, microorganisms produce a large variety of stable pigments such as carotenoids, flavonoids, quinones and rubramines, and fermentation production results in higher yields of pigments and lower residues as compared to that obtained from plants and animals (Durán et al., 2002). Thus, biosynthesis of dyes and pigments via fermentation processes has attracted

*Corresponding author. E-mail: mchandra@cusat.ac.in.

more attention in recent years (Durán et al., 2002; Hobson and Wales, 1998). It may be also noted that prodigiosin is known to have several biological activities such as immunomodulatory, antibacterial, antimycotic, antimalarial (Lazaro et al., 2002; Pandey et al., 2003), anticancer (Montaner and Perez-Tomas, 2001) and algicidal activities (Kim et al., 2007).

Currently, synthetic dyes are used to impart color for rubber products. Polymethyl methacrylate (PMMA) is a transparent thermoplastic, often used as a light or shatter-resistant alternative to glass. These sheets are also given different color shades. Similarly, papers with different colour shades are manufactured. In all these three products, use of prodigiosin as a natural dye has not been reported so far to the best of our knowledge. In this context, we report here the prospects of using prodigiosin, produced by *Serratia* sp. BTWJ8 isolated from marine sediment, as dye for imparting different shades of color to rubber latex, polymethyl methacrylate sheets and paper.

MATERIALS AND METHODS

Purification and characterization of the pigment

The pigment produced by *Serratia* sp. BTWJ8 was purified and characterized as described previously (Jissa et al., 2011). The purified pigment was used for the following application studies:

Dyeing of rubber latex

The colouring dye was prepared by ball milling using the compounding ingredients as mentioned below: Natural rubber latex (60% dry rubber content) 100 (phr), sulphur (50%) 1.5 phr, ZnO (50%) 0.9 phr, accelerator (50%) 0.7 phr, and antioxidant (50%) 0.5 phr. The compounding ingredients were then subjected to sonication for 30 min to make homogenous pigment dispersion. Using this pigment dispersion, four different concentrations viz: 0.0, 0.16, 0.3 and 0.5 phr namely 1, 2, 3 and 4 respectively were prepared to obtain varied color shades. The mixes were then casted onto Petri plates to make required rubber sheets, and kept for 24 h at room temperature (RT) (28 ± 2°C). The rubber sheets with 1 mm thickness were incubated at 70°C for 2 h in hot air oven for proper vulcanization of the rubber. Mix 1 was used as the control.

Dyeing of polymethyl methacrylate sheet

Bacterial pigment in methanol (40 µg/L) was used as the stock solution for imparting colour onto the polymethyl methacrylate sheets prepared using 10% solution of PMMA in chloroform. Towards imparting varied colour shades, four different concentrations of bacterial pigment dispersion viz: 250 µl (0.01 µg; w/v), 500 µl (0.02 µg; w/v), 1 ml (0.04 µg; w/v) and 2 ml (0.08 µg; w/v) were added to PMMA solution from the stock solution separately, mixed well, poured into a watch glass, and kept for 3 h at RT (28 ± 2°C). The watch glasses were covered with a glass plate in order to prevent air contact.

Dyeing of paper

Eight types of paper with different qualities like 'art paper', 'J paper', 'sunlight', '6.9 SPB', '7.8 SPB', '11 Kg JK', '21.3 Kg JK' an '18.6 Kg SPB' commercially available in the market were selecte and used in the present study. All the paper materials were cut in* equal size of 2 cm². Bacterial pigment in methanol (40 µg/L) wa used as the stock solution. An aliquot of 200 µl (0.008 µg; w/v) (the stock solution was applied on to the different paper materials o a warm surface and allowed to dry at RT for 15 min to impa colour. Paper material without pigment was kept as the contro After dyeing, acidic (pH 2.0), neutral (pH 7.0) and alkaline (pH 10.C solutions were spotted over all the paper materials to evaluate th dyed paper as probable pH indicators.

RESULTS AND DISCUSSION

Synthetic dyes made from nonrenewable sources suc as fossil fuels are used extensively in the textile, rubbe paper and plastic industries and the industrial effluent loaded with these dyes has created alarming situatio with respect to environmental health. Wastewater from printing and dyeing units is often rich in color, containin residues of reactive dyes and chemicals. The toxi effects of dyestuffs and other organic compounds, as we as acidic and alkaline contaminants in these dye effluent reach a stage where they are not treated effectivel before their disposal into environment. Considering the i effects of synthetic dyes on human beings an ecosystem, Germany banned the use of numerou specific azo-dyes for their manufacturing and application and most of the countries brought effective laws an regulations related to the customer health and safety an protection of eco-system (Nimkar and Bhajekar, 2006 Premi, 1996). The present trend of work culture, safet and eco-requirements will continue to dominate the trade and the processor will need to understand the change that need to be effected to satisfy these requirement (Burdhan, 2002). As a consequence, there is a renewec interest in the use of natural pigments as dyes, which i normally biodegradable in the environment.

Red pigment isolated from *Serratia* sp. BTWJ8 purifiec and identified as prodigiosin (Jissa et al., 2011) was usec for the application studies. Attempt made to evaluate the probable use of this pigment as a coloring agent showec very promising results with rubber, PMMA and paper Since there was no suitable methodology available ir literature, to the best of our knowledge we employec reliable methodologies standardized in our laboratory fo dyeing experiments.

The impact of rubber and its products is on the rise ir our day-to-day life and the use of rubber is widespread ranging from household to industrial products, entering the production stream at the intermediate stage or as final products. Studies conducted with rubber product show that prodigiosin is an effective dye for inclusion ir rubber products (Figure 1). The maximum color was

Mix 1 (Control)

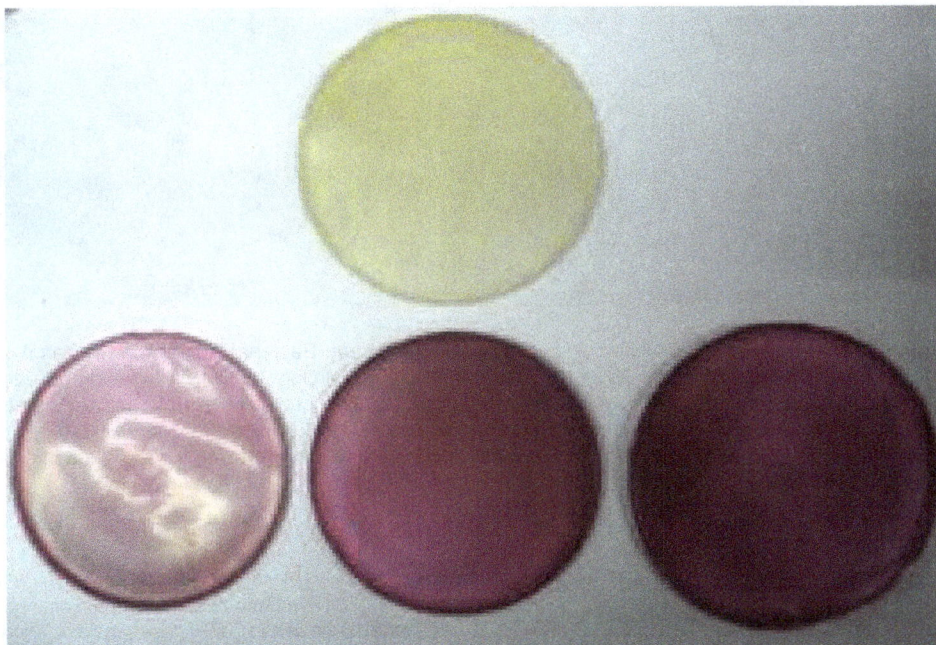

Mix 2: 0.16 phr pigment dispersion   **Mix 3: 0.3 phr pigment dispersion**   **Mix 4: 0.5 phr pigment dispersion**

Figure 1. Rubber sheets dyed with bacterial pigment. Compounding ingredients were prepared by ball milling as follows: Natural rubber latex (60% dry rubber content), 100 parts per hundred gram of rubber (phr), sulphur (50%) 1.5 phr, ZnO (50%) 0.9 phr, accelerator (50%) 0.7 phr, antioxidant (50%) 0.5 phr. From this, 0.0, 0.16, 0.3 and 0.5 phr pigment dispersion, prepared by means of sonication for 30 min, was used to prepare Mix 1, 2, 3 and 4, respectively. Mix 1 was used as the control.

obtained in rubber sheet prepared with Mix 4 that contained 0.5 phr pigment, followed by 0.3 phr and the minimum color shade was obtained with Mix 2 incorporated with 0.16 phr pigment. The results indicate that different color shades can be produced by varying the concentration of pigment in rubber latex. Studies conducted with rubber products show that prodigiosin is an effective dye for inclusion in rubber products.

PMMA or poly methyl 2-methylpropenoate, a synthetic polymer of methyl methacrylate, commonly called acrylic glass or simply acrylic is widely used in the lenses of exterior lights of automobiles and also plastic optical fiber used for short communication. The results obtained for the studies conducted with PMMA (Figure 2) indicated probable scope for exploiting this pigment as a natural dye in synthetic plastic materials. The topmost color intensity was noticed with PMMA sheet incorporated with 0.08 µg pigment followed by those sheets with 0.04 and 0.02 µg pigments. The least color shade was observed with PMMA sheet integrated with 0.01 µg pigment. Hence, it is proposed that different color shades can be produced by varying the concentration of red prodigiosin in PMMA solution in chloroform.

Paper continues to remain as a popular medium for printing, writing and also as a packaging material. So, colorants for paper industry have a bright but challenging prospect. In the present study, we evaluated eight types of paper with different qualities commercially available in the market. All the paper materials were cut into equal sizes of 2 cm^2. Bacterial pigment in methanol (40 µg/L) was used as the stock solution; 200 µl (0.008 µg; w/v) of the stock solution was applied on the different paper materials on a warm surface and allowed to dry at RT for 15 min. Paper material without pigment was kept as control. After dyeing, acidic (pH 2.0), neutral (pH 7.0) and alkaline (pH 10.0) solutions were spotted over all the paper materials. It was observed that the color of all the prodigiosin dyed paper materials recorded a change from white to red, pink and yellowish orange, respectively with acidic (pH 2.0), neutral (pH 7.0) and alkaline (pH 10.0) solutions strongly suggesting that prodigiosin dyed paper can be used as a pH indicator. The results obtained for

0.01 µg   **0.02 µg**   **0.04 µg**   **0.08 µg**

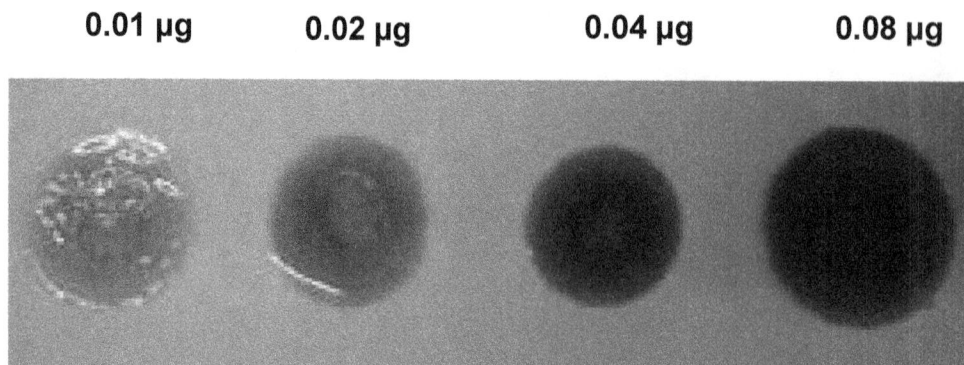

Figure 2. Polymethyl methacrylate dyed with bacterial pigment. Bacterial pigment in methanol (40 µg/L) was used as the stock solution. Aliquots of 10% solution of PMMA was prepared in chloroform 250 µl (0.01 µg), 500 µl (0.02 µg), 1 ml (0.04 µg) and 2 ml (0.08 µg) of bacterial pigment were added to 10% PMMA solution in chloroform separately, mixed well and poured into a watch glass and kept for 3 h at RT (28 ± 2°C). The watch glasses were covered with a glass plate in order to prevent air contact.

the studies conducted with paper indicated that the pigment can be used as a dye for preparation of colored paper.

Based on the results, it is concluded that prodigiosin produced by marine *Serratia* sp. BTWJ8 has the potential for use as a natural dyeing agent for the preparation of colored rubber latex products, PMMA with different color shades and pH indicator paper. Being a natural pigment, it is definitely harmless and would be ecofriendly.

ACKNOWLEDGEMENT

The first author is grateful to Kerala State Council for Science, Technology and Environment (KSCSTE), Kerala, India for the research fellowship.

REFERENCES

Balakrishnan M, Arul AS, Gunasekaran S, Natarajan RK (2008). Impact of dyeing industrial effluents on the groundwater quality in Kancheepuram (India). Indian J. Sci. Technol. 1:1-8.
Burdhan MK (2002). Significance of technical textiles in the context of globalization. Colorage 49:82-86.
Durán N, Teixeira MFS, Conti De R, Esposito E (2002). Ecological-friendly pigments from fungi. Crit. Rev. Food Sci. Nutr. 42:53-66.
Hobson DK, Wales DS (1998). Green colorants. J. Soc. Dyers Color. 114:42-44.
Jissa GK, Soorej MB, Elyas KK, Chandrasekaran M (2011). Prodigiosin from marine bacterium: Production, Characterization and Application as dye in textile industry. Int. J. Biotechnol. Biochem. 7:155-191.
Juailova P, Martinkova LJ, Machet F (1997). The existing genetic diversity in microbes and sophistication of technology has made their choice more feasible. Enzyme Microb. Technol. 16:231-235.

Kim D, Lee JS, Park YK, Kim JF, Jeong H, Oh TK, Kim BS, Lee CH (2007). Biosynthesis of antibiotic prodigiosines in the marine bacterium *Hahella chejuensis* KCTC 2396. J. Appl. Microbiol. 102:937-944.
Lazaro JEH, Nitcheu J, Predicala RZ, Mangalindan GC, Nesslany F, Marzin D, Concepcion GP, Diquet B (2002). Heptyl prodigiosin, a bacterial metabolite, is antimalarial *in vivo* and non-mutagenic *in vitro*. J. Nat. Toxins. 11:367-377.
Mapari SAS, Nielsen KF, Larsen TO, Frisvad JC, Meyer AS, Thrane U (2005). Exploring fungal biodiversity for the production of water-soluble pigments as potential natural food colorants. Curr. Opin. Biotechnol. 16:109-238.
Montaner B, Perez-Tomas R (2001). Prodigiosin-induced apoptosis in human colon cancer cells. Life Sci. 68:2025-2036.
Nimkar U, Bhajekar R (2006). Ecological requirements for the textile industry. Colorage 43:135-142.
Pandey R, Chander R, Sainis KB (2003). A novel prodigiosin-like immunosuppressant from an alkalophilic *Micrococcus* sp. Int. Immunopharmacol. 3:159-167.
Premi GD (1996). Indian textile industry: Emerging eco-friendly standards for exports. Clothline 9:105-106.
Tibor C (2007). Liquid chromatography of natural pigments and synthetic dyes. First ed. Elsevier, UK. Included in series J. Chromatogr. Lib. 71:1-602.

15

Anti-ulcer activity of aqueous leaf extract of *Nauclea latifolia* (rubiaceae) on indomethacin-induced gastric ulcer in rats

Balogun M. E.[1], Oji J.O.[1], Besong E. E.[1], Ajah A. A.[2] and Michael E. M.[3]

[1]Department of Physiology, Faculty of Basic Medical Sciences, College of Health Sciences, Ebonyi State University, Abakaliki, Nigeria.
[2]Department of Physiology, Faculty of Basic Medical Sciences, College of Health Sciences, University of Port Harcourt, Choba, Port Harcourt, Nigeria.
[3]Department of Anatomy, Faculty of Basic Medical Sciences, College of Health Sciences, Ebonyi State University, Abakaliki, Nigeria.

Nauclea latifolia **is known to possess various therapeutic properties. The present study was designed to evaluate the anti-ulcer activity of aqueous leaf extract of** *N. latifolia* **against indomethacin-induced gastric ulcers in rats. Five groups of albino rats were pre-treated orally with: vehicle, distilled water (ulcer control), cimetidine (100 mg/kg, reference control), and 170, 340 and 510 mg/kg** *N. latifolia* **leaf extracts (experimental groups) respectively, 60 min prior to oral administration of indomethacin to generate gastric mucosal injury. Seven hours later, the animals were sacrificed by a blow on the head; their stomachs were removed and examined for ulcer index. The extract produced significant (P<0.05), and dose dependent anti-ulcer activity against indomethacin-induced ulcers in rats. These results suggest that the extract possesses significant anti-ulcer activity against experimentally induced gastric lesions and may justify its use as an anti-ulcerogenic agent.**

Key words: Anti-ulcer activity, *Nauclea latifolia* leaf, mucosal injury, indomethacin, rats.

INTRODUCTION

Peptic ulcer disease is the most common gastrointestinal disorder in clinical practice. It is a chronic disease characterized by ulceration in the regions of upper gastrointestinal tract where parietal cells are found and where they secret Hydrochloric Acid (HCl) and pepsin. The anatomic sites where ulcer occurs commonly are stomach and duodenum, causing gastric and duodenal ulcer, respectively (Rang et al., 2003). Pathophysiology of ulcer is due to an imbalance between aggressive factors (acid, pepsin, helicobacter pylori, and non-

steroidal anti-inflammatory agents) and local mucosal defensive factors (mucus bicarbonate, blood flow and prostaglandins). Integrity of gastro-duodenal mucosa is maintained through a homeostatic balance between these aggressive and defensive factors (Raskin et al., 1995). Herbal medicine deals with plants and plant extracts in treating diseases. These medicines are considered safer because of the natural ingredients with no side effects (Clouatre and Rosenbaum, 1994). Medicinal plants have been shown to possess gastroprotective activity in animal studies (Malairajan et al., 2007; Rao et al., 2008). One of the herbs that have great potential is *Nauclea latifolia*.

N. latifolia (smith) belongs to the family Rubiaceae. It is commonly known as pin cushion tree being a straggling shrub or small tree, native to the tropical Africa and Asia (Gidado et al., 2005). It bears an interesting flower, large red ball fruit with long projecting stamens. The red fruit is edible but not appealing. *N. latifolia* is an evergreen multi stemmed shrub or tree. It grows up to an altitude of 200 m. It is widespread in the humid tropical rainforest zones or in the savannah wood land of West and Central Africa (Burkil, 1985). *N. latifolia* is commonly known as "Ubulu inu" among the Igbo in the Eastern part of Nigeria; as "Tafashiya" among the Hausas in the Northern part of Nigeria; as "Egbesi" among the Yoruba in the Western part of Nigeria and as "Itu" among the Itsekiri (Arise et al., 2012). *N. latifolia* herbal remedies have been commonly seen in various cultures throughout recorded history and still serve as the main means of therapeutic medical treatment. It is found in areas like Abakaliki, Abuja, Enugu, Akwa Ibom, Cross River, Kontagora, Shaki and other parts of Nigeria. The wood of *N. latifolia* is termite resistant and is used as a live stakes in farms. All parts of the plants species are rich source of mono-terpene indole alkaloid. It is used in the treatment of fever, diarrhea and even as an anti-parasitic drug (Deeni and Hussain, 1991). The sticks are used as chewing stick and a remedy against tuberculosis (Burkill, 1985; Esimore et al., 2003). Scientific studies have established hypolipidemic and hypoglycemic effects like most other plants extracts (Schiff, 1970; Chong, 1991; Udoh, 1998; Eno and Owo, 1999). The anti-hypertensive effect and the phyto-chemical screening of this herb have been also documented (Udoh, 1998). Infusion and decoction of the stem bark of the leaves of *N. latifolia* are used for treatment of stomach pain and constipation (Eno and Owo, 1999).

In Kano (Nigeria), the plant is used for tuberculosis (Deeni and Husain, 1991). Abbiw (1990) stated that root infusion of *N. latifolia* is used in Sudan for the treatment of gonorrhea, its roots and leaves are used in Ghana for treating sores. In Nigerian folklores, the fruit are sometimes used in the treatment of piles and dysentery (Reitman et al., 1957). In addition, the plant is used in the treatment of sleeping sickness and to prolong menstrual flow (Elujoba, 1995). Gidado et al. (2005) reported anti-diabetic properties for the root and leaf extracts while Taiwe et al. (2010) reported the anti-depressant and anti-anxiety effects of the root extract of the plant. A decoction of the stem in water has been demonstrated to exhibit a high anti-parasitic potential (Benoit-Vical et al., 1998). However, despite the acclaimed and documented uses, there appears to be a paucity of literature on the anti-ulcerogenic activities of this plant. Therefore, the current study was undertaken to evaluate the anti-ulcer activity of the aqueous leaf extract of *N. latifolia* against indomethacin -induced gastric ulcer in rats.

MATERIALS AND METHODS

Animals

Thirty (30) adult male albino rats (180 to 250 g) obtained from th Central Animal House, Faculty of Basic Medical Sciences, Colleg of Health Sciences, Ebonyi State University, Abakaliki, Nigeria wer used for the experimental study. They were maintained unde standard laboratory conditions and were fed with standard rat pellets (Pfizer Livestock Feeds PLC, Enugu, Nigeria) and tap wate were given ad libitum. They were acclimatized for 2 weeks afte which they were divided randomly into five groups of six animal each coded to prevent observer bias. Animal experimental studie were conducted according to the guidelines of Institutional Anima Ethical Committee (IAEC) of Ebonyi State University, Abakalik Nigeria.

Drugs

Cimetidine was obtained from the Ebonyi State University Medica Centre (EBSUMC) Pharmacy and was used as the reference ant ulcer drug. Cimetidine is H_2-receptor antagonist drug used for th treatment of peptic ulcers. Cimetidine blocks H_2-receptor channel in the wall of the stomach leading to reduction in acid productio allowing the stomach to heal. In this study, the drug wa administered orally to reference control group of rats in a dose c 100 mg/kg suspended in distilled water (5 ml/kg).

Plant material and preparation of aqueous extract

The fresh leaves of *N. latifolia* (family: Rubiaceae) were collecte within the campus of the Ebonyi State University, Abakaliki, Nigeria identified and authenticated by Mr. P. O. Ugwuozo in the herbariur of the Botany Department of University of Nigeria, Nsukka, wit deposition of authenticated voucher specimen (UNH - 303i). Leave were separated from the stalks, washed with distilled water, and ai dried in the shade for 7 to 10 days. The dried leaves wer pulverized to coarse powder. To 200 g of the powdered leaves in container with lid, 1 L of boiling water was added and covered. was allowed to stand for 24 h with intermittent shaking. The mixtur was then filtered with NO. 1 Whatman qualitative filter paper t obtain a pure filtrate. The combined filtrate was concentrated in th required doses of 170, 340 and 510 mg/kg/ 5 ml respectivel (Akpanbiatu et al., 2005). It was stored in the refrigerato throughout the period of the experiment to preserve the prepare extract.

Indomethacin-induced ulcers

Food was withdrawn 24 h and water 1 h before drug treatmer (Ibitoye et al., 2002). Animals in groups 1 and 2 received distille water and cimetidine, respectively, while those in groups 3, 4 and were pre-treated with 170, 340 and 510 mg/kg of the extract. After h, indomethacin (30 mg/kg/5 ml) was administered orally to all th rats. 7 h later, the rats were killed by a blow on the head (Ukwe and Nwafor, 2004). The rats' stomachs were removed and each opene along the greater curvature. After fixing the tissues by immersing i 10% formalin for 24 h, it was rinsed under a stream of water and examined for ulcers. The ulcers were counted by the aid of a han lens (X- magnification) and ulcer score was calculated for eac animals according to the arbitrary scale used by Singh (1997)

Table 1. Effects of N. latifolia aqueous leaf extract on indomethacin- induced ulcers in rats.

Group	Treatment	Dosage (p.o)	Mean ulcer index ± SEM	Percentage protection (%)
1	Distilled Water	5 ml/kg	3.50± 11.18	0.00
2	Cimetidine	100 mg/kg	0.83± 0.69*	76.28
3	Extract	170 mg/kg	1.16± 0.61*	66.85
4	Extract	340 mg/kg	0.66± 0.36*	81.14
5	Extract	510 mg/kg	0.33± 0.02*	90.57

*Significant. All values are expressed as mean ± SEM; n = 6 in each group. *$P<0.05$ as compared with the negative control animal.

where 0 = no lesion, 1 = hyperemia, 2 = one or two slight lesions, 3 = very severe and 4 = mucosal full of lesion. Ulcer index was calculated as mean ulcer scores (Tan et al., 1996).

Histological evaluation of gastric lesions

Immediately after macroscopic evaluation, the stomachs were washed with saline and fixed in 10% buffered formalin solution for histo-pathological studies. Sections of the gastric walls were made at a thickness of 5 to 6 μm, stained with hematoxylin and eosin (H&E), were assessed for histo-pathological changes such as congestion, edema, necroses and heamorrhage (Shahl and Khan, 1997). The microscopic slides were photographed.

Statistical analysis

Results were expressed as mean ± S.E.M. The significance of the data was calculated at the 95% confidence interval using the Student's 't' test.

RESULTS

Gross evaluation of gastric lesions

As shown in Table 1, indomethacin induced ulcers in 100% of the animals in the negative control (distilled water; 5 ml/kg) group. The ulcer index was 3.5±11.18, which is characterized with severe disruption of surface epithelium of gastric mucosa (Figure 1). Pre-treatment with cimetidine significantly (P< 0.05) reduced the severity of indomethacin-induced ulcers compared to rats pre-treated with distilled water (ulcer control). The N. latifolia leaf extracts were also shown to exert cytoprotective effects in a dose-dependent manner (Table 1).

Histological evaluation of gastric lesions

The cytoprotective effect was confirmed by histological examination. The rats pre-treated with distilled water (negative control) before administration of ulcer-inducing indomethacin showed markedly extensive damage to the gastric mucosa, with lesions extending deep into the mucosal layer, and edema and leucocytes infiltration of the sub-mucosa layer (Figure 1). Rats pre-treated with 340 or 510 mg/kg N. latifolia leaf extracts had comparatively better gastric mucosal protection compared to rats pre-treated with cimetidine or 170 mg/kg plant extract as evidence by the marked reduction in ulcer index, inhibition of edema and leucocytes infiltration of the submucosa layer (Figures 2, 3 and 4). Percentage inhibition to ulcer formation in rats by the extract was calculated as follows:

$$\% \text{ Inhibition of ulceration} = \frac{\text{Ulcer index }_{Control} - \text{Ulcer index }_{Test}}{\text{Ulcer index }_{Control}} \times 100\%$$

DISCUSSION

This research work was designed to investigate the anti-ulcerogenic activity of aqueous leaf extract of N. latifolia against indomethacin-induced gastric ulceration in albino rats. The finding of the present study demonstrate that aqueous extract of N. latifolia significantly protected against mucosal damage induced by indomethacin and curative ratios of plant extracts 170, 340 and 510 mg/kg were 66.85, 81.14 and 90.57%, respectively. It is remarkable that the leaf extract at 340 and 510 mg/kg doses produced a greater protection than cimetidine (100 mg/kg) against the indomethacin. The effect of the extract compared favorable to cimetidine 100 mg/kg (positive control). However, the mechanism by which the aqueous leaf extract on N. latifolia produced its gastroprotective effects in rats is not clear. It has been established that indomethacin is an ulcerogenic agent especially when administered on an empty stomach (Blaargawa et al., 1993). The ulcerogenic activity of indomethacin and other non-steroidal anti-inflammatory agents as postulated might be due to their ability to inhibit prostaglandin synthesis (Vane, 1971). Several lines of evidence suggest that prostaglandins inhibit gastric secretion and are important to normal gastric physiology and mucosal integrity (Robert et al., 1968; Jacobson, 1970; Main and White, 1976). Some of the mechanisms suggested for their effect include tightening of the gastric mucosal barrier (Bolton and Cohen, 1979) and stimulation of the gastric sodium pump (Robert, 1979).

The protective effect of the extract on indomethacin

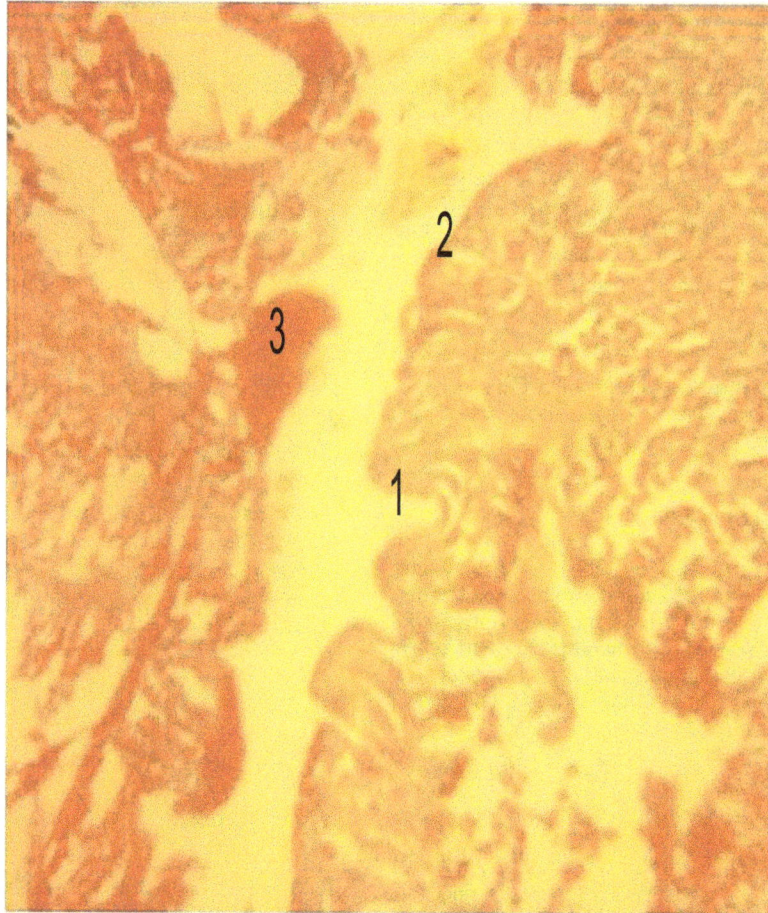

Figure 1. Histological section of the gastric mucosa in a rat pre-treated with distilled water (ulcer control). 1, gastric pit; 2, columnar epithelium; 3, ulcerated section of the epithelium. There is severe disruption of the surface epithelium, deep penetration of necrotic lesions into mucosa and edema of submucosa layer with leukocyte infiltration of ulcerative tissues (H&E stain, 10x).

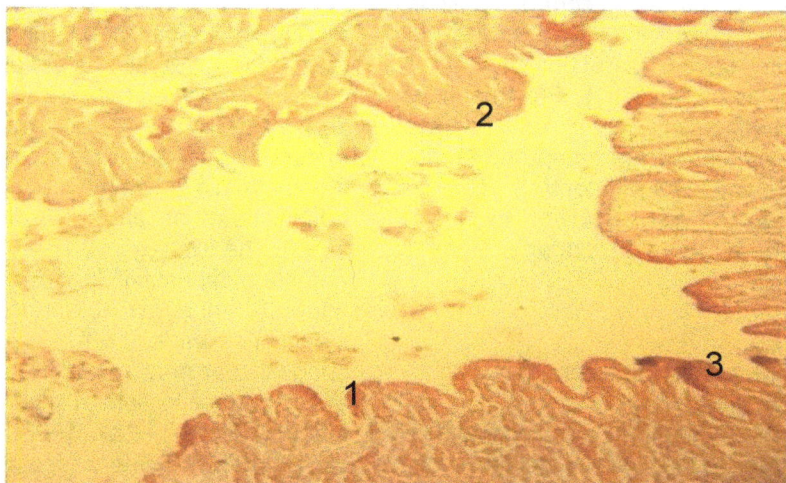

Figure 2. Histological section of the gastric mucosa in a rat pre-treated with cimetidine (100 mg/kg). 1, gastric pit; 2, columnar epithelium; 3, ulcerated section of the epithelium. There is mild disruption of the surface epithelium and edema of submucosa layer with leukocyte infiltration of ulcerative tissues (H&E stain, 10x).

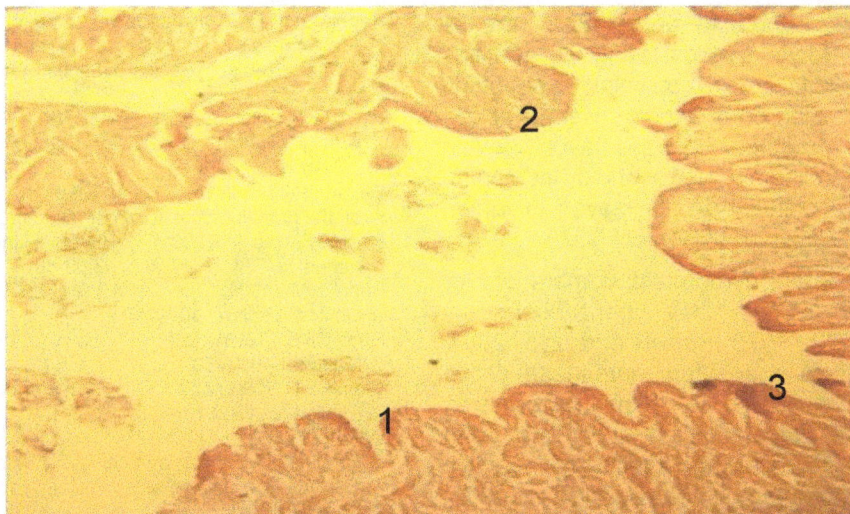

Figure 3. Histological section of the gastric mucosa in a rat pre-treated with *N. latifolia* (170 mg/kg). 1, gastric pit; 2, columnar epithelium; 3, ulcerated section of the epithelium. There is mild disruption of the surface epithelium and edema of submucosa layer with leukocyte infiltration of ulcerative tissues (H&E stain, 10x).

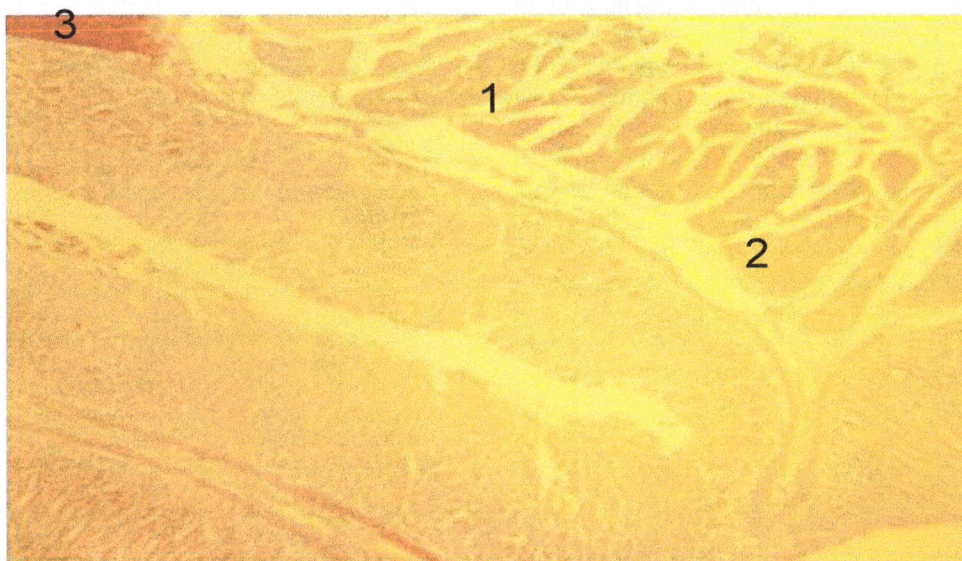

Figure 4. Histological section of the gastric mucosa in a rat pre-treated with *N. latifolia* (510 mg/kg). 1, gastric pit; 2, columnar epithelium; 3, ulcerated section of the epithelium. There is mild disruption of the surface epithelium and there is no submucosal edema and no leukocyte infiltration (H&E stain, 10x).

induced-ulcers in rats might be related to any of the mechanism suggested. Infection of the stomach mucosa with helicobacter pylori a Gram-negative spiral-shaped bacterium is now generally considered to be a major cause of gastro-duodenal ulcer (Rang et al., 2003). Although, a number of anti-ulcer drugs such as H_2 receptor antagonists, proton pump inhibitors and cytoprotectants are available for ulceration, all these

drugs have side effects and limitations. Hot aqueous extract also showed effectiveness against chloroquine resistance strains of *Plasmodium falciparum* (Benoit-Vical et al., 1998). Hot aqueous and ethanolic extracts were demonstrated to exhibit strong anti-bacterial property (Okiei et al., 2011). Alkaloid rich extract of *N. latifolia* can react *in vitro* with mammalian DNA, leading to G2-M cell cycle arrest and heritable DNA- damage. In the liver, kidney and blood cells, it induces single strand breaks (Traore et al., 2000). Phytochemical analysis identifies indole-quinolizidine, alkaloids (glycoalkaloid), saponins and tannins as the major components (Karou et al., 2011). It has also been reported that, the flavonoids like flavones, glycosides, tannins and isoflavonoid (Indicanine B and C) have been isolated from its leaves (Yadava, 1999; Borrelli and Izzo, 2001).

Flavonoids are among the cytoprotective materials for which anti-ulcerogenic efficacy has been extensively confirmed (Di Carlo et al., 1999; Borrelli and Izzo, 2001; Galati et al., 2001). It is suggested that, these active compounds would be able to stimulate mucus, bicarbonate and the prostaglandin secretion and counteract with the deteriorating effects of reactive oxidants in gastrointestinal lumen (Salvayre et al., 1982; Asuzu and Onu, 1990; Suja et al., 2002). So the antiulcer activity of the leaf may be attributed to its flavonoids content.

Conclusion

The present study showed that pre-treatment with the leaf extract of *N. latifolia* caused a beneficial effect on indomethacin-induced gastric ulcers in rats as evident by the reduction in the ulcer index. The gastroprotective effect of the leaf extract is dose dependent and this may justify its use as an anti-ulcerogenic agent. The exact mechanism of action by which the extract protects laboratory animals from experimentally induced gastric ulcers as studied with indomethacin has not been elucidated. Further studies are necessary to isolate the responsible active compound(s) and elucidate its mechanism of actions. If these findings are extrapolated to man, aqueous leaf extract of *N. latifolia* may be beneficial to peptic ulcer-prone individuals.

ACKNOWLEDGEMENT

The authors are thankful to Mr. Christian Chukwu Bright for his active participation and contribution during the course of this work.

REFERENCES

Abbiw KD (1990). Useful plants of Ghana, West Africa; Uses of Wild and cultivated plants, Intermediate Technology publication, London. pp. 98-212.

Akpanabiatu MI, Umoh IB, Eyong EU, Udoh FV (2005). Influence of *Nauclea latifolia* Leaf Extracts on Some Hepatic Enzymes of Rats fed on Coconut Oil and Non-Coconut Oil Meals. Pharm. Biolo. 43:153-157.

Arise RO, Akintola AA, Olarinoye JB, Balogun EA (2012). Effects of *Nauclea latifolia* stem on lipid profile and some Enzymes of Rat Liver and Kidney. Int. J. Pharmacol.10(3):23-39.

Asuzu IU, Onu OU (1990). Antiulcer activity of the ethanolic extract of Combretum dolichopetalum root. Int. J. Crude Drug Res. 28:27-32.

Benoit-Vical FA, Valentin V, Cournac Y, Pellisier M, Mallie JM (1998). Antidepressant, myorelaxant and anti anxiety like effects of *Nauclea latifolia* (Rubiaceae) root extractin murine models. Int. J. Pharmaco 6:364-361.

Blaargawa KP, Gupta MB, Tangri KK (1993). Mechanism of ulcerogen activity of indomethacin and oxyphenbutazone. Eur. J. Pharmaco 22:191-195.

Bolton JP, Cohen MM (1979). Effect of 16,16-dimethyl prostglandin E on the gastric mucosal barrier. Gut 20(6):513-517.

Borrelli F, Izzo AA (2001). The plant kingdom as a source of anti-ulce remedies. Phytother. Res. 53:82-88.

Burkil HM (1985).The useful plants of West Africa- Whiferrers Pres Limited, London. 401-415.

Chong YH (1991). Effects of palm oil on cardiovascular risk. Med. Malaysia 46:41-50.

Clouatre D, Rosenbaum M (1994). The diet and benefits of HCA: Kea Publishing: New York. pp. 23-32.

Deeni YY, Hussain HSN (1991). Screening for antimicrobial activity an for alkaloids of Nauclea latifolia J. Ethnopharmacol. 35:91-96.

Di Carlo G, Mascolo N, Izzo AA, Capasso F (1999). Flavonoid: Old an new aspects of a class of natural therapeutic drugs. Life Scienc 64:337-357.

Elujoba AAA (1995). Female infertility in the Hands of traditional Birth Attendants in South West Nigeria. Fitoterapia 66(3):239-248.

Eno AE, Owo OI (1999). Cardiovascular effects of extract from the roo of a shrub Elaeophorbia drupifera. phytother. Research 13:549-554.

Esimore CO, Ebebe IM, Chan KF (2003). Comparative antibacteri effect of Psidiu guajava aqueous extract. J. Trop. Med. Plants 4:185 189.

Galati EM, Monforte MT, Tripodo MM (2001). Anti ulcer activity c Opuntia ficus indica L. (Cactaceae): Ultrastructural study. Ethnopharcol. pp. 76:19.

Gidado A, Ameh DA, Atawodi SE (2005). Effect of *Nauclea latifoli* Leaves of aqueous extract on the blood glucose levels of normal an alloxane-induced diabetic rats. Afri. J. Biotech. 4(1):91-93.

Ibitoye SF, Ogunleye D, Fagboun AO, Ekor M (2002). An investigatio of the anti-ulcer and gastroprotective effects of Annona muricat (Family-Anonaceae) leaf extract in rats. Nigerian J. Pharm.Res 1(1):75-77.

Jacobson EP (1970). Comparison of prostaglandins EE2 an norepinephrine on the gastric mucosal circulation. Proceeding of th society of Experimental Biology and Medicine.133: 516.

Karou SD, Tchacondo T, Iboudo DP, Simpore J (2011). SubSahara Rubiaciaea: A review of their traditional uses, phytochemistry an biological activities. Pak. J. Biol. Sci.149 -169.

Main IHM, White BWR (1976). The effects of E and A prostaglandins o the gastric mucosal blood flow and the acid secretion in the ra British J. Pharmacol. pp. 49: 428.

Malairajan P, Gopalakrishnan G, Narasimhan S (2007). Anti-ulce activity of crude alcoholic extract of Toona ciliata Roeme (heartwood). J. Ethopharmacol. 110 (2): 348-351.

Okiei WM, Ogunlesi EA, Osibote MK, Binutu, Ademoye MA (2011) Comparative studies of the antimicrobial activity of components o different polarities from the leaves of Nauclea latifolia. Res. J. Med Plant 5:321-329.

Rang HP, Dale MM, Ritter M, Moore PK (2003). Pharmacology, 5 edition. Churchill, Livingstones, Edinburgh, p. 797.

Rao CV, Verma AR, Vijayakumar M, Rastogi S (2008). Gastroprotectiv effect of standardized extract of Ficus glomerata fruit on experimenta gastric ulcer in rats. J. Ethnopharmacol. 15: 323-326.

Raskin JB, White RH, Jackson JE, Weaver AL, Tindall EA, Lies RB Stanton DS (1995). Misoprostol dosage in the prevention non steroidal anti-inflammatory drug-induced gastric and duodenal ulcers A comparison of three regimens. Ann. Int. Med. 123:344-350.

Reitman S, FrankelS (1957). A colourimetric method for determinatio of Serum glutamate-oxaloacete and pyruvate transaminases. Am. J Clin. Pathol. 28:56-63.

Roberts A (1979). Cytoprotection by prostaglandins. Gastroenterolog 77:761-767.

Roberts A, Phillips JP, Nezamis JE (1968). Effect of prostaglandin E o gastric secretion and ulcer formation in rats. Gatroenterology 55:481.

Salvayre R, Braquet P, Perochot L, Douste-Blazy L (1982). Comparison of the scavenger effect of bilberry anthocyanosides with various flavonoids. Flavonoids Bioflavonoids 11:437-442.

Schiff PL (1970). Thalictrum alkaloids Lloydia 33:403-452.

Shah AH, Khan ZA (1997). Gastroprotective effects of pretreatment with Zizyphus sativa fruits against toxic damage in the rats. Fitoterapia 3:226-234.

Singh GB (1997). Anti-inflammatory activity of Lupeol. Fitoterapia 68:9-16.

Suja PR, Anuradha CV, Viswanathan P (2002). Gastroprotective effect of fenugreek seeds (Trigonella foenum graecum) on experimental gastric ulcer in rats. J Ethnopharmacol. 8: 393-397.

Taiwe GS, Bum EN, Dimo T, Talla, Weiss N (2010). The Effects of aqueous extract of Nauclea latifolia leaf in glucose concentration in rat. A paper biochemistry and molecular biology (Calabar),1-5.

Tan PV, Nditafon NG, Yewah MP, Dimo T, Ayafor FI (1996). Eremomoatax speciosa: effect of leaf aqueous extract on ulcer formation and gastric secretion in rats. J. Ethno.Pharmacol. 54: 139-142.

Tan PV, Nyasse B, Dimo T, Mezui C (2002). Gastric cytoprotective anti-ulcer effects of leaf methanolic extract ocimum suave (Lamiaceae) in rats. J. Ethnopharmacol. 82:69-74.

Traore FM, Gasquet M, Laget H, Guirand, Di-Giorgio C (2000). Toxicity and genotoxicity of antimalarial alkaloidrich extracts Derived from Mytragy na inermis O. kuntze and Nauclea latifolia. Phytother.Res.14:608-611.

Udoh FV (1998). Effect of leaf and root extract of Nauclea latifolia on ca rdiovascular antagonist on radiation induced system. Fitoterapia 69: 141-145.

Ukwe CV, Nwafor SV (2004). Anti- ulcer activity of leaf extract of Persea Americana. Nig. J. Pharm. Res. 3 (1):91-95.

Vane JR (1971). L inhibition of prostaglandin synthesis as a mechanism of action for Aspirin like drugs. Nature New Biol. 231-232.

Yadava RN, Reddy KIS (1999). A novel prenylated flavone glycosides from the seeds of Erythrina indica. Fitoterapia 70: 357-360.

Studies on some active components and antimicrobial activities of the fermentation broth of endophytic fungi DZY16 Isolated from *Eucommia ulmoides* Oliv.

Ding Ting[1], Sun Wei-Wei[1], Qi Yong- Xia[1] and Jiang Hai-Yang[2]

[1]School of Plant Protection, Anhui Agricultural University, Hefei 230036, People's Republic of China.
[2]Anhui Provincial Key Laboratory of Crop Biology, Hefei 230036, People's Republic of China.

Research into plant-derived endophytic fungi has grown in recent decades. Endophytic fungi still have enormous potential to inspire and influence modern agriculture. In this study, the endophytic fungi DZY16 isolated from *Eucommia ulmoides* Oliv. was tested for its bioactive components and antimicrobial activities using phenol-sulfuric acid method, high performance liquid chromatography method and growth inhibition measurements. The results show that variation trend of extracellular polysaccharide content at different growth stages of the strain DZY16 and the maximum content of extracellular polysaccharide was 2.02 g/L at the sixth day. Moreover, the fermentation broth of the DZY16 contained guanosine, uridine and adenosine; the contents were 1.54 mg/g, 1.07 mg/g and 1.36 mg/g respectively. On the other hand, the strongest antimicrobial activity was exhibited by the acetylacetate extract of strain DZY16 against *Rhizoctonia solani* and *Gibberella zeae*, showing 59.84 and 70.86% respectively. The strain DZY16 was identified by internal transcribed spacer (ITS) sequence as belonging to *Nigrospora*. The results indicate that the endophytic fungi DZY16 of the plant *E. ulmoides* Oliv. is a promising source of novel bioactive compounds.

Key words: *Eucommia ulmoides* Oliv., endophytic fungi, extracellular polysaccharide, nucleotides, antimicrobial activity.

INTRODUCTION

Endophytic fungi are microorganisms that reside in living plant tissues, apparently without inflicting negative effects. They are quite ubiquitous and have been found in all plant species examined to date (Arnold et al., 2000). In recent years, many reports have showed that endophytic fungi can be capable of synthesizing active compounds produced in host plants, and the compounds are potentially useful for modern medicine and agriculture (Hung et al., 2006; Strobel, 2003). Hence, screening with activity of endophytic fungi as alternative sources of medicinal plant are crucial for conservation and utilization of fungal resources in plants.

Eucommia ulmoides Oliv. is one of the most valuable timber resources in China, and the only species both in its genus and in its family (Ma et al., 2007). This natural resource is now in short supply because of the over-collection of the wild plant. Therefore, it is important to find an alternative way to produce its active constituents to satisfy the demand. On the basis of the above studies of endophytic fungi, it is feasible to study the products of

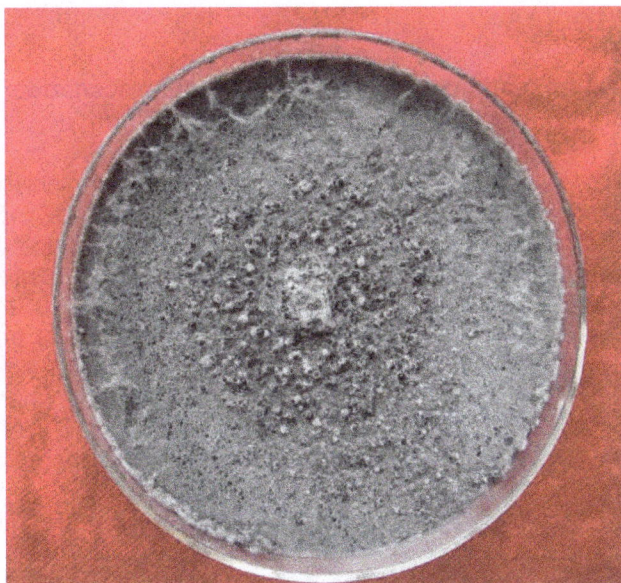

Figure 1. The colony characteristics of the strain DZY16.

secondary metabolism of the endophytic fungi isolated from *E. ulmoides* Oliv. During the initial stage of the researches on the endophytic fungi of *E. ulmoides* Oliv., a total of 78 strains were obtained from the leaves, stems and fruits of *E. ulmoides* Oliv., and the strain DZY16 isolated from the leave was found to have better antimicrobial activity using growth inhibition measurements ; the colony morphology has shown in Figure 1. In order to make an intensive study of the strain DZY16, this study was carried out to analyze the extracellular polysaccharide and nucleosides of the fermentation broth of DZY16 and to detect the anti-phytopathogen activity of the extracts of the fermentation broth. The research may contribute to the development and utilization of endophytic fungi DZY16.

MATERIALS AND METHODS

The test microorganisms were obtained from the School of Plant Protection, Anhui Agricultural University (31°86′ N and 117°25′ E), AnHui Province, Southeast China.

Culture of the shaking flask seeding

The endophytic fungi DZY16 taken from the slope were added to potato dextrose liquid medium in Erlenmeyer flasks and incubated at 28°C and 160 rpm with normal daily light and dark periods for six days, and the shaking flask seeding of DZY16 was obtained.

Preparation of the crude extracellular polysaccharides

The shaking flask seeding culture of DZY16 were added to 60 mL potato dextrose liquid medium with 10% (V/V) inoculation quantity in Erlenmeyer flasks and incubated at 28°C and 160 rpm with nor-

mal daily light and dark periods for ten days. Then, the flasks was sampled at special time every day during incubation period of DZY16, and biomass was separated by filtering with Whatman filter paper to be dried in a freeze-drying system (FreeZone12, Labconco Ltd., U.S.), so that the supernatant (the fermentation broth) was obtained.

The crude extracellular polysaccharides were precipitated from the supernatant for 12 h at 4°C by a three fold volume of 95% ethanol alcohol, after the precipitation, the crude extracellular polysaccharides were washed three times with 100% ethanol, acetone, ethyl ether successively, and finally, they were dissolved in water and freeze-dried.

Determination of extracellular polysaccharide

The contents of extracellular polysaccharides consisting in the crude extracellular polysaccharides were determined by phenol-sulfuric acid method.

Fermentation and treatment of the fermentation broth of DZY16

The strain DZY16 was added to 200 mL potato dextrose liquid medium in Erlenmeyer flasks and incubated at 28°C and 160 rpm with normal daily light and dark periods for ten days. The mycelia of the DZY16 were separated by centrifugation (5000 r. min^{-1}, 10 min), so that the fermentation broth of DZY16 was obtained, and finally the fermentation broth of DZY16 was freeze-dried in freeze-drying system.

Preparation of the nucleotides of the fermentation broth of DZY16

The freeze-dried fermentation broth of DZY16 (1.0 g) was dissolved in 70% ethyl alcohol in 5 mL volumetric flask to the mark, ultrasoniction (40 KHZ, 400 W) was performed at room temperature under optimized conditions: 70% ethyl alcohol solvent; 5 mL volume of solvent, and 30 min extract time. After centrifugation (10000 r. min^{-1}, 5 min), the supernatant was filtered with 0.22 μm Millipore filters before HPLC analysis.

High-performance liquid chromatography (HPLC) analysis

High performance liquid chromatography method was used for the determination of the contents of nucleoside in the samples (Acme9000HPLC, YOUNGLIN Ltd., Korea); detection was performed on Waters Spherisorb ODS$_2$ (4.6×250mm, 5μm) column by UV730D detector at 254 nm with methanol and KH$_2$PO$_4$ (0.01mol·L^{-1}) as the mobile phases at a flow rate of 1.0 ml/min, and the volume ratio of the solution methanol and KH$_2$PO$_4$ was 10:90.

Calibration curves

Adenosine (0.2500 g), uridine (0.2455 g), guanine (0.2480 g) and inosine (0.2550 g) were dissolved in 70% ethyl alcohol in 50 mL volumetric flask to the mark, to get four standard solutions, then the stock solution of standards was prepared and diluted with 70% ethyl alcohol to appropriate concentrations for the establishment of calibration curves. At least five concentrations of the four standards mixture were injected in triplicate, respectively and then the calibration curves were constructed by plotting the peak areas versus the concentration of each standard.

Limits of detection and quantification

The stock solutions containing reference compounds were diluted with 70% ethyl alcohol to appropriate concentrations, and an aliquot of the diluted solutions was injected into the HPLC for analysis. Limits of detection (LOD) and quantification (LOQ) for each standard were determined at a signal to noise ratio (S/N)of about 3 and 10 respectively.

Accuracy

The recovery was determined by adding a known amount of individual standards to the freeze-dried fermentation broth of DZY16 (1.0 g). The mixture was extracted and analyzed using the methods mentioned above. The quantity of each analyte was subsequently obtained from the corresponding calibration curve.

Preparation of three extracts of the fermentation broth of DZY16

With a solid to liquid ratio of 1:15 (w/V), the freeze-dried fermentation broth of DZY16 was extracted two times for 5 h by petroleum ether, ethyl acetate and methanol in sequence at 45°C. Three extracts were evaporated by vacuum concentration, and were dissolved in sterilized water at a concentration of 1 mg/ mL, then the solutions were sterilized by filtration with 0.22 μm Millipore filters.

Growth inhibition measurements

The antifungal activity of three extracts of the fermentation broth was determined in two pathogenic fungi: *Rhizoctonia solani* and *Gibberella zeae*. Prior to testing, indicator organisms were cultured in PDA medium at 28°C

0.5 mL samples were poured into sterile Petri dishes containing 15 mL PDA, followed by adequate mixing. A negative control was prepared using 0.5 mL sterilized water. For the positive controls, 100 mg/L carbendazim was used for two pathogenic fungi.

A 5 mm diameter plug of the actively growing mycelium of the pathogenic fungi was placed in the center of the each plate. The plates were incubated at 28°C in the dark (3 plates per treatment). The diameters of the inhibition zones were measured by vernier caliper. According to the growth rate of each fungus, colony diameter data taken after 2 days (*R. solani*) and 5 days (*G. zeae*) were used. The inhibitory activity of each treatment was carried out using the following formula, where DC = diameter of control, and DT = diameter of fungal colony with treatment. The experiments were repeated twice and the data presented here are the averages of two experiments.

$$\text{Growth inhibition (\%)} = \left[\frac{DC - DT}{DC} \right] \times 100\%$$

DNA extraction, PCR amplification, and sequencing

The strain DZY16 was identified using molecular tools. Genomic DNA was extracted from ground mycelium (Lee et al, 1988). Primers ITS5 5′ TCCGTAGGTGAACCTGCGC 3′ and ITS4 5′ TCCTCCGCTTATTGATATGC 3′ were used to amplify the 5.8S and flanking ITS regions of the orphospecies. The DNA fragment was amplified and sequenced according to previously described methods (Guo et al, 2000). The sequence data from this study have been submitted to GenBank under accession No. JQ359020.

RESULTS

Determination of extracellular polysaccharide a different growth stages of the strain DZY16

The variation trend of content of extracellular polysac charide at different growth stages of the strain DZY16 i shown in Figure 2. Firstly, the content of extracellula polysaccharide of DZY16 was slowly increased at la phase and logarithmic phase of the strain DZY16, the the content of extracellular polysaccharides had a tren of rapid growth at stationary phase of the strain DZY16 and the maximum content of extracellular polysaccharid was 2.02 g/L at the sixth day. Subsequently, the conter of extracellular polysaccharide significantly decreased until the eighth day; downtrend of the content of extra cellular polysaccharides tended stable. On the othe hand, the trend of the biomass of the strain DZY16 slowl increased and eventually was stable; at the eighth da the maximum biomass reached 14.30 g/L.

Identification and quantification of the investigate nucleosides in the fermentation broth of DZY16

Linearity, regression, and the linear ranges of the fou analytes were determined using the HPLC method. Th correlation coefficient values indicated appropriate corre lations between the investigated compound concentra tions and their peak areas within the test ranges. Th limits of detection and quantification were less than 1. and 2.0 ng respectively (Table 1) and the method ha good accuracy with the overall recovery of 94.6 t 101.4% for the analytes (Table 2). The results indicate that this HPLC method was accurate and sensitive fo quantitative determination of the four compounds.

Chromatograms of the mix-standard and ultrasonic 70% ethyl alcohol extract from the fermentation broth of DZY1 are shown in Figures 3 and 4. The peaks were identifie by two means: (1) comparing the retention times of th unknown peaks with those of the standards eluted unde the same conditions and (2) spiking the sample wit stock standard solutions of analytes. By the calibratio curve of each investigated compound, the contents of th four analytes in the fermentation broth of DZY16 wer determined. The data is summarized in Table 3. In brie the amounts of the inosine was undetectable, further more, the content of guanosine was higher than those o uridine and adenosine.

The antifungal activities of the fermentation broth o DZY16

The antifungal activities of the three extracts of the fer mentation broth of DZY16 were detected in two pathogeni fungi, and the results show that the antifungal activitie

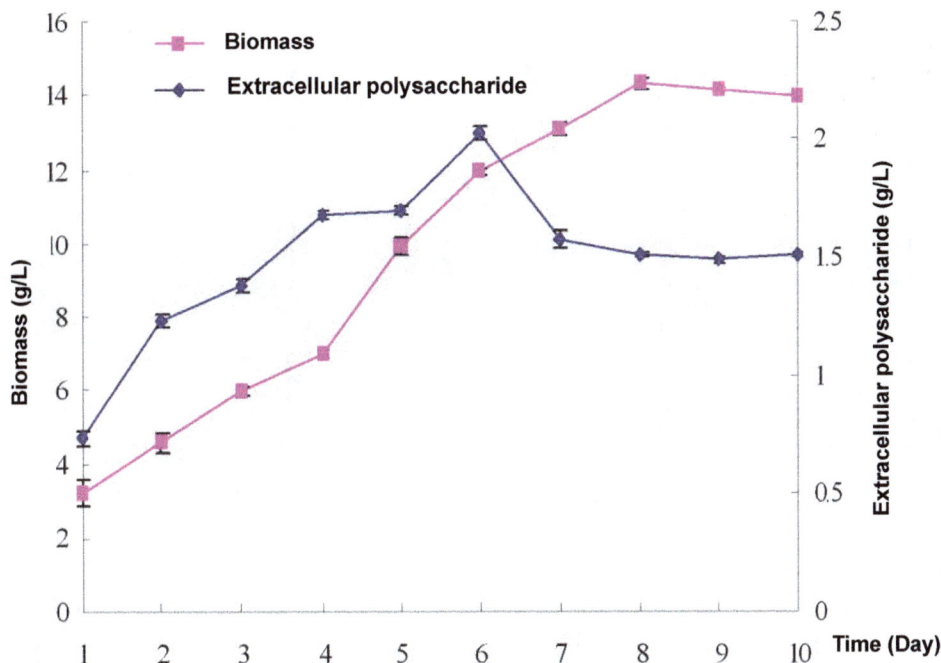

Figure 2. Time profiles of extracellular polysaccharide and biomass.

Table 1. Calibration data, LOD and LOQ of the four analytes.

Analyte	Linear regression data			LOD/ng	LOQ/ng
	Regressive equation	R^2	Test range/ mg/ml		
Uridine	Y=3227.2X-170.57	0.9975	0.0491-0.2500	0.8	1.1
Inosine	Y=3399.3X-137.15	0.9976	0.0510-0.2550	1.0	2.0
Guanosine	Y=2520.1X-125.41	0.9985	0.0496-0.2480	0.8	1.1
Adenosine	Y=5696.5X-315.05	0.9965	0.0500-0.2500	0.6	1.0

Table 2. Recoveries for the assay of the investigated components in the fermentation broth of DZY16.

Analyte[a]	Original (µg)	Spiked (µg)	Found (µg)	RSD (%)	Recovery[b] (%)
Uridine	10.7	50.0	61.4	0.6	101.4
Guanosine	15.4	50.0	62.7	1.2	94.6
Adenosine	13.6	50.0	63.1	1.3	99.0

[a] the inosine was absence in the fermentation broth of DZY16, so the recovery of the inosine was not analyzed . [b] Recovery(%) = 100 × amount found - original amount)/amount spiked.

of the three extracts were significantly different. With *R. solani* and *G. zeae*, the inhibition rate of the acetyl acetate extract was significantly higher than those of petroleum ether extract and methanol extract (Table 4). With *R. Solani,* the inhibition rate of the acetyl acetate extract on the growth of *R. solani* was 59.84%, which was lower compared to that of the fungicide carbendazim on *R. solani*, whereas the inhibition rates of the petroleum ether extract and methanol extract were only 15.44 and 48.00% respectively (Table 4). The acetyl acetate extract showed higher growth inhibitory effect on the growth of *G. zeae,* where the acetyl acetate extract at 1 mg/mL caused 70.86% inhibition, compared to 63.94% for 100 mg/L carbendazim, used as the positive control, whereas the inhibition rates of the petroleum ether extract and methanol extract were 16.19 and 55.40% respectively (Table 4).

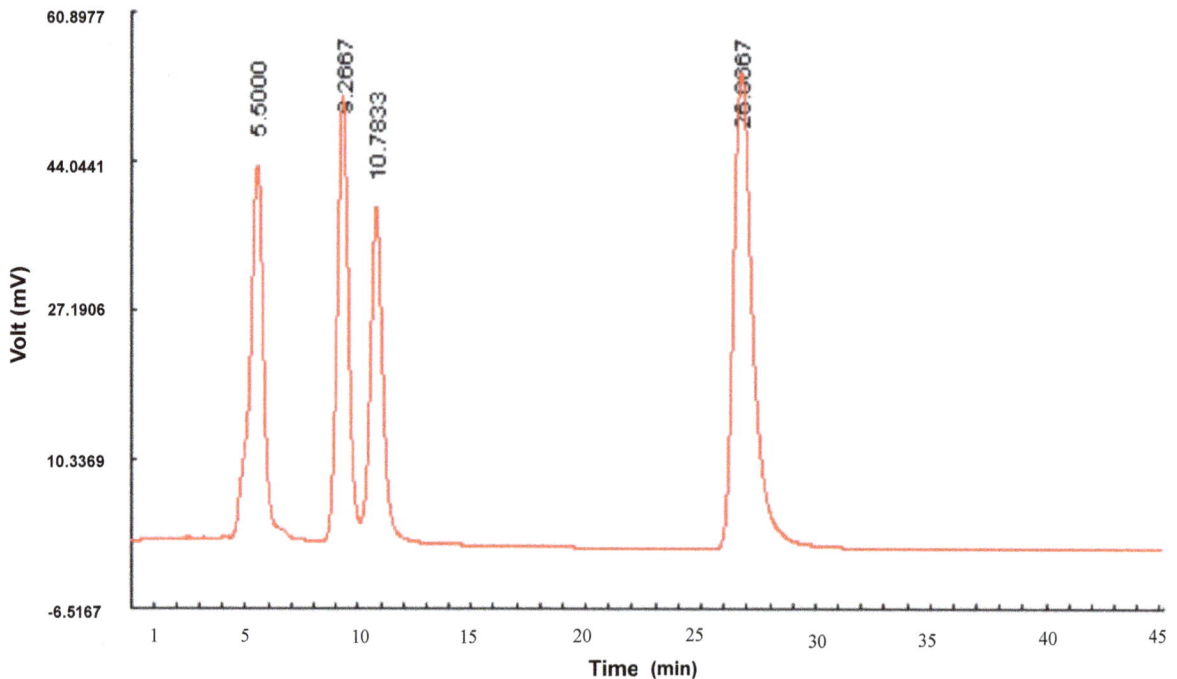

Figure 3. HPLC Diagram of mixture of nucleoside standard sample (i) uridine (ii) inosine (iii) guanosine (iv) adenosine.

Table 3. kinds and contents of nucleosides in the fermentation broth of DZY16 [a]Undetectable.

Sample	Adenosine (mg/g）	Uridine (mg/g)	Guanosine (mg/g)	Inosine[a] (mg/g)
The fermentation broth of DZY16	1.36±0.09	1.07±0.07	1.54±0.02	

Table 4 Antifungi activities at different extracting solvents of DZY16.

Sample	Growth inhibition (%)	
	Gibberella zeae	*Rhizoctonia solani*
Petroleum ether extract	16.19 ± 0.29[d]	15.44 ± 0.72[d]
Acetylacetate extract	70.86 ± 0.55[a]	59.84 ± 0.72[b]
Methanol extract	55.40 ± 0.63[c]	48.00 ± 0.62[c]
Positive control	63.94 ± 0.25[b]	71.19 ± 0.67[a]

The different lowercase show significant difference among treatments at 95% confidence level using the Duncan multiple comparisons.

Molecular phylogenetics

ITS of DZY16 strains were sequenced, the sequence long was 776 bp, including 18 S, 5.8 S, 28 S and interval region. The sequences were aligned by ClustalW first, and then the alignment sequences were subjected to the construction of the phylogenetic trees based on the Bootstrap neighbor-joining (NJ) method with Kimura 2-parameter model by using MEGA version 3.1 (Figure 5). The sequence similarity can also be demonstrated in tree graphs. DZY16 and *Nigrospora* had high homology; one

of the largest homology was up to 98%, so DZY16 strain is theoretically a member of *Nigrospora*.

DISCUSSION

During this studies, the fermentation broth of DZY16 was found to contain extracellular polysaccharide, and the maximum content of extracellular polysaccharide reached 2.02 g/L. Qualitative HPLC analysis showed that the fermentation broth of the strain DZY16 had uridine, guanosine and adenosine; the type of nucleoside was more

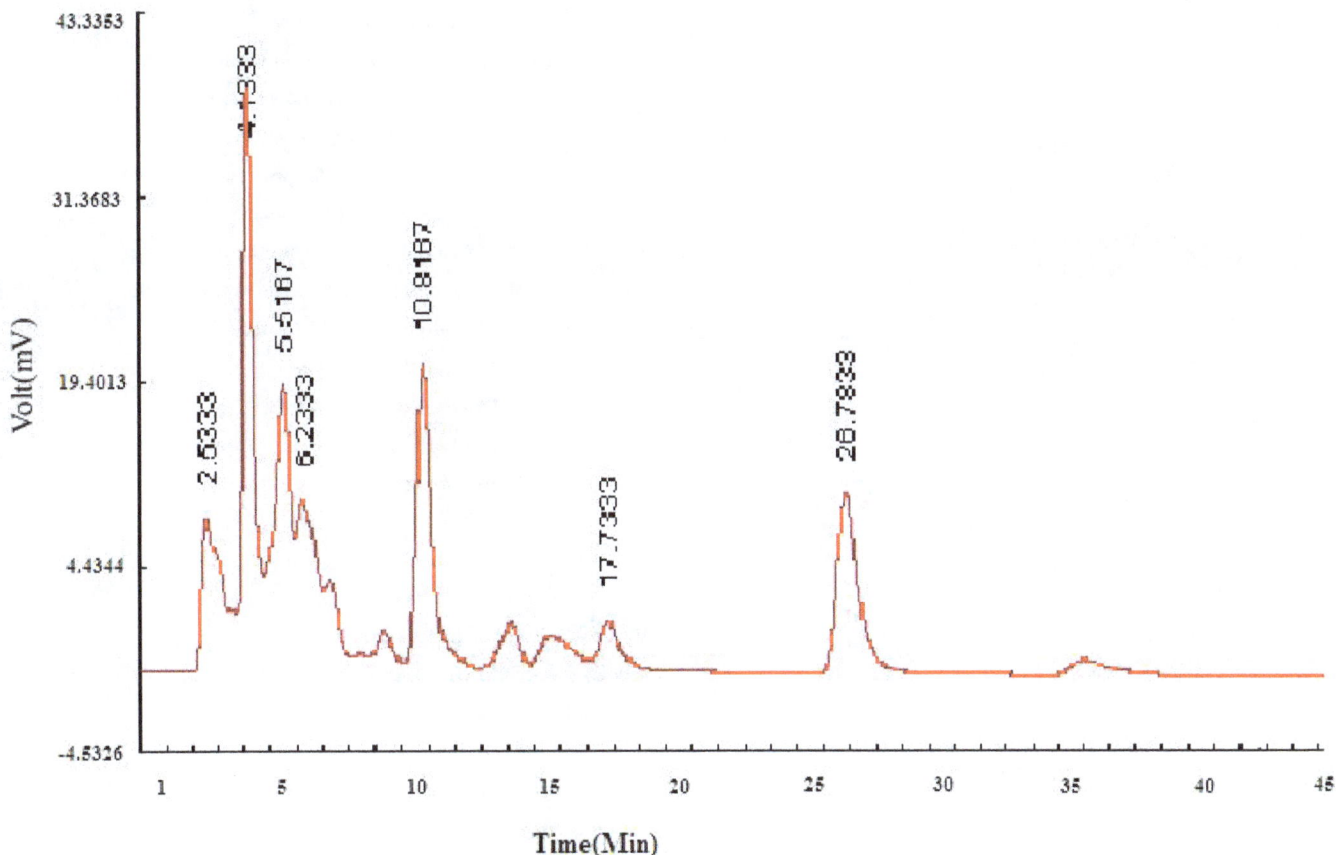

Figure 4. HPLC digram of the fermentation broth of DZY16

than that of the reported fungi *Phomopsis sp.* from *Azadirachta indica* A. Juss (Wu et al., 2008), at the same time, quantitative HPLC analysis showed the contents of uridine, guanosine and adenosine were 1.07 mg/g, 1.54 mg/g and 1.36 mg/g respectively. Thus, the above results indicated the strain DZY16 could be used as a source of extracellular polysaccharide and nucleosides, and these biological substances could be obtained by liquid fermentation easily without influence of external conditions.

Antifungal activities of plant endophytic fungi have been reported by a few groups (Liu et al., 2001; Li et al., 2005). In the present study, three extracts of the fermentation broths of endophytic fungi DZY16 were screened for antifungal activities in two pathogenic fungi, as an indication of its capability to produce secondary metabolites of potential therapeutic interest. The results show that the effect of the three extracts on the growth of *R. solani* was lower compared to the effect on *G. zeae*. Among the three extracts , the acetyl acetate extract exhibited strong antifungal activity to two pathogenic fungi; the antifungal activity of methanol extract was secondary, and the antifungal activity of petroleum ether extract was poor (Table 4). Moreover, the inhibition rate of DZY16 to *G. zeae* was 70.86% and significantly higher

than that of the fungicide carbendazim (100 mg/L); it could be another potential source of bioactive antifungal agents, and their active metabolites are worth further research.

In recently years, many investigations of the *E. ulmoides Oliver.* indicated that the principal components of *E. ulmoides* Oliver. were geniposidic acid and chlorogenic acid, moreover, chlorogenic acid was widely used for its antimicrobial, anti-inxammatory, antioxidant, anticancer, and anti-hepatitis B virus activities (Riitta et al., 2005; Wu et al., 2007). Therefore, the next goal is to isolate and purify the acetyl acetate extract in order to confirm whether chlorogenic acid exists in the acetyl acetate extract or obtain other compounds with antifungal activities. In this way, we hope to lay a stable foundation for a new method for producing these important bioactive compounds.

ACKNOWLEDGMENTS

Research supported by the National Natural Science Foundation of China (Grant No.31000871) and Projects of University Science Foundation for Outstanding Young

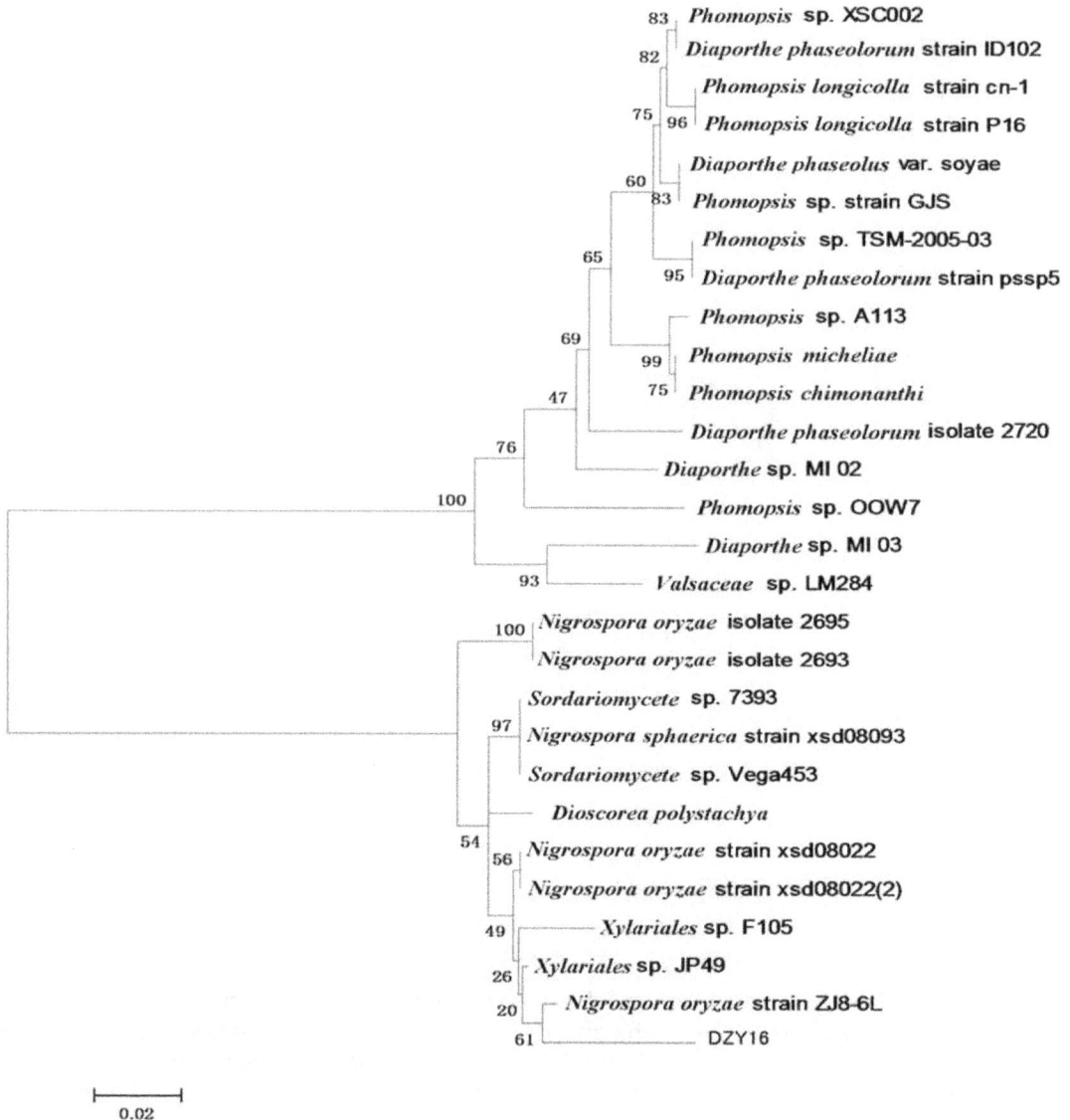

Figure 5. Strain DZY16 ITS sequence blast analysis.

Scholars of Anhui Province (Grant No. 2010SQRL063).

REFERENCES

Arnold AE, Maynard Z, Gilbert GS, Coley PD, Kursat TA (2000). Are tropical fungal endophytes hyperdiverse? Ecology Lett. 3: 267-274.

Guo LD, Hyde KD, Liew ECY (2000). Identification of endophytic fungi from Livistona chinensis based on morphology and rDNA sequences. New Phytol. 147: 617-630.

Hung MY, Fu TYC, Shih PH, Lee CP, Yen GC (2006). Du-Zhong (Eucommia ulmoides Oliv) leaves inhibit CCl4-induced hepatic damage in rats. Food Chem. Toxicol. 44:1424-1431.

Lee SB, Milgroom MG, Taylor JW (1988). A rapid, high yield mini prep method for isolation of total genomic DNA from fungi. Fungal Genet. Newslett. 35: 23-24.

Liu CH, Zou WX, Lu H, Tan RX (2001). Antifungal activity of Artemisia annua endophyte cultures against phytopathogenic fungi. J. Biotechnol. 88:277-282.

Li HY, Chen Q, Zhang YL, Zhao ZW(2005).Screening for endophytic

fungi with antitumour and antifungal activities from Chinese medicina plants. World J. Microbiol. Biotechnol. 21:1515-1519.

Ma XH, Zhang KJ, Qin WQ, Wang L (1997). Identification of th structure of chlorogenic acid from the leaves of Eucommia ulmoides Beijing: China Forestry Publishing House. 42-47.

Riitta PP, Nohynek L, Alakomi HL, Oksman-Caldentey KM (2005) Bioactive berry compounds-novel tools against human pathogens Appl Microbiol Biot. 67:8-18.

Strobel G, Daisy B (2003). Bioprospecting for microbial endophytes an their natural products. Microbiol Mol Biol Rev. 67: 491-502.

Wu SH, Chen YW, Li ZY, Yang LY, Li SL , Huang R (2008). Studies o the metabolites of endophytic fungus Phomopsis sp . fror Azadirachta indica A . Juss. Nat. Prod. Res. Dev. 20: 1014 1015,1030.

Wu SD, Jiang XY, Chen QY, Chen QX (2007) Comparison o techniques for the extraction of the hypotensive drugs geniposidi acid and geniposide from Eucommia Ulmoides. J. Iran Chem. Soc 4(2): 205-214.

17

Characterization of lactic acid bacteria isolated from indigenous dahi samples for potential source of starter culture

Shabana Maqsood[1], Fariha Hasan[1] and Tariq Masud[2]

[1]Department of Microbiology, Quaid –e- Azam University, Islamabad, Pakistan.
[2]Department of Food, Technology, PMAS-Arid Agriculture University, Rawalpindi, Pakistan.

Diversity and density of lactic acid bacteria from indigenous dahi were studied by the determination of morphological, cultural, physiological and biochemical characteristics. A total of 143 isolates were identified phenotypically and divided into three genera: *Lactobacillus*, *Lactococcus* and *Streptococcus*. The microorganisms isolated were *Lactobacillus bulgaricus* (23.77%), *Streptococcus thermophilus* (26.57%), *Lactococcus lactis* (13.9%), *Lactobacillus acidophilus* (9.79%), *Lactobacillus lactis* (9.79%), *Lactobacillus delbreuckii* (4.89%), *Lactobacillus helveticus* (2.79%), *Lactobacillus casei* (1.39%), *Lactobacillus casei* ssp. *psuedoplantarum* (2.79%), *Streptococcus cremoris* (2.09%) and *Streptococcus lactis* sub. *diacetylactis* (2.09%).

Key words: Lactic acid bacteria, fermented milk, dahi, characterization.

INTRODUCTION

Dahi analogue to yoghurt is a popular fermented milk product of Indo-Pak subcontinent. It's consumption stands next to whole milk especially during summer. Dahi has been reported to contain a mixture of lactic acid bacteria (LAB) in addition to *Lactobacillus bulgaricus* and *Streptococcus thermophilus* mostly used cultures for yoghurt making (Masud et al., 1991).

The developing interest in microorganisms found in food is primarily due to the biotechnological potential of new bacterial species and strains (Leisner et al., 1999). LAB widely distributed in nature and found naturally as indigenous micro flora in raw milk are Gram positive, catalase negative microorganisms that play an important role in many food and feed fermentation. For many centuries, LAB has served to provide an effective form of natural preservation. In addition, they strongly determine the flavor, texture and frequently, the nutritional value of the food and feed products.

LAB is also capable of producing inhibitory substances known as bacteriocins that are antagonistic towards other microorganisms. In several, fermented dairy products, LAB contribute to the characteristic structure of the product by producing exo-polysaccharides (Talarico and Dobrogosz, 1989).

In Pakistan, 70% of the dairy products viz dahi, butter and cheese available in the local market are prepared from buffalo milk in addition to cow, sheep, camel and goat's milk. In these products, the species combination of LAB is more varying and inconsistent as compared to those of the trade products. In biotechnological aspects, the wild strains of the LAB are prospective bacteriocins producers (Padmanabha-Reddy et al., 1994; Park et al., 2003) and probiotics (Rinkinen et al., 2003).

The aim of this work was the isolation and taxonomic determination of a large number of LAB from dahi samples in order to develop an indigenous culture collection of LAB strains and to use them as starter for improving the quality of different traditional fermented milk products such as yoghurt, cheese and allied products.

MATERIALS AND METHODS

Isolation of bacterial strains

The LAB was isolated from indigenous dahi samples. The isolation was performed by the routine microbiological procedures and inoculation on a solid medium. Selective media for LAB used were MRS and M17 agar plates. The samples were inoculated on the media and then incubated for 48 h at 37°C. The colonies that showed different morphological characteristics were then identified by using various biochemical tests as described by Collins and Lyne (1980).

Physiological and biochemical tests

All strains were initially tested for Gram reaction and catalase production (Harrigan and McCance, 1976; Sharpe 1979). Cell morphology and colony characteristics on MRS and M17 agar plates were also examined and separation into phenotypic groups was undertaken. Only the Gram positive, catalase negative isolates were further identified. Growth at different temperatures was ob-served in MRS and M17 broth after incubation for five days at 15, 37 and 45°C. Supplemented test was performed and resistance at 63°C for 30 min was done in order to discard enterococcus bacteria. Growth in the presence of 4 and 6.5% NaCl was performed in MRS and M17 broth for five days. Production of CO_2 from glucose was done in MRS broth containing inverted Durham tubes.

Growth at different temperatures

Growth at 15 and 45°C are most frequently used criteria for the classification of bacilli (Hammes and Vogel, 1995). *Lactobacilli* cannot grow at 15°C, however can grow at 45°C. For the classification of cocci isolates, the growth temperatures of 10 and 45°C were used. *S. thermophilus* cannot grow at 10°C, but can grow well at 45°C. In order to determine the growth at given temperatures, the modified MRS and M17 media given in the Appendix B were used. Basically, all ingredients were the same, except for Bromocresol purple. Bromocresol purple was used to determine the color change in acidity with from purple to yellow, indicating lactic acid production and cell growth. Fifty microliters of overnight activated cultures were inoculated into 5 ml test media and the incubation period at the given temperatures was observed for seven days to determine the growth and color change. For performing the biochemical tests, all strains were tested for fermentation of the following seven sugars: Lactose, Glucose, Sucrose, Maltose, Mannitol, Galactose, and Fructose. To ensure anaerobic conditions; two drops of sterile liquid paraffin were placed in each tube after inoculation. For further identification of the LAB, API 50 CH tests (bioMerieux) were also used.

Growth at different NaCl concentration

Unlike other bacilli and cocci, *Lb. delbrueckii* ssp. *bulgaricus Lb. acidophilus* and *S. thermophilus* are highly sensitive to NaCl. *S. thermophilus* does not grow even at 2% NaCl concentration, but there is no data available for *Lb.* delbrueckii ssp. bulgaricus. The most frequently used NaCl concentration for the identification of bacilli are 4 and 6.5% salt concentrations. Hence growth at 2 and 4% NaCl concentrations were used for cocci isolates and the

growth at 4 and 6.5% NaCl concentrations were used for bacill isolates in NaCl test medium (Appendix B). Fifty microliters o overnight activated cultures were inoculated into 5 ml NaCl tes media and the incubation at 42°C was observed for 7 days to determine the growth with the indication of color change from purple to yellow.

Unlike other bacilli and cocci *Lb. bulgaricus* and *S. thermophilus* are highly sensitive to NaCl. *S. thermophillus* does not grow even a 2% NaCl concentration, but there is no data available for *Lb bulgaricus.* The most frequently used NaCl concentration used fo identification of bacteria is 4 and 6.5% salt concentration. Hence growth at 2 and 4% NaCl concontration were used for cocci isolates and growth at 4 and 6.5% NaCl concentration for bacilli isolates in Nacl test medium. (50 µl of overnight activated culture was inoculated at 5 ml NaCl test medium and incubation at 42°C was observed for seven days so as to determine growth with indication of color change from purple to yellow).

RESULTS AND DISCUSSION

The lactic acid bacteria were isolated from indigenous dahi samples. The isolation was performed by the routine microbiological procedure and inoculation on a solid medium. Selective media were used for LAB were M17 and MRS plates. In most of the cases more than one colony was observed on the surface of M17 and MRS plates. In most of the cases more than one colony was observed. The cultural and morphological characteristics were examined within the help of microscope. Different types of microorganisms were observed, majority of them belonged to Gram positive rods and cocci shaped bacteria.

The Gram positive rods and cocci shaped bacteria were specifically transferred to the plates of selective media MRS and M17 respectively to purify the isolated. Subculturing of the isolate was done until pure isolates were obtained. Once pure colonies were obtained they were cultured in MRS and M17 and stored at 4°C in refrigerator until it was used.

From fifty samples, a total of 143 isolates were Gram positive and catalase negative bacteria were recorded. The organisms identified in this study are listed in Table 1. The Majority of the isolates identified belongs to genus *Lactobacillus* and the rest were referred to genus *Lactococcus* respectively. Results reveal that LAB dominated the microbial flora of dahi. It might be due to the reason that two specific media MRS and M17 agar were used to study the morphological characteristics of rods and cocci isolates respectively. This selective media allows only specific type of microorganisms to grow therefore the ability of bacterial species to grow on specific media is regarded as an important characteristic in identification. MRS and M17 media are the best suitable media for the isolation of LAB as reported earlier by Ghoddusi (2002).

All the isolates lactobacilli grew at 37 and 45°C but non at 15°C and also produced no gas from glucose and no growth at 4 and 6.5% concentration except *Lb helvitus*. None of the *Lactobacilli* isolates grew at 15°C and all characteristics of LAB, attention will be given to the

Table 1. Incidence of Lactic acid bacteria isolated from indigenous dahi samples.

Isolated specie	Number of strains/(percentage)
Lb. delbrueckii ssp. bulgaricus	34 (23.77)
Lb. delbrueckii ssp. lactis	14 (9.79)
Lb. acidophilus	14 (9.79)
Lb. helveticus	4 (2.79)
Lb. delbrueckii	7 (4.89)
Lb. casei	2 (1.39)
Lb. casei ssp. psuedoplantarum	4 (2.79)
L. lactis	20 (13.9)
S. thermophilus	38 (26.57)
S. cremoris	3 (2.09)
S. lactis ssp. diacetylactis	3 (2.09)
Total number of strains	143

distinguish sugars for the purpose of indicating which were more relevant to the identification. All the dahi isolates fermented Lactose and glucose except four isolates of Lb. acidophilus which were unable to ferment glucose, whereas all the isolates of Lb. acidophilus ferment sucrose, maltose, galactose and fructose (Table 2).

The presence of such bacteria has been reported in earlier studies (Samelis et al., 1994; Masud et al., 1993; Naeem and Rizvi, 1983). Moreover, it was observed that all the isolated bacteria from indigenous dahi were thermophilic and mesophilic in nature (Kosikowski, 1982). This diversity of species is relative and dependent primarily on the nature of the material isolated and different criteria used for each study as reported by Fitzsimmons et al. (1999) and Bissonnette et al. (2000).

Out of 143 isolates, 38 (26.57%) were of S. thermophilus followed by 34 (23.77%) Lb. bulgaricus, 20 (13.9%) L. lactis, 14 (9.79%) Lb. acidophilus, 14 (9.79%) Lb. delbreuckii sub lactis, 7 (4.89%) Lb. delbreuckii, 4 (2.79%) Lb. helveticus, 4 (2.79%) Lb. casei ssp. psuedoplantarum, 3 (2.09%) S. cremoris, 3 (2.09%) S. lactis ssp. diacetylactis and 2 (1.39%) Lb. casei.

It was observed that S. thermophilus and Lb. bulgaricus along with L. lactis constituted the dominant micro flora of dahi. Thus these species play an important role for the preparation of fermented milk products, as also noted by Warsey (1983).

S. thermophilus is mostly used in the manufacture of yoghurt, Mozzarella cheese and in some other cheeses such as cheddar cheese etc. It is further reported by Hitchener et al. (1982) that it provide protection against phages when used in combination with lactococci and produce acid rapidly during scalding.

The presence of large number of Lb. bulgaricus as recorded in the present investigation could be attributed to the presence of old inoculums containing large number of Lb. bulgaricus. Similar views are expressed by Mohanan et al. (1983). They are mainly used for acid and flavor production in yoghurt making (Kosikowski, 1982). For the thermophilic lactobacilli, the production of flavor is

largely due to acetaldehyde, which is regarded as being the most characteristic flavoring compound in dairy products. Indeed, Lb. bulgaricus plays an active part in the production of flavor resulting from threonine (Zourari et al., 1991). The data about the phenotypic characterization, identification and biochemical characteristics of isolates is presented in Table 2.

The presence of Lactococcus lactis and Streptococcus cremoris in our study is of great importance. Several studies reported that L. lactis was more frequently isolated from raw milk samples (Weerkamp et al., 1996; Badis et al., 2004), raw milk cheese (Centeno et al., 1996), Pecornio Sardo cheese (Mannu et al., 2000), Moroccan traditional fermented milk Raib (Hamama, 1992), Dahi and buttermilk samples from India (Padmanabha-Reddy et al., 1994) and Amazi, fermented milk in Zimbabwe (Mutukumira, 1996), similar findings have been reported by Guessas and Kihal (2004), who on the basis of morphological, cultural, physiological and biochemical characteristics found that Algerian raw goat milk carried different LAB with majority of Lactococcus lactis. These results are also in agreement with the findings of El Shafei et al. (2002).

Kosikowski (1982) reported that L. lactis and S. cremoris are used for the preparation of different types of cheese. However, they may be used for the preparation of yoghurt where multi-strains starter culture is used to produce the desired acidity. The role of these strains have not been yet characterized, however, they may play a role in the preparation of this product in winter season. Different studies elsewhere reported that these strains have ability to produce slime characteristics, which may be used to improve the quality of the final product.

The low incidence of S. cremoris strains in this study might be due to the reason that the incubation temperature was 37°C during isolation and identification of the bacterial strains and S. cremoris strains are mesophilic in nature. In addition, the study demonstrates the problem in attempting to classify the LAB found in dahi samples using the classical methods, which were developed largely

Table 2. Phenotypical characteristics of lactic acid bacteria isolated from indigenous Dahi.

Characteristics	Lactobacilli							Cocci			
	Lb. delbrueckii ssp bulgaricus	*Lb. delbrueckii ssp lactis*	*Lb. acidophilus*	*Lb. helveticus*	*Lb. delbrueckii*	*Lb. casei*	*Lb. casei ssp. psuedoplantarum*	*S. thermophilus*	*L. lactis*	*S. cremoris*	*S. lactis ssp diacetylactis*
	Number of isolates										
	34	14	14	4	7	2	4	38	20	3	3
Gram stain reaction	+	+	+	+	+	+	+	+	+	+	+
Catalase reaction	-	-	-	-	-	-	-	-	-	-	-
CO_2 from glucose	-	-	-	-	-	-	-	-	-	-	-
Growth at											
10°C	-	-	-	-	-	-	-	-	+	+	+
37°C	+	+	+	+	+	v	+	+	+	-	
45°C	+	+	+	+	+	v	-	+	-	-	-
Growth in medium											
4% NaCl	-	-	-	+	-	+	-	-	+	-	+
6.5%NaCl	-	-	-	+	-	-	-	-	-	-	-
Sugar fermentation											
Lactose	+	+	+	+	+	+	+	+	+	+	+
Glucose	+	+	10	+	+	+	+	+	+	+	+
Mannitol	-	-	-	-	-	+	+	-	-	-	+
Sucrose	-	+	+	-	+	+	+	+	+	-	+
Maltose	-	+	+	+	d	d	+	+	+	+	
Galactose	+	+	+	+	w	+	+	-	+	-	+
Fructose	+	+	+	-	+	+	+	-	+	-	+

Positive Reaction (+), Negative Reaction (-), Weak reaction (w), delayed Reaction (d), varying reaction (v) the number designate the amount o positive strains.

for bacteria of dairy origin. In this regard, genotypic tests should confirm the phenotypic results. Identification of *S. cremoris* is becoming scarce in industrial countries but some cases have been described in milk from Morocco or Eastern Europe (Salama, 1995).

Fourteen (14) strains of *Lactobacillus acidophilus* are recorded in the present study. These strains are considered to produce higher titrable acidity and result in the production of low pH that may be considered objectionable (Naeem and Rizvi, 1983). However, the results of these studies report that these strains have the ability to produce bacteriocins and widely used as a probiotic. Its importance is well documented by Isani et al. (1986). Due to the fact that *Lb. acidophilus* produces D-lactate,

there have been concerns about its use in infant nutrition Therefore, we must have to look its role in dahi making with special reference to public health.

Four strains of *Lb. helveticus* and three strains of *S lactis* ssp. *diacetylactis* are reported in this study. *Lb helveticus* strains are capable of producing high acidity o 2% or more of lactic acid. They can be used for the production of yoghurt in combination with other strains as reported by Badis et al. (2004). These strains could be used as starters, in the manufacture of dairy products with organoleptic qualities liked by tasters. *S. lactis* ssp *diacetylactis* are mesophilic strains, which might be the reason for their low incidence in dahi samples. They have low acidifying activity, high diacetyl content and mostly involved in flavor and aroma production. They are mostly used for cheese and buttermilk production (Badis et al., 2004).

Only two strains of *Lb. casei* have been reported in the present investigation. *Lb. casei* is used as a probiotic although it is found in some starter cultures and is commonly one of the numbers of non-starter lactic acid bacteria found in cheddar cheese (Gomes and Malcata, 1998).

The difference in the microbial composition observed among the tested samples may be contributed to the starter culture used in their preparation, processing techniques and also to the duration of fermentation. The different microbial composition may also be due to the competitive difference between the different microorganisms. The result of this study also reveals that selected LAB can be implicated in the fermentation of dahi. Therefore, there is a need in the selection of most suitable strains for controlled fermentation of dahi. The potential use of our strains as starter culture will depend on further studies on selected isolate and assessment on the effect of the quality of yoghurt.

The identified isolates will undergo tests for lactic acid production and selected for further tests (production of bacteriocins, organic acids and volatile compounds) to asses their potential as starter culture in yoghurt making. The lactic acid bacterial starter culture can greatly contribute in solving the problem of inconsistent quality and short shelf life of dairy product in Pakistan.

On the basis of these results, it can be concluded that indigenous dahi contains a mixture of lactic acid bacteria, and the quality of dahi varies with the type of species predominant in the starter culture. Therefore, efforts should be made to select suitable indigenous strains of LAB in order to produce high quality fermented dahi and its allied products.

ACKNOWLEDGMENTS

This material is based upon works supported by Higher Education Commission of Pakistan under Project R and D/HEC/03/684.

REFERENCES

Badis AD, Guetarni, Moussa-Boudjemaa B, Henni DE, Tornadijo ME, and Kihal M (2004). Identification of cultivable lactic acid bacteria isolated from Algerian raw goat's milk and evaluation of their technological properties. Food Microbiol. 21:343-349.

Bissonnette F, Labrie S, Deveau H, Lamoureux M, Moineau S (2000). Characterization of mesophilic mixed starter cultures used for the manufacture of aged Cheddar cheese. J. Dairy Sci. 83:620-627.

Centeno JA, Cepeda AA, Rodriguez-Olero JL (1996). LAB isolated from Arzua cow's milk. Int. Dairy J. 6:65-78.

Collins CH, Lyne PM (1980). Micobiological Methods. Vol 4. Butterworths, London. pp. 215-217.

El Shafei HJ, El Sabour HA, Ibrahim N, Mostafa YA (2002). Isolation, screening and characterization of bacteriocin producing LAB isolated from traditional ermented food. *Microbiol. Res.* 154:321-331.

Fitzsimmons NA, Cogan TM, Condon S, Beresford T (1999). Phenotypic and genotypic characterization of non-starter lactic acid bacteria in mature cheddar cheese. Appl. Environ. Microbiol. 65:3418-3426.

Ghoddusi HB (2002). A comparative study on suitability of culture media for yogurt starter culture. Agri. Sci. Technol. 16(1):153-160.

Gomes AMP, Malcata FX (1998). Development of probiotic cheese manufactured from goat milk: response surface analysis via technological manufacture. J. Dairy Sci. 81:1492-1507.

Guessas B, Kihal M (2004). Characterization of LAB isolated from Algerian arid zone raw goat's milk. Afr. J. Biotechnol. 3(6):339-342.

Hamama A (1992). Morrocan traditional fermented dairy products. In: Ruskin, F. R. (Ed.), Applications of Biotechnology to Traditional Fermented Foods. National Academy Press, Washington, DC. 75-79pp.

Harrigan WF, McCance ME (1976). *Laboratory methods in food and dairy microbiology.* Academic Press, New York.

Hitchener BJ, Egan AF, Roger PJS (1982). Characteristics of lactic acid bacteria isolated from vaccum-packaged beef. J. Appl. Bacteriol. 52:31-37.

Isani GB, Arain MA, Sheikh BA (1986). Study of lactic acid producing bacteria isolated from sour milk and dahi. Presented at the 2nd World Congress on Food borne Infections and Intoxications, Berlin (West), May, 1986 (Proceeding).

Jay JM (1978). Modern Food Microbiology. 2nd Ed. New York.

Kosikowski VF (1982). Cheese and fermented milk foods. 2nd Ed. New York.

Leisner JJ, Pot B, Christensen H, Rusul G, Olsen JO, Wee BW, Muhammad K, Ghazali HM (1999). Identification of lactic acid bacteria from Chili Bo, a Malaysian food ingredient. Appl. Environ. Microbiol. 65(2):599-605.

Mannu L, Paba A, Pes M, Scintu MF (2000). Genotypic and phenotypic heterogenity among lactococci isolated from traditional Pecorino Sardo cheese. J. Appl. Microbiol. 89:191-197.

Masud T, Sultana K, Shah MA (1991). Incidence of lactic acid bacteria isolated from indigenous dahi. Aust. J. Anim. Sci. 4:329-331.

Mohanan KR, Shankar PA, Laxminarayana H (1983). Growth and acid production of dahi starter cultures at sub-optimum temperature. Indian J. Dairy Sci. 3:177-181.

Mutukumira AN (1996). *Investigation of some prospects for the development of starter cultures for industrial production of traditional fermented milk in Zimbabwe.* Doctor Scientiarum Thesis, Dept. of Food Science, Agriculture University of Norway.

Naeem K, Rizvi SA (1983). Studies on the physiochemical and bacteriological aspects of dahi manufacturing in Lahore city. J. Animal Health Pro. 4:50-67.

Niku-Paavola ML, Laitila AT, Mattila-Sandholm Haikara A (1999). New types of antimicrobial compounds produced by *Lactobacillus plantarum*. J. Appl. Microbiol. 86:29-35.

Padmanabha-Reddy V, Habibulla-Khad MM, Purushothaman MM (1994). Plasmid linked starter characteristics in *Lactococci* isolated from Dahi and buttermilk. Cult. Dairy Prod. J. 29:25-26, 28-30.

Park SH, Itoh K, Fujisawa T (2003). Characteristics and identification of enterocins produced by *Enterococcus faecium* JCM 5804T. J. Appl. Microbiol. 95(2):294-300.

Rinkinen M, Jalava K, Westermarck E, Salminen E, Ouwehand AC (2003). Interaction between probiotic lactic acid bacteria and canine enteric pathogens: a risk factor for intestinal *Enterococcus faecium* colonization. Vet. Microbiol. Mar. 92(1-2):111-9.

Salama SM, Musafija-Jeknic T, Sandine EN, Giovannoni JS (1995). An ecological study of LAB. Isolation of new strains of *Lactococcus* including *Lactococcus lactis* subsp. cremoris. J. Dairy Sci. 78:1004-1007.

Samelis J, Maurogenakis F, Metaxapoulos J (1994). Characterization of lactic acid bacteria isolated from naturally fermented Greek dry salami. Int. J. Food Microbiol. 23:179-196.

Sharpe ME (1979). Identification of the lactic acid bacteria. In: F/A/Skinner and DW lovelock (editors), *Identification methods for Microbiologists. Academic Press*, London. 233-259 pp.

Talarico TL, Dobrogosz WJ (1989). Chemical characterization of an antimicrobial substance produced by *Lactobacillus reuteri. Antimicrob.* Agents Chemother. 33:674-679.

Warsy JD (1983). Production of volatile aroma compounds in dahi. J. Agric. Res. 21(1):31-36.

Weerkamp AH, Klijn K, Neeter R, Smit G (1996). Properties of mesophilic lactic acid bacteria from raw milk and naturally fermented raw milk products.Neth. Milk Dairy J. 50:319-332.

Zourari A, Roger S, Chabanet C, Desmazeaud M (1991). Characterisation de bacteries lactiques thermophiles isolees de yaourts artisanaux grecs.Souches de *Streptococcus salivarius ssp thermophilus*. Lait. 71:445-461.

18

Screening of verotoxin-producing *Escherichia coli* (VETC) O104-2011 from Egyptian market in 2011

Nashwa A. Ezzeldeen[1], Khaled F. Al-Amary[1], Mohamed M. Abdalla[2] and Sherein I Abd El-Moez[3,4]

[1]Department of Microbiology, Faculty of Veterinary Medicine, Cairo University, Giza, Egypt.
[2]Department of Microbiology, Central Laboratory of Residue Analysis of Pesticides and Heavy Metals in Foods (QCAP), Dokki, Giza, Egypt.
[3]Department of Microbiology and Immunology, National Research Center (NRC), Giza, Egypt.
[4]Food Risk Analysis Group- Center of Excellence for Advanced Sciences, NRC, Giza, Egypt.

***Escherichia coli* strains are important causes of diarrheal disease in the world and remain a major public health problem of animals and human. Sixty seven samples from different kind of food, water, soil and composit were screened for the detection of verotoxin-producing *E.coli* (VTEC) in the Egyptian markets from different location. Result showed that none of the samples gives positive results for VTEC (O104) (*vt1* and *vt2*) detection. All samples from the Egyptian markets are negative for Vero cyto toxin producing *E. coli* (O104) with percentage of 100.**

Key words: Verotoxin-producing *Escherichia coli*, Shiga toxins.

INTRODUCTION

Escherichia coli are bacterial population of the gastro-intestinal tract of humans and animals (Gross, 1994). They are commensals or pathogenic, enterotoxigenic, enteropathogenic, enteroinvasive, or enterohaemorrhagic according to the presence of specific virulence factors (Nataro and Kaper, 1998). Some isolates are shiga-like toxin producers which are zoonotic agents causing serious diseases like diarrhoea, haemorrhagic colitis (HC) and haemolytic-uraemic syndrome (HUS). The emergence of strains showing multi resistance to several antimicrobial drugs is a public health concern (White et al., 2002). Some strains of this species are agents of colibacillosis, an increasingly challenging disease in animal production, resulting in significant economic losses to the poultry industries (White, 2005). Shiga toxin–producing or verotoxin-producing *E. coli* (VETC) are the most important recently emerged groups of foodborne pathogens (Beutin et al., 2002). These *E. coli* produce either one or two cytotoxins called Shiga toxins *(stx1* and *stx2)* or verotoxins *(vt1 and vt2)* (Paton and Paton, 1998). Currently, available epidemiological information on this Shiga - toxin producing *E. coli* bacteria (STEC) outbreak in Germany suggests that STEC-contaminated food is the vehicle of infection.

A case control study carried out in Hamburg identified consumption of contaminated raw tomatoes, cucumbers and /or leafy salad as significant risk factors (Frank et al., 2011). The present study aims to screen the Egyptian

Table 1. Dye settings for multiplex reaction.

Parameter	VT1 Probe	IPC
TaqMan® VT1 (stx1) Assay	Reporter = FAM™	Reporter = VIC®
TaqMan® VT1 (stx2) Assay	Reporter = FAM™	Reporter = VIC®
TaqMan® E. coli O104 Assay	Reporter = FAM™	Reporter = VIC®
TaqMan® E. coli O157 Assay	Reporter = FAM™	Reporter = VIC®

IPC, Internal positive control.

markets for detection of the main serotyping of VTEC (O104:H4).

On the 21st of May 2011, Germany reported an ongoing outbreak of Shiga-toxin producing *E. coli* (STEC45), serotype O104:H4 (Frank et al., 2011). In Germany, between the 1st of May and the 28th of June 2011, 838 Haemolytic Uremic Syndrome (HUS) cases and 3 091 STECcases with diarrhea was reported, of which 47 persons died (RKI, 2011). On Friday the 24th of June, France reported a cluster of patients with bloody diarrhoea, after having participated in an event in the Commune of Bègles near Bordeaux on the 8th of June. As of 28 June, eight cases of bloody diarrhoea and a further eight cases with HUS have been identified. Eleven (11) of these patients, seven women and four men, between 31 and 64 years of age, had attended the same event in Bègles.

Infection with *E. coli* O104:H4 was confirmed for four patients with HUS. Six of the cases reported having eaten sprouts at the event on the 8th of June, and leftovers were analysed. Outbreak investigation revealed that the suspected sprouts of fenugreek, rocket and mustard was privately produced in small quantities by the organiser of the event from seeds bought at an approved garden centre, and were not imported from the sprout producer implicated in the outbreak in Germany (INVS, 2011).

An analytical epidemiological study went on with the persons that attended the event on 8th of June. Local trace back investigations in France suggested that the seeds for sprouting were distributed to the approved garden centre by a UK based company. European food safety authority (EFSA) was urgently requested by the Commission to initiate a comprehensive tracing back exercise (followed by tracing forward) to identify the source of the two outbreaks and identify appropriate risk mitigating measures regarding potential further outbreaks. These further investigations particularly aimed at determining whether the origin of the suspected sprout-seeds from the French cluster was linked to the large outbreak in northern Germany. This report documents the steps taken in the trace back process. Any activities already undertaken by the Task Force with regard to tracing forward are also described.

A trace back investigation is the method used to determine and document the distribution and production chain, and the source(s) of a product that has bee implicated in a food-borne illness investigation. A trac forward investigation aims to find the distribution of th suspected food products along the food chain from th origin in the direction of the consumer. Using thi approach for this investigation, at each step of th delivery/production chain identified in the trace back further investigation was initiated to try and account for a seeds in any suspect lots.

The objective was to identify critical lots and the current location. To this end, detailed information on eac lot of seeds was established for each step of th delivery/production chain back to the importation into th EU. The comparison of the back tracing information fror the French and German outbreaks leads to th conclusion that a lot of fenugreek seeds imported by th Importer, from Egypt, is the most likely common link although it cannot be excluded that other lots may b implicated.

MATERIALS AND METHODS

Samples

Sixty-seven (67) samples from different kinds of food, water, so and composit was screened for the detection of VTEC in th Egyptian markets from different location.

Screening of VTEC

VTEC screening was carried after overnight incubation of th sample in Buffer peptone water at 37°C for 24 h, using Prep ma ultra for extraction of DNA, TaqMan Environmental Master Mi TaqMan *E. coli* 2011 O104:H4 Assay, Custom TaqMan *VT2 (stx2* Assay, Custom TaqMan *VT1 (stx1)* Assay (Applied Bio system Custom TaqMan O157 Assay and 7500 real time polymerase chai reaction (PCR) (Applied Bio system). The Master Mix Set-up wa prepared as follow: 15 µL of 2X EMM 2.0 and 3 µL of 10X Targe Assay Mix and 18 µL of total volume master mix per reaction. Th dye settings for multiplex reaction were prepared as shown in Tabl 1 and the thermo cycler settings was carried out in 2 steps, enzym activation and template denaturation step which occurred at 95°C /10 min and amplification step which is repeated 45 times, includin stage 1 at 95°C /15 min and stage 2 at 60°C /45 min. Positiv reference strains included within the run and purchased fron reference laboratory of *E. coli* in Rome Italy incude *E. coli* O15 strain C210-03 genotype (*eae+, VTx2+, VTx1+*) and *E. coli* O104:H strain 11 2027 genotype (*eae-, VTx2+, VTx1-*).

Detection of VTEC from different types of food, water, soil and composit in the Egyptian markets (EU RL Method, 2011)

The procedure includes three main steps:

Enrichment

Food sample was enriched by adding 25 g sample to 225 ml of enrichment broth (buffer peptone water) then incubated at 37°C for 24 h. Water sample was enriched by filtration of 100 ml of water samples and the filterate added to 50 ml buffer peptone water.

Extraction of DNA

1 ml of enrichment was transfered in micro centrifuge tube, centrifuged at maximum speed (15000rpm/3 min) to spin down the contents, and the supernatant removed. The pellets were resuspended in 100 µl of PrepMan® Ultra Sample preparation reagent and vortex to mix the contents. The tube was heated in a heat block at 95 to 100°C/10 mins, then centrifuged at 15000rpm/3 min to spin down the contents. 10 µl of the supernatant (sample DNA) was transfered to a new tube containing 90 µl of water then vortexed to mix the contents, and then the sample DNA was ready for PCR.

Preparing the sample for PCR

According to the number of samples, the premix solution of master mix and assay was calculated and added in external screw capped tube. Both the samples and negative controls require 15 µl of master mix and 3 µl assays. Premix solution (18 µl) was transfered into each well, gently pipetting at the bottom of the well. 12 µl of unknown sample was transfered into each well, gently pipetted to mix the solution. 12 µl of negative and positive control were transferred and the tubes closed.

Preparation of PCR run

The samples runned on real-time PCR System, plate were loaded into the instrument, and then the cycle was adjusted as follow holding stage where the temperature was gradually raised to 95.0°C/10min followed by cycling stages which include 40 cycles; 15 min at 95.0°C and 1 h at 60°C, then the run started.

Data analysis and documentation

Data analysis was carried out according to Flow-diagram of the screening procedure of VTEC according to EU RL (2011) method. 25 g of sample was added to 225 ml BPW and enriched at 37°C, 18 to 24 h followed by extraction of DNA for screening and detection of the presence of Verotoxin genes. Negative result to vtx gene will be reported as absence of VTEC. Positive samples to vtx genes will undergo test for O104 and 0157 gene, followed by isolation onto Maconkey agar or sorbitol Maconkey agar, then the isolated colonies were tested for vtx producing genes by real time PCR to confirm positive results by O104 and/or O157. Negative results indicate non VTEC producing O104 or O157 while positive VTEC, O104 and /or O157.

The interprition of results was carried out according to the analysis of verotxin where samples showing positive VT1 and /or VT2 are VTEC positive while samples showing negative VT1 and /or VT2 are VTEC negative.

RESULTS AND DISCUSSION

Sixty seven (67) samples from different kind of food, water, soil and composit were screened for the detection of VTEC in different places in the Egyptian markets. The result shows that none of the samples gave positive results for VTEC (vt1 and vt2) as shown in Table 2.

E. coli is a major pathogen of worldwide importance in commercially raised, contributing significantly to economic losses in both turkeys and chickens. E. coli has been associated with a variety of diseases in birds including enteritis, arthritis, omphalitis, coligranuloma, septicemia, salphingitis and complicated air sacculitis about 10 to 15% of intestinal coliform are pathogenic serotypes (Roy et al., 2006). Although normally commensal in nature, certain strains of E. coli are associated with a variety of infections in human and animals. E. coli are present on most uncooked foods and in the inanimate environment. The major route of E. coli transmission is through the consumption of contaminated food and water, person to person and animal to person contact (Heuvelink et al., 1995; Leyer et al., 1995; Reilly, 1998). Recent food borne outbreaks caused by E. coli had heighlighted the demand for rapid detection of this organism in food (Heuvelink et al., 1995; Takeda, 1995).

On the 21st of May 2011, Germany reported an ongoing outbreak of STEC, serotype O104:H4 (Frank et al., 2011). In the past, STEC O104:H4 had been isolated in humans twice in Germany in 2001 (Mellmann et al., 2008) and once in Korea in 2005 (Bae et al., 2006). In addition, according to the information reported to the European Centre for Disease Prevention and Control (ECDC), a total of 10 persons were infected with other STEC O104 types in the European Union (EU) Member States from 2004 to 2009 (EFSA 2011). In Germany, between the 1st of May and the 28th of June 2011, 838 HUS cases and 3,091 STEC cases with diarrhea were reported, of which 47 persons died (RKI, 2011). The last date of onset of disease reported from Germany was on the 23rd of June for all EHEC or HUS cases reported, while for confirmed STEC O104:H4 cases, the last date of disease onset was the 12th of June. Up to the 29th of June, 13 EU/EEA9 countries reported cases associated with the outbreak in Germany for a total of 885 HUS and 3,170 non-HUS STEC cases (ECDC 2011). Until a recent outbreak in the Bordeaux area in France, with a rare exception, these cases in other European countries had all been linked to travel to northern Germany, where the outbreak had occurred.

The German outbreak strain is a STEC that belongs to serotype O104:H4, and has been microbiologically characterised in detail (EFSA, 2011). Preliminary information on the microbiological characterisation of the isolates implicated in the French outbreak indicate that many characteristics (stx2 positive, eae negative, hlyA negative, multi-resistance pattern to antimicrobials) are common with the German outbreak strain. In addition, the two molecular techniques (repetitive sequence based polymerase chain reaction (Rep-PCR) and pulsed-field gel electrophoresis (PFGE) used to fully characterise and compare the outbreak strains in France and Germany

Table 2. Result of screening of 67 samples from the Egyptian markets for the detection of VTEC.

Sample	Number of sample	Place	Results	
			VT1	VT2
Okra	1	Alexandria	Negative	Negative
Cheese	1	Cairo	Negative	Negative
Composite	1	Cairo	Negative	Negative
Corn	1	Cairo	Negative	Negative
Dehydrated leek	1	Cairo	Negative	Negative
Drinking Water	1	Cairo	Negative	Negative
Fenugreek	2	Cairo	Negative	Negative
Green beans	4	Cairo	Negative	Negative
Juice	1	Cairo	Negative	Negative
Nigella sativa	1	Cairo	Negative	Negative
Soil	1	Cairo	Negative	Negative
Strawberry	1	Cairo	Negative	Negative
Water irrigation	1	Cairo	Negative	Negative
Majoram	1	El fayoum	Negative	Negative
Carawy	1	El fayoum	Negative	Negative
Drinking water	1	El fayoum	Negative	Negative
Irrigation water	8	El fayoum	Negative	Negative
Fenugreek	1	El fayoum	Negative	Negative
Dry mint	2	El fayoum	Negative	Negative
Basil	1	El fayoum	Negative	Negative
Soil	3	El fayoum	Negative	Negative
Fym	1	El fayoum	Negative	Negative
Pepper	1	El fayoum	Negative	Negative
Water well	1	El fayoum	Negative	Negative
Fennel	3	El fayoum	Negative	Negative
Chamomile	3	El fayoum	Negative	Negative
Composit	1	El fayoum	Negative	Negative
Fenugreek	1	EL Minya	Negative	Negative
Compsite	2	EL Minya	Negative	Negative
Irrigation water	2	EL Minya	Negative	Negative
Waste water	1	El Minya	Negative	Negative
Soil	1	El Minya	Negative	Negative
Fenugreek	1	Ismailia	Negative	Negative
Water bore hole	3	Ismailia	Negative	Negative
Nigella sativa	1	Ismailia	Negative	Negative
Sweet potatoes	1	Ismailia	Negative	Negative
Canal water	2	Ismailia	Negative	Negative
Well water	1	Ismailia	Negative	Negative
Green beans	1	Minufia	Negative	Negative
Dehydrated beans	1	Minufia	Negative	Negative
Fenugreek	2	Minufia	Negative	Negative
Dry mint	1	Minufia	Negative	Negative

showed the genetic relatedness of the strains (Gault et al., 2011).

The analysis and discussion in EFSA (2011) focuses primarily on data obtained from the back tracing process to identify the source of the seeds suspected of causing the STEC O104:H4 outbreaks. The German EHEC Task Force trace back methodology was successfully extended to support the investigations involving five other European Member States. The comparison of the back tracing information from the French and German outbreaks leads to the conclusion that fenugreek seeds imported from Egypt are the common link for these two outbreaks. The implication is that the seeds became contaminated with STEC O104:H4 at some point prior to leaving the importer. Such contamination typically reflects a production or distribution process which allowed contamination by faecal material of human and/or animal origin. The results show that from this importer, seeds were sold to many businesses in Germany and many other countries in Europe.

For these reasons, invistgation on 67 samples from different kinds of food, water, soil and composit was carried for the detection of VTEC in the Egyptian markets. The result show that all analyzed samples from the Egyptian markets were negative for Vero cyto toxin producing E. coli VTEC (vt1 and vt2) with 100% and there is no such serotype of VTEC (O104:H4) detected in any kind of food especially fenugreek and also water, soil and composite. Moreover, It is recommended that all laboratories for VTEC analysis use harmonised methods to define VTEC seropathotypes from human and non-human sources to allow more effective monitoring by comparison of isolates from food and animals with those from humans.

REFERENCES

Gross WG (1994). Diseases due to Escherichia coli in poultry. In C.L. Gyles (ed.), Escherichia coli in Domestic Animals and Humans. CAB International, Wallingford, UK.: 237–259.

Nataro JP, Kaper, JB (1998). Diarrheagenic Escherichia coli. Clinical Microbiology Reviews. 11: 142–201.

White DG, Zhao S, Simjee S (2002). Antimicrobial resistance of foodborne pathogens. Microbes Infect. 4: 405–412.

White DG (2005). Antimicrobial resistance in pathogenic Escherichia coli from animals. In F.M. Aarestrup (ed.), Antimicrobial Resistance in Bacteria of Animal Origin. ASM Press, Washington, DC.: 145–166.

Beutin L, Kaulfuss T, Cheasty B, Brandenburg S, Zimmerman K, Gleier GA, Willshaw S, Smith HR (2002). Characteristics and associations with disease of two major subclones of Shiga toxin (verotoxin) producing strains of Escherichia coli (STEC) O157 that are present among isolates from patients in Germany. Diagn. Microbiol. Infect. Dis. 44: 337–346.

Paton AW, Paton JC (1998). Detection and characterization of Shiga toxigenic Escherichia coli by using multiplex PCR assays for stx1, stx2, eaeA, enterohemorrhagic E. coli hlyA, rfbO111, and rfbO157. J Clin Microbiol. 36: 598-602.

Frank C, Faber MS, Askar M, Bernard H, Fruth A, Gilsdorf A, Höhle M, Karch H, Krause G, Prager R, Spode A, Stark K, Werber D (2011). Large and ongoing outbreak of haemolytic uraemic syndrome, Germany. Euro Surveill. 16(21): 19878.

RKI (Robert Kock Institut) (2011). Final presentation and evaluation of epidemiological findings in the EHEC O104:H4 Outbreak, Germany.

Institut de Veille Sanitaire (INVS). Cas groupés de syndrome hémolytique et urémique (SHU) en Gironde - Point au 24 juin 2011. 2011 [27 June 2011] , Available from: http://www.invs.sante.fr/Dossiers-thematiques/Maladies-infectieuses/Risques-infectieux-dorigine-alimentaire/Syndrome-hemolytique-et-uremique/Actualites/Cas-groupes-d-infectionsa-Escherichia-coli-entero-hemorragique-EHEC-en-Gironde-Point-au-24-juin-2011.

EU RL Method (European reference lab) (2011). Detection and identification of Verocytotoxin-producing Escherichia coli (VTEC) O104:H4 in food by Real Time PCR. EU RL Method_food_2 Rev. 2.

Roy P, Purushothaman V, Koteeswaran A, Dhillon AS (2006). Isolation, characterization, and antimicrobial drug resistance pattern of Escherichia coli isolated from Japanese quail and their environment. J. Appl. Poult. Res. 15(3): 442-446.

Heuvelink AE, Van de kar NCAJ, Meis JFGM, Monnes LAH, Melchers WHG (1995). Characterization of verocytotoxin producing Escherichia coli O157 isolates from patients with haemolytic uremic syndrome in western Europe." Epidemiol. Infect., 115,1-14.

Leyer GJ, Wand LL, Jonshon EA (1995). Pidiateric diarrhea. J. Paediatr. Infect. Dis. 5(1): 29-43.

Reilly A (1998). Prevention and control of enterohaemorrhagic Esherichia coli (EHEC) infection. Memorandum form a WHO meeting, Bull. WHO. 76: 245-255.

Mellmann A, Bielaszewska M, Kock R, Friedrich AW, Fruth A, Middendorf B, Harmsen D, Schmidt MA, Karch H (2008). Analysis of collection of hemolytic uremic syndrome-associated entero-hemorrhagic Escherichia coli. Emerg. Infect. Dis. 14: 1287-1290.

Bae WK, Lee YK, Cho MS, Ma SK, Kim SW, Kim NH, Choi KC (2006). A case of hemolytic uremic syndrome caused by Escherichia coli O104:H4. Yonsei Med. J. 47: 437-439.

EFSA (European food safety authority) (2011). Tracing seeds, in particular fenugreek (Trigonella foenum-graecum) seeds, in relation to the Shiga toxin-producing E. coli (STEC) O104:H4 Outbreaks in Germany and France.

ECDC (European Centre for Disease Prevention and Control and European Food Safety Authority) (2011). Shiga toxin/verotoxin-producing Escherichia coli in humans, food and animals in the EU/EEA, with special reference to the German outbreak strain STEC O104. Stockholm.

Gault G, Weill FX, Mariani-Kurkdjian, Jourdan-da Silva N, King L, Aldabe B, Charron M, Ong N, Castor C, Macé M, Bingen E, Noël H, Vaillant V, Bone A, Vendrely B, Delmas Y, Combe C, Bercion R, d'Andigné E, Desjardin M, de Valk H, Rolland P (2011). Outbreak of haemolytic uraemic syndrome and bloody diarrhoea due to Escherichia coli O104:H4, south-west France. Eur. Surveill. 16(26): 19-905.

19

Effect of light and aeration on the growth of *Sclerotium rolfsii in vitro*

Muthukumar, A. and Venkatesh, A.

Department of Plant Pathology, Faculty of Agriculture, Annamalai University, Annamalainagar-608 002, Chidambaram, Tamil Nadu, India.

***Sclerotium rolfsii* is one of the devastating soil-borne plant pathogens which cause severe loss at the time of seedling development. It also causes leaf spots in several crops and wild plants. In this experiment, exposure of pathogen to different light period and aeration in order to assess the mycelial growth, biomass production, weight and number of sclerotia of *S. rolfsii* was done. Three-fourth area of three plates, 50% area of three plates and 100% area of three plates were sealed with cellophane tape. The other three plates were not sealed. Two sets of such plates were prepared. All the plates were incubated at 28±2°C. One set was incubated in light whereas the other set was incubated in the dark. The results reveal that there was no significant difference in mycelial growth and number of sclerotia among them but significant difference was observed when compared with the control, that is, the plates which were not sealed. Sclerotial formations were directly influenced by air as completely sealed plates failed to produce sclerotia. Generally, the light condition induces the production of more number of sclerotia than dark condition. In another study, the exposure of pathogen to different light periods revealed that alternative cycles of 12 h light and 12 h darkness for ten days resulted in the maximum mycelial growth and dry weight, more number of sclerotia and weight of sclerotia was also seen when compared with other treatments.**

Key words: *Sclerotium rolfsii*, aeration, light, peppermint.

INTRODUCTION

Peppermint (*Mentha piperita* L.) is an important aromatic perennial herb grown throughout the world; it belongs to the family Lamiaceae. It is extensively cultivated in India and about 70% of the international annual requirement is met from crops raised in the central region of the Indo-Gangetic plains (Singh et al., 1999). *Mentha* is cultivated in Himalaya-hills, Haryana, Uttar Pradesh, Punjab and Bihar. Of these, Uttar Pradesh is the largest producing state in the country contributing 80-90% of the total production followed by Punjab, Haryana, Bihar and Himachal Pradesh. Cultivated peppermint, serves as a source of menthol, menthone, isomenthone, menthofuran, linanool, linalyl acetate, methyl acetate, terpenes, carvone, piperitenone

oxide and other aromatic compounds. In India, pepper mint is grown throughout the year (Shukla et al., 1998 and it is affected by several fungal diseases caused by *Rhizoctonia solani* (Kumar et al., 1997; Merin, 2002) *Verticillium dahliae* (Johnson and Santo, 2001) *Collectotrichum cocodes* (Johnson et al., 2002) and *Sclerotium rolfsii* (Anand and Harikesh Bahadur Bahadur 2004) of which, collar rot caused by *S. rolfsii* is a majo constraint in the peppermint cultivation in Tamil Nadu. *S rolfsii* is a soil borne plant pathogen causing root rot stem rot, collar rot, wilt and foot rot diseases on more than 500 plant species of agricultural and horticultural crops throughout the world (Aycock, 1966). The pathogen causes a great economic loss in various crops. It has been reported that *S. rolfsii* caused about 25% seedling mortality in the groundnut cultivar JL-24 (Ingale and Mayee, 1986). In tomato, this pathogen was responsible for a crop loss of 30% (Thiribhuvanamala et al., 1999). Its occurrence on crossandra has been observed to be about 40-50% mortality of plants. In peppermint, this pathogen caused about 5 to 20% of crop loss under field condition (Anand and Harikesh, 2004). Diseases caused by *S. rolfsii* are initiated either directly from soil-borne sclerotia which germinate to form fine cottony hyphae infecting the collar region of host plants or sclerotia sticking on the lower/upper surfaces of the leaves by rain splashes where they germinate and cause leaf spots (Singh and Pavgi, 1965). Soil temperature of 25-30°C and soil moisture 90% play significant role in disease development (Gupta et al., 2002). Various biotic and abiotic factors which directly or indirectly influence the development of sclerotia were discussed in literature (Ellil, 1999; Sarma, 2002). The objectives of the present study were i) to isolate and identify the pathogen ii) to study the pathogenicity test iii) to study the role of air in the growth of *S. rolfsii* and influence of light on the growth of pathogen.

MATERIALS AND METHODS

Isolation, identification and maintenance of pathogen

The collar rot symptoms were collected from the Department of Plant Pathology, Faculty of Agriculture, Annamalai University, Chidambaram, Tamil Nadu, India. The infected plant materials brought back from the field were washed, cut into 5 mm segments including the advancing margins of infection. The segments were surface sterilized in 0.5% sodium hypochlorite solution for 5 min. and rinsed in three changes of sterile distilled water. The segments were separately dried in between sheets of sterile filter paper and placed (3 pieces per plate) on fresh potato dextrose agar (PDA) medium (Ainsworth, 1961) impregnated with streptomycin, and incubated for seven days at 28 ± 2°C. The fungal growth on 5th day, which arose through the sclerotial bodies was cut by inoculation loop and transferred aseptically to the PDA slants and allowed to grow at room (28 ± 2°C) temperature to obtain the pure culture of the fungus. The culture thus obtained was stored in refrigerator at 5°C for further studies and was sub cultured periodically. The purified isolate was identified as *S. rolfsii* based on morphological and colony characteristics (Punja and Damini, 1996; Sarma et al., 2002; Watanabe, 2002).

Assessing the pathogenicity of *S. rolfsii* isolate

The pot mixture was prepared by thoroughly mixing clay loam soil, sand and farm yard manure at the ratio of 1:1:1. The inoculum of *S. rolfsii* isolate was grown on sand-corn meal medium (twenty days old) mixed thoroughly at five percent (w/w basis) level, and applied to top two centimeter of the soil (Abeygunawardena and Wood, 1957). Then, apparently healthy surface sterilized mint cuttings were planted in inoculated pots. The cuttings planted in pots without inoculum served as control. Soil moisture was maintained at moisture holding capacity of soil by adding sterilized water on weight basis throughout the period. After 20 days of inoculation, the plants showing the typical wilting symptoms were observed. Re-isolation

was made from such affected portion of the plant tissue and compared with that of original isolate for conformity.

Effect of air on sclerotial development of *S. rolfsii* in potato dextrose agar medium

Fifteen milliliters of molten PDA medium was dispensed into 12 sterile Petri plate. Mycelial discs taken from the advancing margins of seven days old culture of respective *S. rolfsii* isolate by the aid of cork borer were separately placed at the centre of the plate containing PDA medium. Three-fourth area of three plates, 50% area of three plates and 100% area of three plates were sealed with cellophane tape. The other three plates were not sealed. Two sets of such plates were prepared. All the plates were incubated at 28±2°C. One set was incubated in light whereas the other set was incubated in the dark. In this experiment, there were four treatments and each treatment consists of three plates and each treatment is repeated three times. The inoculated plates were sealed with the help of lab seal in the following manner, that is, no sealing (control), half sealed, 3/4th and complete sealing. Each set contained three plates. After inoculation and sealing, Petri plates were incubated at 28 ± 2°C (light and dark) and the other sealed plates were wrapped with black paper and incubated as above. Visual observations were periodically made for sclerotial initiation, sclerotial development and number of sclerotia per plate.

Effect of light on the growth of *S. rolfsii*

Potato dextrose broth and agar were used in this experiment. Conical flasks of 250 ml capacity and each containing 100 ml of liquid broth were inoculated and exposed to different length of light hours viz., alternate cycles of twelve hours light and twelve hours darkness, continuous light and continuous darkness in an environmental conditions. Flasks were inoculated with 6 mm mycelial disc obtained from the periphery of seven days old culture of *S. rolfsii* and incubated at different light intensities. All the inoculated plates were incubated for ten days under different length of light hours. The number of sclerotia/flask and weight of sclerotia was recorded at the end of the incubation period. There were three treatments and each treatment consists of three plates and each treatment was repeated three times. Then the mycelial mat was filtered through Whatman No. 41 filter paper discs of 12.50 cm diameter dried to a constant weight at 60°C prior to filtration. The mycelial mat on the filter paper was washed thoroughly with distilled water to remove any salts likely to be associated with the mycelium and dried to a constant weight in an electrical oven at 60°C, cooled in a dessicator and weighed immediately on an analytical electrical balance. The weight of dry mycelium was recorded and the data were statistically analyzed.

To carryout study on solid media, 15 ml of potato dextrose agar was poured in 90 mm sterile Petri plate. Such plates were inoculated with six mm mycelial disc obtained from the periphery of seven days old culture of *S. rolfsii* and incubated at different light intensities. Each treatment consists of three plates and each treatment was repeated three times. All the inoculated plates were incubated for ten days under different length of light hours. The mycelial growth, number of sclerotia/plate and weight of sclerotia was recorded at the end of the incubation period.

Statistical analysis

The data on effect of the treatments on the growth of pathogens was analyzed by analysis of variance (ANOVA) and treatment means were compared by Duncan's multiple range test (DMRT) and by least significance difference (LSD) at P = 0.05. The package used for analysis was IRRISTAT version 92-1 developed by the

Table 1. Effect of air on sclerotial development of *S. rolfsii* in potato dextrose agar medium.

Treatment	Observation									
	In dark visual observation after (days)				Average number of sclerotia/plate	In light visual observation after (days)				Average number of sclerotia/plate
	6	8	10	12		6	8	10	12	
No sealing (control)	+	++	++	+++	213[a]*	+	++	++	+++	276[a]*
1/2 sealing	+[F]	++	+++	+++	168[b]	+[F]	++	+++	+++	185[b]
3/4 sealing	+[F]	++	+++	+++	157[c]	+[F]	++	+++	+++	178[c]
Complete sealing	-	-	-	-	0[d]	-	-	-	-	0[d]

+- Sclerotial initial; ++- white sclerotia; +[F] -fewer sclerotia initials; +++ - dark brown sclerotia; - = no sclerotial initials; *Values in each column followed by the same letter are not significantly different according to the DMRT method (P = 0.05).

Table 2. Effect of light on the growth of *S. rolfsii* in potato dextrose agar medium and potato dextrose broth.

Treatments	Mycelial growth (mm)	Mycelial dry weight (mg)	Average number of sclerotia/plate	Weight (mg)/100 sclerotia
Continuous light	68.66[b]*	280.33[b]*	225[b]*	74[b]*
Continuous dark	47.33[c]	130.00[c]	156[c]	54[c]
Alternate cycle of 12 h light and 12 h darkness	89.66[a]	382.66[a]	263[a]	85[a]

*Values in each column followed by the same letter are not significantly different according to the DMRT method (P = 0.05).

Biometrics Unit of the International Rice Research Institute, The Philippines (Gomez and Gomez, 1984).

RESULTS AND DISCUSSION

The results of the present study reveal that the number of sclerotia in 3/4 and 1/2 sealed plates placed in light and darkness affected the mycelial growth and number of sclerotia significantly as compared to the control (unwrapped plates). In the control plates, sclerotia initials were observed after 6 days of inoculation as whitish, tiny, pinhead-like structures and after 6-8 days exudation commenced. In completely sealed plates, the fungal growth was relatively very slow, compact and profusely growing mycelium was observed after 6 to 8 days as compared to the control. In all completely sealed plates, there was no sclerotium formation even after 12 days after inoculation. In 3/4 and 1/2 sealed plates, the number of sclerotia were less but they were bigger in size as compared to the control. In control plates, mature sclerotia became brownish at 3/4th day after inoculation but in 1/2 and 3/4 sealed plates, such sclerotia were seen after 10 days (Table 1). Sclerotia are the asexual structures formed due to the aggregation of fungal mycelium. Several biotic and abiotic factors influence the aggregation of fungal hyphae in the culture medium. Punja and Damini (1996) and Singh et al. (2002) reported that sclerotial exudates directly influenced the development and maturation of sclerotia. The number and sclerotial weight were affected drastically due to improper aeration as average numbers of sclerotia were more in unsealed plates (Sudarshan et al., 2010). Bhoraniya et al. (2002) reported that due to patho-

genesis, the level of oxalic acid increases in the infected plants and the increase of oxalic acid induces formation of sclerotial initiation at the collar region. It was reported that depletion of exudate inhibits the development of sclerotia of *S. sclerotiorum* (Singh et al., unpublished observation).

The exposure of the pathogen to alternative cycles of 12 h light and 12 h darkness for ten days resulted in the maximum mycelial growth and dry weight (89.66 mm; 382.66 mg, respectively) with more number of sclerotia/plate and weight of sclerotia of *S. rolfsii* which was significantly superior over other treatments tested (Table 2). The mycelial growth of pathogen exposed to continuous light resulted in moderate growth (68.66 mm; 280.33 mg) and continuous darkness resulted in minimum mycelial growth and dry weight of *S. rolfsii* (47.33 mm; 130.00 mg) and less number of sclerotia/plate and weight of sclerotia was also very less (156; 54, respectively). Similarly, Basamma (2008) reported that, *S. rolfsii* was exposed to alternate cycles of 12 h light and 12 h darkness recorded more number of sclerotia of *S. rolfsii*. This is in agreement with the findings of Chung and Kim (1977) and Punja (1985).

In the present experiment, we found that a proper aeration and light is essential for the mycelial growth anddevelopment of sclerotia of *S. rolfsii*.

REFERENCES

Abeygunawardena DVW, Wood RKS (1957). Factors affecting the germination of sclerotia and mycelial growth of *Sclerotium rolfsii* Sacc. Trans. Br. Mycol. Soc. 40:221-231.

Ainsworth GC (1961). Dictionary of fungi. Common Wealth Mycological Institute., Kew Burrey, England. p. 547.

Anand S, Harikesh Bahadur S (2004). Control of collar rot in mint (*Mentha* spp.) caused by *Sclerotium rolfsii* using biological means. Curr. Sci. 87:362-366.

Aycock R (1966). Stem rot and other diseases caused by *Sclerotium rolfsii* or the status of Rolf's fungus after 70 years. N.C. Agric. Exp. Stn. Tech. Bull. 174.

Basamma (2008). Integrated management of *Sclerotium* wilt of potato caused by *Sclerotium rolfsii* Sacc. M.Sc. Thesis, University of Agricultural Sciences, Dharwad.

Bhoraniya MF, Khandar RR, Khunti JP (2002). Estimation of oxalic acid in chillies infected with *Sclerotium rolfsii*. Plant Dis. Res. 17:325.

Chung HS Kim HK (1977). Effect of light on *Sclerotium* formation of *Sclerotium rolfsii*. Sacc. on agar media. Korean J. Mycol. 5:21-23.

Ellil AHAA (1999). Oxidative stress in relation to lipid peroxidation, sclerotial development and melanin production by *S. rolfsii*. J. Phytopathol.147:561-566.

Gomez KA, Gomez AA (1984). Statistical Procedure for Agricultural Research. John Wiley Sons, New York, NY, USA. p. 680 .

Gupta SK, Sharma A, Shyam KR, Sharma JC (2002). Role of soil temperature and moisture on the development of crown rot (*Sclerotium rolfsii*) of French bean. Plant Dis. Res. 17:366-368.

Ingale RV, Mayee CD (1986). Efficacy and economics of some management practices of fungal diseases of groundnut. J. Oilseeds Res. 3: 201-204.

Johnson DA, Douhan LI Greary B (2002). Report of *Colletotrichum cocodes* associated with mentha. Plant Dis. 86:695.

Johnson DA, Santo GS (2001). Development of wilt in mint in response to infection by two pathotypes of *Verticillium dahliae* and co infection by *Prataylenchus penetrans*. Plant Dis. 85: 1189-1192.

Kumar S, Khol AP, Pata DD, Rar TM, Sinqii S Tyagi BR (1997). Cultivation of menthol mint in India. Farm Bull. 4:13-20.

Merin B (2002). Studies on epidemiology and stolon rot of mentha. M.Sc. Thesis, Department of Plant Pathology, Tamil Nadu Agricultural University, Coimbatore.

Punja Z (1985). The biology, ecology and control of *Sclerotium rolfsii*. Ann. Rev. Phytopathol. 23: 97-127.

Punja ZK, Damini A (1996). Comparative growth, morphology and physiology of three *Sclerotium* species. Mycologia 88:694-706.

Sarma BK (2002). Studies of variability, sexual sage production and control of *S. rolfsii* Sacc., the causal agent of collar rot of chickpea (*Cicer arietinum*). Ph.D., Thesis, Department of Mycology and Plant Pathology, Institute of Agricultural Sciences, Banaras Hindu University, Varanasi, India.

Sarma BK, Singh UP, Singh KP (2002). Variability in Indian isolates of *Sclerotium rolfsii*. Mycologia 94:1051-1058.

Shukla PK, Haseeb AS Sharma S (1998). Soil texture, root lesion nematodes and yield of peppermint (*Mentha piperita*). J. Herbs Spices Med. Plants 6:1-8.

Singh AK, Srivastav RK Kumar S (1999). Production and trade of menthol mint in India. Curr. Sci. 87.

Singh UP, Pavgi MS (1965). Spotted leaf rots of plants- a new sclerotial disease. Plant Dis. Rep. 49:58–59.

Singh UP, Sarma BK, Singh DP, Bahadur A (2002). Studies on exudate-depleted sclerotial development in *Sclerotium rolfsii* and the effect of oxalic acid sclerotial exudates and culture filtrate of phenolic acid induction in chickpea (*Cicer arietinum*). Can. J. Microbiol. 48:443-448.

Sudarshan Maurya Udai Pratap S, Rashmi S, Amitabh S, Harikesh BS (2010). Role of air and light in sclerotial development and basidiospore formation in *Sclerotium rolfsii*. J. Plant Prot. Res. 50:206-209.

Thiribhuvanamala G, Rajeswar, E, Sabitha Doraiswamy (1999). Inoculum levels of *Sclerotium rolfsii* on the incidence of stem rot in tomato. Madras Agric. J. 86: 334.

Watanabe T (2002). *Sclerotium* sp. morphologies of cultured fungi and key species: Pictorial Atlas of Soil and Seed fungi. 2nd Edn., CRC Press, New York.

Evaluation of antimicrobial and antioxidant properties of leaves of *Emex spinosa* and fruits of *Citrillus colocynthis* from Saudi Arabia

Mona A. Aldamegh[1], Emad M. Abdallah[2] and Anis Ben Hsouna[2]

[1]Departmet of Biology, College of Sciences and Arts at Onaizah, Qassim University, Saudi Arabia.
[2]Department of Laboratory Sciences, College of Sciences and Arts at Al-Rass, Qassim University, P. O. Box 53, Saudi Arabia.

The crude methanol extract of *Citrullus colocynthis* fruit and *Emex spinosa* leaves were examined for antimicrobial and antioxidant potentialities. The phytochemical analysis revealed presence of some bioactive principles, such as alkaloids, flavonoids and anthraquinones for *E. spinosa* and saponin, flavonoids, terpenoids and alkaloids for *C. colocynthis*. The antimicrobial activities were determined against seven bacterial strains (*Proteus vulgaris* NCTC 8196, *Escherichia coli* ATCC 25922, *Pseudomonas aeruginosa* ATCC 27853, *Klebsiella pneumonia* ATCC 53651, *Salmonella typhi* NCTC 0650, *Staphylococcus aureus* ATCC 25923 and *Bacillus cereus* NCTC 8236) and one fungal strain (*Candida albicans* ATCC 7596). *E. spinosa* leaf methanol extract was most active against fungus, while *C. colocynthis* fruit methanol extract was most active against bacteria, particularly *E. coli* ATCC 25922 and *P. aeruginosa* ATCC 27853. The antioxidant properties of extracts were investigated *in vitro* using 1,1-diphenyl, 2-picryl hydrazyl (DPPH) radical scavenging assay and *in vivo* in rats using serological and enzymatic tests. Both plant extracts showed considerable antioxidant activities. The promising findings of this investigation could be used as a novel natural antimicrobial and antioxidant agents.

Key words: *Emex spinosa*, *Citrullus colocynthis*, antimicrobial, antioxidant activity.

INTRODUCTION

The medicinal properties of plants were an issue of human interest since times immemorial. Since modern medicine has increasingly benefitted from exploring compounds present in traditional medicine, researches should focus on medicinal plants which are rich sources of natural remedies (Abdallah and El-Ghazali, 2013). Saudi Arabia which is located in the Arabian Peninsula has an arid hot desert climate and rainfall is mostly scarce.

Based on these harsh conditions, the studies on the pharmaceutical properties of the flora of Saudi Arabia have been neglected for a long time (El-Ghazali et al. 2010). However, arid climate may activate the production of secondary phytochemical compounds in high concentrations which are presumably more capable of fighting its natural enemies such as insects, herbivores and diseases. *Citrulluscolocynthis*(L.)Schraderfromfamily Cucurbitacea and *Emex spinosa* (L.) Campd. from family Polygona-ceae are among the plants of folk medicinal applications growing in the deserts of Arabian Peninsula. The fruits of *C. colocynthis* have many folk medicinal applications, its juice with sugar are prescribed for dropsy (Anonymous, 1970), against tumors of gastrointestinal tract, joint pains, anti-leukemia and as anti-cancerous drug (Rodge and Biradar, 2012). The leaves of *E. spinosa* are edible by the local inhabitants and Bedouin in eastern Saudi Arabia as well as its carrot-like tap root (Mandaville, 1990). It is used to relief dyspepsia, appetite stimulant, diuretic and also used for gastrointestinal disorders (Watt and Breyer-Brandwjk, 1962). The present investigation describes the antimicrobial and antioxidant properties of two medicinal plants, *C. colocynthis* and *E. spinosa* growing in the arid desert regions in Saudi Arabia.

MATERIALS AND METHODS

Plant collection

E. spinosa (leaves) and *C. colocynthis* (fruits) were collected manually from Qassim district, Saudi Arabia. The botanical identi-fication was confirmed by Gamal E. El-Ghazali (Taxonomist). The fresh plant samples were washed and dried in shade for up to 15 days for leaves and 30 days for fruits. Then, they were crushed into fine powders using crushing machine and kept in dark well tight bottles for further investigations.

Plant extraction

Extraction was performed as described by Samie et al. (2005) with minor modifications. 50 g of each ground plant material was soaked in 500 ml of methanol for up to 72 h with frequent shaking. Then, samples were filtered twice using Whattman No.1 filter paper and evaporated to dry (semi-solid residues) under reduced pressure at 40°C. The semi-solid residues were left in Incubator at 40°C until totally dried (about two days). Dry extracts were reconstituted with methanol 70% (50 and 100 mg/ml) and kept in refrigerator in a dark well tight bottles.

Tested microorganisms

Seven reference bacterial strains representing the gram negatives (*Proteus vulgaris* NCTC 8196, *Escherichia coli* ATCC 25922, *Pseudomonas aeruginosa* ATCC 27853, *Klebsiella pneumonia* ATCC 53651 and *Salmonella typhi* NCTC 0650) and gram positives (*Staphylococcus aureus* ATCC 25923 and *Bacillus cereus* NCTC 8236,), and one reference fungal strain (*Candida albicans* ATCC 7596) were used in this study.

Phytochemical screening

Some phytochemical compounds claimed to have many bioactive properties were investigated; those compounds were tannins, saponins, flavonoids, terpenoids, phenolic compounds, alkaloids and anthraquinones (Edeoga et al., 2005; Krishnaiah et al., 2009; Abdallah et al., 2009).

Determination of antimicrobial activity

The antimicrobial activity of the methanol extract of leaves of *E. spinosa* and fruits of *C. colocynthis* were investigated using agar-

well diffusion method as mentioned by Güven et al. (2006), with minor modifications. 15 ml of molten Mueller-Hinton for bacterial isolates or potato dextrose agar (PDA) for fungal isolates (Oxoid Ltd, UK) was poured in a sterile Petri-dish and left until it solidied. Fresh cell suspensions were prepared and adjusted to 0.5 McFarland's standard. Then, 100 µl was spread onto the surface of the plates of Mueller-Hinton agar or potatoes dextrose agar. 6 mm wells were punched into the agar with a sterile cork borer. The methanol extracts of leaves of *E.spinosa* and fruits of *C. colocynthis* were dissolved in 70% methanol to a final concentration of 50 and 100 mg/ml. 80 µl (8000 µg/well) from each concentration was loaded into the wells and incubated for 24 h at 37°C for bacterial strains and 72 h at 28°C for fungal strain. Reference antibiotics, chloramphenicol (5 mg/ml, 50 µg/wells) and clotrimazole (5 mg/ml, 100 µg/wells) were used as antibacterial and antifungal agents, respectively. Tests were repeated two times and the mean inhibition zone was recorded.

1,1-Diphenyl, 2-picryl hydrazyl (DPPH) radical scavenging activity

Radical scavenging activity of tested extracts was evaluated using DPPH as a reagent as reported by Kirby and Schmidt (1997) with minor modifications. 1 ml of a 4 % (w/v) solution of DPPH radical in methanol was mixed with 500 µl of sample solutions in ethanol at different concentrations. Then, the mixture was incubated for 20 min in the dark at room temperature. The scavenging capacity was determined spectrophotometrically by monitoring the decrease in absorbance at 517 nm against a blank using a spectrophotometer (Bio-Rad Smart SpecTM plus). Lower absorbance of the reaction mixture indicated a higher free radical scavenging activity. Ascorbic acid was used as positive control. The percent DPPH scavenging effect was calculated using the following equation:

DPPH scavenging effect (%) = (A $_{control}$- A $_{sample}$ / A $_{control}$) ×100

Where, A$_{control}$ is the absorbance of the control reaction containing all reagents except the tested compound and A$_{sample}$ is the absorbance of the tested compound.

The extract concentrations providing 50% inhibition (IC$_{50}$) was calculated from the graph plotting inhibition percentage against the extract concentration. Tests were repeated three times.

In vivo antioxidant properties

This test was performed as described by Parthasarathy et al. (2006) with some modifications. Male albino *Wistar* rats (weight of 200-220 g) were used in this study. The animals were purchased from the Central Pharmacy of Tunisia (SIPHAT, Tunisia). They were housed at standard conditions and fed with a commercial balanced diet (SICO, Sfax, Tunisia). The drinking water was offered *ad libitum*. Our Institutional animal Care Committee approved the protocols for the animal study; the animals were cared for in accordance with the institutional ethical guidelines.

After two weeks of acclimatization, the rats were allocated ran-domly to four experimental groups of eight animals each with free access to food and water. Based on the preliminary experiments, the hepato protective dose of the methanolic extract of leaves of *E. spinosa* and fruits of *C. colocynthis* were decided. In multiple dose pretreatment experiment, methanolic extract was administered at 250 mg/kg bw by intraperitoneal injection. Group I (control rats) received the vehicle (olive oil, 1 mL/kg orally) at day 8; Group II received CCl$_4$ in olive oil (1 mL/kg, i.p) at day 8; Group III received the methanol extract of tested plant (250 mg/kg BW) daily by i.p injection for 8 days followed by a single dose of CCl4 in olive oil at a dose of 1 ml/kg using an intragastric tube twenty-four hours after

Table 1. The phytochemical analysis of leaves of *Emex spinosa* and fruits of *Citrillus colocynthis*.

Plant	Tannin	Saponin	Flavonoid	Terpenoid	Phenolic compound	Alkaloid	Anthraquinone
Emex spinosa	–	–	+	–	–	+	+
Citrillus colocynthis	±	+	+	+	–	+	–
Negative control (D.W.)	–	–	–	–	–	–	–

+, Presence; –, absence; ±, weak positive reaction; D.W., distilled water.

Table 2. Antimicrobial activities of leaves of *Emex spinosa* and fruits of *Citrillus colocynthis*.

Test *	Concentration (mg/ml)	Mean zone of inhibition (mm) of microorganisms (Mean ± SEM)**							
		Pr	Ec	Bc	Sal	Kp	Ps	Sa	Cand
Emex spinosa	50	6.0 ± 0.0	6.0 ± 0.0	6.0 ± 0.0	6.0 ± 0.0	6.0 ± 0.0	6.0 ± 0.0	6.0 ± 0.0	12.0 ± 1.0
	100	6.0 ± 0.0	6.0 ± 0.0	6.0 ± 0.0	6.0 ± 0.0	6.0 ± 0.0	6.0 ± 0.0	6.0 ± 0.0	14.5 ± 0.5
Citrillus colocynthis	50	8.5 ± 0.5	19.5 ± 1.5	12.5 ± 0.5	13.0 ± 1.0	8.25 ± 0.25	17.0 ± 0.0	10.75 ± 0.75	6.0 ± 0.0
	100	9.0 ± 1.0	22.5 ± 0.5	16.5 ± 1.5	17.0 ± 0.0	10.0 ± 1.0	22.0 ± 2.0	15.0 ± 0.0	9.0 ± 0.5
Chloramphenicol	5	26.0 ± 1.0	13.7 ± 1.05	21.0 ± 1.0	13.0 ± 1.0	22.0 ± 0.2	24.0 ± 1.0	16.5 ± 1.5	–
Clotrimazole	10	–	–	–	–	–	–	–	13.0 ± 1.0

*Plants tested are as methanol extracts at 100 mg/ml; chloramphenicol as antibacterial at 5 mg/ml and clotrimazole as antifungal at 10 g/ml. **Mean ± standard error of means (SEM), mm=millimeter; Pr, *Proteus vulgaris* NCTC 8196; Ec, *Escherichia coli* ATCC 25922; Bc, *Bacillus cereus* NCTC 8236; Sal, *Salmonella typhi* NCTC 0650; Kp, *Klebsiella pneumonia* ATCC 53651; Ps, *Pseudomonas aeruginosa* ATCC 27853; Sa, *Staphylococcus aureus* ATCC 25923; C and, *Candida albicans* ATCC 7596.

the last dosing; Group IV received methanol extract (250 mg/kg BW) daily by i.p injection for 8 days.

The animals were killed on day 9 by cervical decapitation. Blood samples were collected, allowed to clot at room temperature and serum separated by centrifuging at 4000 r.p.m. for 15 min for various biochemical parameters. The liver and the kidney were quickly excised, minced with ice cold saline, blotted on filter paper and homogenized (Ultra Turrax T25, Germany) (1:2, w/v) in 50 mmol/l phosphate buffer (pH 7.4). The supernatant and serum were frozen at -30°C in aliquots until analysis.

Blood samples were collected, allowed to clot at room temperature and serum separated by centrifuging at 4000 rpm for 15 min for various biochemical parameters. The supernatant and serum were frozen at -30°C in aliquots until analysis. The levels of serum alanine aminotransferase (ALT), aspartate aminotransferase (AST), alkaline phosphate (ALP), lactate dehydrogenase (LDH), γ-Gluta-

myl transpeptidase (γ-GT), urea and creatinine were measured using Chemistry Automatic analyzer 911, with the appropriate reagents purchased from Hitachi LTD., Japan.

RESULTS AND DISCUSSION

The phytochemical testing on the studied plant extracts revealed presence of many compounds of biological activities as presented in Table 1. Leaves of *E. spinosa* showed presence of alkaloids, flavonoids and anthraquinones. Similar study confirms the current results (Rizk, 1986). Fruits of *C. colocynthis* revealed presence of saponin, flavonoids, terpenoids and alkaloids.

This is in partial agreement with the results of Rodge and Biradar (2012) who found that *C. colocynthis* growing in India consists of tannins, saponins, flavanoids, terpenoids, alkaloids, steroids and cardic glycoloids. The current study did not detect tannins. Variation of season of harvesting or different geographical localities may have effect on the production of some phytochemical compounds. Such phytochemical compounds are interesting, as they are claimed to have many biological activities on animals and human including antioxidant, anti-inflammatory and anti-cholinesterase effects (Loizzo et al., 2007), in addition to antimicrobial properties (Cowan, 1999).

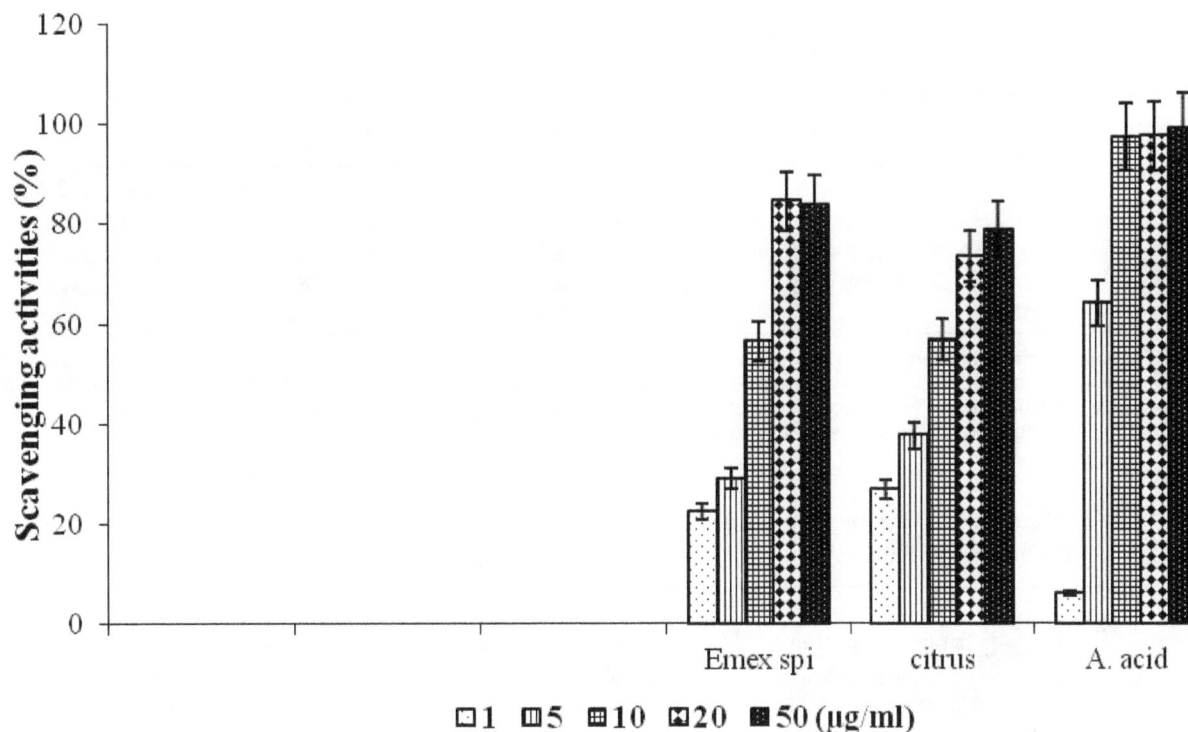

Figure 1. The DPPH free radical-scavenging activity of leaves of *Emex spinosa* and fruits of *Citrillus colocynthis* at different concentrations. A. acid, Ascorbic acid; Emex spi, *Emex spinosa,* Citrus, *Citrillus colocynthis.* Each value represents the mean ± SD of three experiments.

Table 2 shows the antimicrobial testing results of methanol extracts. It is clear that leaves of *E. spinosa* did not show any antibacterial activity against tested bacteria, while it revealed a considerable antifungal activity against *C. albicans* ATCC 7596 particularly at concentration 100 mg/ml, which was 14.5 ± 0.5 mm, higher than clotrimazole (10 mg/ml) (13.0 ± 1.0 mm). This interesting result requires more investigation on the chemical constituents of leaves of *E. spinosa* in order to categorize the active antifungal compound(s). However, different result was published by Abd El-kader et al. (2006) who reported that ethyl acetate extract of areal parts of *E. spinosa* from Egypt showed weak antifungal and variable antibacterial activity. Fruits of *C. colocynthis* showed weak antifungal activity and significant antibacterial activity at 100 mg/ml, the highest susceptible bacterium was *E. coli* ATCC 25922 (22.5 ± 0.5 mm), followed by *P. aeruginosa* ATCC 27853 (22.0 ± 2.0 mm), *S. typhi* NCTC 0650 (17.0 ± 0.0 mm), *B. cereus* NCTC 8236 (16.5 ± 1.5 mm), *S. aureus* ATCC 25923 (15.0 ± 0.0 mm), *K. pneumonia* ATCC 53651 (10.0 ± 1.0 mm) and the least susceptible bacterium was *P. vulgaris* NCTC 8196 (9.0 ± 1.0) compared to the antibiotic chloramphenicol 5 mg/ml (Table 2). Similar study was conducted by Rodge and Biradar (2012) who mentioned that the methanol, ethanol and water fruit extracts of *C. colocynthis* showed signi-ficant antibacterial and antifungal activities.

Figure 1 presents the *in vitro* antioxidant activities of

leaves of *E. spinosa* and fruits of *C. colocynthis.* As shown in the figure, the tested plant extracts and the standard for the *in vitro* antioxidant activity using DPPH method revealed significant antioxidant activity. The DPPH radical is used for assessment of the free radical-scavenging activity in plant extracts (Brand-Williams et al., 1995). These antioxidants generally include flavonoids, tannins, phenolic compounds, coumarins, anthocyans and essential oils (Ennajar et al., 2009; Bettaieb et al., 2010).

Table 3 shows the effects of CCl_4, the methanol extracts of examined plants and their combination (EACs /CCl_4) on hepatic markers in serum of control and experimental rats. The positive effects of these extracts on serum parameters are clearly seen. Table 4 reveals the effects of CCl_4, the methanolic extract of tested plants and their combination on the activities of enzymatic antioxidants in liver of control and experimental rats, which possesses significant protective effect against hepatotoxicity induced by CCl_4. Moreover, the phytochemical analysis of the studied plants revealed presence of some compounds known to have hepatoprotective activities such as alkaloids, flavonoids and phenolic compounds (Ranawat et al., 2010).

Conclusion

The medicinal plants mentioned in folk medicine are rich

Table 3. Effects of CCl$_4$, the methanolic extract *of* studied plants and their combination (Emex/CCl4 and Citrillus/CCl4) on hepatic markers in serum of control and experimental rats

Enzyme (U/L)	Experimental group			
	C	CCl4	Emex/CCl4	Citrillus/CCl4
AST	61.7 ± 2.5	125 ± 2.5	55 ± 2.4	65.15 ± 3.4
ALT	16.5 ± 1.25	34.5 ± 4.5	20.1 ± 4.4	18.7 ± 2.4
ALP	32.2 ± 0.75	44.5 ± 2.3	33.3 ± 1.5	31.2 ± 1.9
LDH	450 ± 45	520 ± 45	525 ± 46	523 ± 47
γGT	1.25 ± 0.3	2.23 ± 0.15	1.5 ± 0.25	1.84 ± 0.20

*AST, Aspartate aminotransferase; ALT, alanine aminotransferase; ALP, alkaline phosphatase; LDH, lacatate dehydrogenase; γGT, gamma glutamyl transferase. Values are mean ± SEM for eight rats in each group; CCl4, rat group treated with carbon tetrachloride; Emex/CCl4, rat group treated with a combination of Emex spinosamethanol extract and carbon tetrachloride,Citrillus/CCl4, rat group treated with a combination of Citrillus methanol extract and carbon tetrachloride.

Table 4. Effects of CCl$_4$, the methanol extracts and their combination on the activities of enzymatic antioxidants in liver of control and experimental rats.

Treatment	SOD (Units/mg protein)	CAT (µmol H$_2$O$_2$/mg protein)	GPx (µmol GSH/min/mg protein)
C	9.1 ± 1.15	267.5 ± 8.4	5.2 ± 0.25
CCl$_4$	6.65 ± 0.25	96 ± 1.33	3.3 ± 0. 45
Emex/CCl$_4$	7.96 ± 1.25	243.25 ± 14.5	4.4 ± 0.3
Citrus/CCl$_4$	8.82 ± 1.4	244 ± 3.15	4.3 ± 0.828

Emex, *Emex spinosa;* Citrus, *Citrillus colocynthis.* C, control group; CCL4, rat group treated with carbon tetrachloride; Emex/CCl4, rat group treated with a combination of Emex spinosamethanol extract and carbon tetrachloride,Citrus/CCl4, rat group treated with a combination of Citrillus colocynthis methanol extract and carbon tetrachloride. Values are mean ± SEM for eight rats in each group.

sources for new therapeutics. Leaves of *E. spinosa* revealed significant antifungal activities and fruits of *C. colocynthis* showed significant antibacterial activities. Both plants are rich sources for antioxidant compounds. These promising plants require more interest and deep research in order to explore their medicinal properties.

ACKNOWLEDGEMENT

We wish to thank the Deanship of Scientific Research, Qassim University, Saudi Arabia, for the financial support, grant No. SR-D-1787.

REFERENCES

Abdallah EM, Khalid AS, Ibrahim N (2009). Antibacterial activity of oleogum resins of *Commiphora molmol* and *Boswellia papyrifera* against methicillin resistant *Staphylococcus aureus* (MRSA). Sci. Res. Essay. 4(4):351–356.

Abdallah EM, El-Ghazali GE (2013). Screening for antimicrobial activity of some plants from Saudi Folk medicine. Global J. Res. Med. Plants Indig. Med. 2(4):210-218.

Abd El-kader AM, Abd El- Mawla AMA, Mohamed MH, Ibrahim ZZ (2006). Phytochemical and biological studies of *Emex spinosa* (L.) Campd. growing in Egypt. Bull. Pharma. Sci. Assiut University 29(2):328-347.

Anonymous (1970). Hamdard pharmacopoeia of Eastern Medicine, Hamdard National foundation, Pakistan 2nd Impression. p. 373.

Bettaieb I, Bourgou S, Wannes WA, Hamrouni I, Limam F, Marzouk (2010). Essential oils, phenolics, and antioxidant activities of differen parts of cumin (*Cuminum cyminum* L.). J. Agric. Food Chem 58:10410-10418.

Brand-Williams W, Cuvelier ME, Berset C (1995). Use of a free radica method to evaluate antioxidant activity. Lebensmittel-Wissenscha und Technologie. 28: 25-30.

Cowan MM (1999). Plant Products as Antimicrobial Agents. Clir Microbiol. Rev. 12(4):564-582.

Edeoga HO, Okwu DE, Mbaebie BO (2005). Phytochemical con stituents of some Nigerian medicinal Plants. Afr. J. Biotechno 4(7):685–688.

El-Ghazali GE, Al-Khalifa KS, Saleem GA, Abdallah EM (2010) Traditional medicinal plants indigenous to Al-Rass province, Sauc Arabia. J. Med. Plants Res. 4(24): 2680-2683.

Ennajar M, Bouajila J, Lebrihi A, Mathieu F, Abderraba M, Raies A Romdhane M (2009). Chemical composition and antimicrobial and antioxidant activities of essential oils and various extracts c *Juniperus phoenicea* L. (Cupressacees). J. Food Sci. 74:364-371.

Kirby AJ, Schmidt RJ (1997). The antioxidant activity of Chinese herb for eczema and of placebo herbs. J. Ethnopharma. 56(2):103-108.

Krishnaiah D, Devi T, Bono A, Sarbatly R (2009). Studies on phytochemical constituents of six Malaysian medicinal plants. J. Med Plants Res. 3(2):67–72.

Loizzo MR, Tundis R, Conforti F, Saab AM, Statti GA, Menichini F (2007). Composition and alpha-amylase inhibitory effect of essentia oils from Cedrus libani. Fitoterapia, 78:323-326.

Mandaville JP (1990). Flora of Eastern Saudi Arabia. Kegan Pau International, London. p. 482.

Parthasarathy NJ, Sri Kumar R, Manikandan S, Sheela Devi R (2006) Methanol induced oxidative stress in rat lymphoid organs. J Occupational Health. 48(1):20-27.

Ranawat L, Bhatt J, Patel J (2010). Hepatoprotective activity of ethanolic extracts of bark of *Zanthoxylum armatum* DC in CCl₄ induced hepatic damage in rats. J. Ethnopharma. 127(3):777–780.

Rizk AM (1986). The phytochemistry of the flora of Qatar, Scientific and Applied Research Center, Qatar Univ. Qatar. p.318.

Rodge SV, Biradar SD (2012). Preliminary Phytochemical screening and antimicrobial activity of *Citrullus colocynthis* (Linn.) Shared, Indian J. Plant Sci. 2(1):19-23.

Samie A, Obi CL, Bessong PO, Namrita L (2005). Activity profiles of fourteen selected medicinal plants from Rural Venda communities in South Africa against fifteen clinical bacterial species. Afri. J. Biotechnol. 4(12):1443–1451.

Watt JM, Breyer-Brandwijk MG (1962). The Medicinal and poisonous plants of southern and eastern Africa, 2nd ed., E & S Levingstone, Ltd., Edinburgh, London.

21

Emerging *Acinetobacter schindleri* in red eye infection of *Pangasius sutchi*

M. Radha Krishna REDDY[1] and S. A. MASTAN[2]

[1]Department of Biotechnology, Krishna University, Machilipatnam-521001, Andhra Pradesh, India.
[2]P.G Department of Biotechnology, PG Courses, Research Center, DNR College, Bhimavaram- 534202, Andhra Pradesh, India.

This communication provides an insight into the emerging of new infection "red eye" in *Pangasius sutchi* and aimed to screen the prime pathogens involved in disease. The pathogen was isolated from diseased *P. sutchi* and characterized by morphological, biochemical and molecular approach, which includes 16s r RNA gene sequencing. Polymerase chain reaction (PCR) amplified 16s RNA was separated using agarose gel electrophoresis, eluted product was sequenced and BLAST analysis was carried out to identify the pathogens. Identified virulent bacterial strain *Acinetobacter schindleri* with LD_{50} $10^{8.35}$ initiated re-infection in experimentally in infected *Pangasius* fingerlings. This study provided the evidence of *A. schindleri* which is true causative agents in red eye disease in *P. sutchi*. To the best of knowledge of this study, there was no track record of *A. schindleri* eye infection in fishes till date around the globe.

Key words: *Pangasius sutchi*, 16s r- RNA gene sequencing, *Acinetobacter schindleri*, LD_{50}.

INTRODUCTION

Pangasius sutchi is the exotic fish introduced in India from Thailand because of its high commercial value. The farmers of Janardhanapuram, Nandivada (Md), Krishna (Dist), Andhra Pradesh, culturing *Pangasius* in fresh water as intensive, monoculture with stock density of 50,000 per hectare, fed with floating feed having 20 to 23% protein, feeding rate up to 1.2 to 1.6% 10^3 kg body mass of fishes. *P. sutchi* is highly resistant species, and it is voracious feeder shown good food conversion rates (FCR) and give maximum sustainable yields (MSY) in short period. Nutrient rich feeds leave higher concentrations of ammonia and nitrite in culture waters and these stresses the fish in enormous rate and make them susceptible to different diseases. Culture waters with high organic matter pollute not only the tank, but also surroundings, and support the growth of many pathogens. For *Pangasius,* the most important bacterial diseases are bacillary necrosis; red spot have been reported by (Tu et al., 2008). These pollution problems may support the growth of *Acinetobacter* members in culture waters.

Acinetobacter members are found in water and act as common flora (José Américo, 2001; Marian, 1990). The alimentary tract of fresh water trout has *Acinetobacter* members (Trust, 1974). A significant increase in the microbial load of *Acinetobacter* members in ponds treated with different chemotherapeutics has been reported by Andreas Petersen (2002) and their entrance into culture waters along with contaminated feed has been reported by Trevors et al. (1977). Although *Acinetobacter spp* acts as a severe human pathogen, there are only few studies to date that report it as a pathogen for fish. The genus *Acinetobacter* show wide range of distribution, recovered from soil, water, living organisms. The bacteria very quickly became important member in bacteria landscape in hospitals, responsible for number of nosocomial infections in humans like surgical wound, urinary tract, respiratory tract (Wolff et al., 1997), pneumonia, secondary meningitis (Bukhary et al., 2005), endocarditis (Levi and Rubinstein, 1996), peritonitis, skin and soft tissue infections (Fierobe et al., 2001). Kalidas Rit and Rajdeep (2012) have reported *Acinetobacter* sp and their member's cause nosocomial infections and susceptibility patterns for different antibiotics. The *Acinetobacter* members show resistance to wide range of antibiotics like ampicillin, carbapenems (Mussi et al., 2005), carbenicillin, cephalosporin's (Heritier et al., 2006), amino glycosides, fluoroquinolones (Vila et al., 1993), carboxy pencillins (Joly and Guillov et al., 1995). They produce a wide range of amino glycoside inactivating enzymes (Buisson et al., 1990). Nemec (2001) reported *Acinetobacter schindleri* infections in human nosocomial infections, and Bouvet and Grimont (1986) firstly reported *A. haemolyticus* infections in humans. Emerging of new multi drug resistant bacterial pathogen, *Acinetobacter baumannii* associated with snake head *Channa striatus* eye infection has been reported by Rauta et al. (2011). The present study aimed to identify pathogens at molecular level from diseased *P. sutchi* suffering from red eye infection and to be proved as primary agents in disease.

MATERIALS AND METHODS

Collection of water and diseased fish samples

Diseased moribund fish samples (10) were collected from above said locality ponds, and brought to the laboratory. The fishes show different symptoms like gill impairment, erythro dermatitis, petechiae at lateral line, red mouth, redness at fin bases, swollen red colour anus, pop eye, red arched region around eye, swollen enlarged liver in light yellow colour, shrunken gastro intestinal tract and spleen, and hemorrhages on internal body cavity. Three water samples were collected from sequential days of 15 for 45 days to be checked the parameters like water temperature, pH, ammonia, nitrite, calcium, magnesium, alkalinity, hardness, chlorides, total dissolved solids, conductivity, and dissolved oxygen (APHA, 1988).

Isolation and identification of bacteria

A loop full of sample was collected with the help of inoculation loop from eye transferred on to Rimler Shots agar medium (Hi media, Mumbai). The plates were incubated at 37°C for 24 h. The nature of the cell wall of isolate was tested by gram staining method. For further differentiation, the culture was tested for biochemical characteristics with the Enterobacteriaceae kit (Hi media, Mumbai) as per manufacturer instructions. Later organisms were subjected for molecular characterization, to differentiate organisms up to species level.

DNA extraction

Extraction of genomic DNA and polymerase chain reaction (PCR) mediated amplification of the 16s r RNA gene of bacterial strain was carried out as per the method described by Neal Stewart et al. (1993). DNA from saturated bacteria liquid cultures was extracted by above said methodology, includes collection of bacterial cell pellet by centrifugation, lysis of cell pellet were attained by suspending in TE buffer with 100 µg of proteinase K and 0.5% sodium dodecyl sulfate (SDS) final concentrations. After 1 h of incubation at 37°C the lysate was treated with 80 µl of 5 M NaCl and 100 µl of 10% cetyl trimethyl ammonium bromide (CTAB) solution. Cell lysate was incubated at 60°C for 10 min. Degraded proteins from the cell lysate were removed by precipitation with phenol, phenol\ chloroform and chloroform treatment, respectively. Followed by protein precipitation, bacterial genomic DNA was recovered from the resulting supernatant by iso-propanol precipitation. Precipitated DNA pellet was washed with 70% alcohol for removal of salts. The DNA pellet was allowed for air drying and re suspended in 50 µl of deionized water with 1 µl 10 mg ml⁻L RNA ase A enzyme for the removal of RNA. Quality of the isolated DNA was analyzed by resolving on 1% agarose gel electrophoresis with 1X TAE buffer.

PCR amplification

The variable V3 region of DNA coding for 16s RNA was amplified by PCR with primers F- 5'- AGAGTTTGATCCTGGCTCAG –3' and R-5'- GGTTACCTTGTTACGACTT–3'. All the PCR amplifications were conducted in 50 µl volume containing 2 µl of total DNA having 54 ng per µl concentration, 200 M each of the four de oxy nucleotide tri phosphates, 1.5 µl $MgCl_2$, 5 µl of individual primers and 1 IU of Taq polymerase. The PCR amplification, used for gene amplification was consisted of initial denaturation at 95°C for 3 min, followed by 39 cycles of denaturation for 1 min at 95°C, annealing for 30 s at 56°C, and extension for 1 min at 72°C and a final extension at 72°C for 10 min. Finally, the amplified PCR product was stored at 4°C. The samples were verified on 1% agarose gel (Lonza, USA) to know Ribo print pattern. The separated bands were excised from the gel (Figure 1) by using surgical blade for elution of DNA. The elution of DNA from agarose gel was carried out as per manufacturer instructions (Real Biotech DNA/PCR purification kit CAT NO 36105).

DNA sequence and phylogenetic analysis

For sequencing analysis, amplified PCR product was sent to EUROFIN Company. All the 16s r RNA partial sequence were aligned with those of the reference micro organisms in the same region of the closet relative strains available in the Gen Bank data base by using the BLAST N facility (http://www.ncbi.nlm.nih.gov\BLAST) and were also tested for possible chimera formation with the CHECK CHIMERA program(http://www.35.8.164.52\cgis\chimera.cgi? Su: SSU). The sequences were further analyzed by using Clustal Omega (www.ebi.ac.uk/Tools/msa/clustalo/). Neighbor joining phylogenetic tree (Figure 2) was constructed with the Molecular Evolutionary Genetic Analysis Package (MEGA VERSION 5.1) (Tamura K et al., 2011). A boot strap analysis with 500 replicates was carried out to check the robustness of the tree. Boot strap re-sampling analysis, for the replicates was performed to estimate the confidence of the tree topologies.

Artificial challenge studies

Bacterial suspension was prepared by culturing the isolates on trypticase soy agar (TSA) plates at 30°C for 24 h and harvesting them with 50 ml of 0.85% physiological saline. Colony forming unit (CFU) per mL of this solution was determined by plating 10 fold dilu-

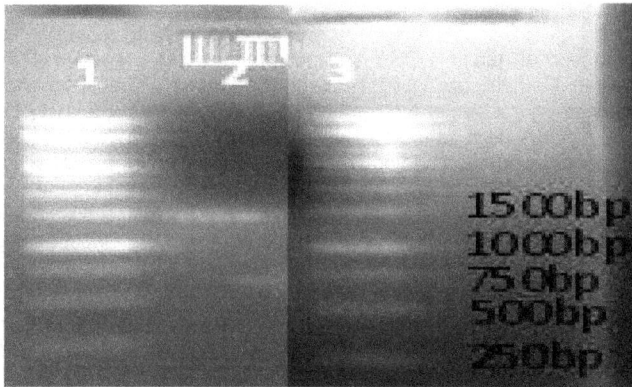

Figure 1. Ribo print pattern of isolate DNA on agarose gel.

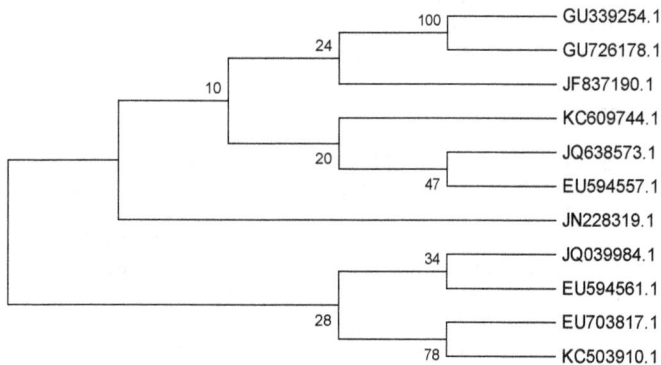

Figure 2. Neighbor-joining tree constructed using Mega 5.1 showing phylogenetic relationships of 16s RNA sequences from diseases fish to closely related sequences from Gen Bank.

Table 1. Physico- chemical parameters of water.

Parameter	Mean value+ Sd
Water temperature	28.5+0.5°C
pH	8.2+0.458
Ammonia	1+0.416 mg\L
Nitrite	0.8+0.2 mg\L
Calcium	54.6+7.02 mg\L
Magnesium	91.3+6.42 mg\L
Alkalinity	473.3+30.55 mg\L
Hardness	155+5.56 mg\L
Chlorides	175.6+4.5 mg\L
TDS	1710+36 mg\L
Conductivity	1.1457+0.024 ms\cm^2
DO	1.76+0.25 mg\L

dilution series. For this purpose, the solution was diluted with distilled water. Apparently active healthy, fingerlings of *P. sutchi* (50+ 10 g) were taken from the fish farm of Kaikalur, AP, India. They were stocked in 500 L cement tanks filled with fresh water and acclimatized in the laboratory condition for two weeks before starting the experiment. They were fed with standard diet in 2 divi-

ded doses daily during the experiment. Water was exchange partially to remove left out feed and fecal matter. The lethal dos LD_{50} of the isolate was estimated according to Reed and Muenc (1938). Five groups (Group1- 5) with 6 fish in each group were cha lenged with a series of dilutions of bacteria. The bacterial susper sion prepared in phosphate buffered saline (0.15 M, pH 7.4) wa injected to each fish intraperitoneally with 0.1 ml of different dilu tions of bacteria. The final concentration of the bacteria injected t each was 10^5 - 10^8 CFU/mL. Control fish was injected with 0.1 m phosphate buffered saline without bacteria. Mortality was observe till 5 days, and pathogenicity was confirmed by re-isolating th bacteria from experimentally infected fishes.

RESULTS AND DISCUSSION

The result of physico-chemical parameters of waters ar presented in Table 1. Water chemistry results indicate that variation in ranges of pH, ammonia, nitrite and tota dissolved solids, alkalinity, biological oxygen deman (BOD) and chemical oxygen demand (COD) show grea impact on aquatic biota including fish. Inoculated fis isolates on RS medium resulted in green colour colonie are 1.5 to 2.5 mm in diameter, circular, convex, smoot and slightly opaque with entire margins. The nature of the cell wall composition of isolate was tested by gran staining method; the results confirmed the organisms a: gram negative rods. The green colour colonies are positiv for ONPG, lysine, ornithine, citrate, H_2S, voge: proskaver, melonate, trehalose and negative for urease phenyl alanine, nitrate reduction, methyl red, indole esculin, melibiose, and glucose, respectively. The isolate showed good growth on nutrient agar, brain heart infu sion agar, and tryptone soya agar. The isolate grew bes in the temperature range of 30 to 37°C and pH 6 to 8 Biochemical results of green colonies are given in the Table 2. Sequencing analysis revealed a 100% identity with the sequence corresponding to the 16s r RNA gene of *A. schindleri YNB 103 strain* (Gen Bank accessior number JQ 039984.1). Experimental infection stud confirmed the pathogenicity of *A. schindleri* to *P. sutch* The LD_{50} of *A. schindleri* $10^{8.35}$ CFU per fish, which indi cates the isolated strain, was highly virulent and capabl of causing re-infection in *P. sutchi* and cause death ir experimentally infected *Pangasius* fingerlings and showec similar signs even in the collected fishes from the tank outbreak (Table 3).

For effective cultivation of the fish, good quality water is needed; due to lack of sustainable management prac tices in water quality, fishes are prone to stress and sus ceptible to different diseases. All living organisms have optimum range of pH where growth is best. Water witr high alkalinity not more buffered and the degree of pH fluctuation is high. Alkalinity changes can affect the pri mary productivity in cultured ponds. Dissolved oxygen is not at all problem to *P. sutchi* because it is air breathinc fish. Elevated levels of ammonia causes gill damage anc reduce the growth of fishes. Water temperature show direct impact on metabolism, feeding rates, respiratory rates of aquatic biota, and influence the solubility of oxy-

Table 2. Physical and biochemical characteristics of *A. schindleri* YNB 103.

Character	*A. schindleri*
Colony colour	Green
Gram reaction	Negative
Shape (R/C)	Rod
Motility	Non motile
Growth at different temp (°C)	
20	Negative
25	Negative
30	Positive
35	Positive
42	Positive
Growth on different media	
Nutrient agar	Positive
BHIA	Positive
Rimler-Shots agar medium	Positive
Tryptone soy agar	Positive
Growth in NaCl (w/v)	
2	Negative
4	Negative
6	Positive
8	Positive
10	Negative
Oxidative/Fermentative	Oxidative
Acid-fast test	Negative
Oxidase reaction	Negative
ONPG	Positive
Lysine	Positive
Ornithine Decarboxylase	Positive
Urease	Negative
Phenylalanine	Negative
Nitrate reduction	Negative
H_2S	Positive
Citrate	Positive
VP	Positive
MR	Negative
Indole	Negative
Production of acid from	
Melonate	Positive
Esculin	Negative
Arabinose	Variable
Xylose	Variable
Adonitol	Variable
Rhamnose	Variable
Cellobiose	Variable
Melibiose	Variable
Saccharose	Variable
Raffinose	Variable
Trehalose	Positive
Glucose	Negative
Lactose	Variable

Table 3. Lethal dose value CFU per mL of *A. schindleri*.

Group	Log dose	Death	Survived	Death	Cumulative survival	Total	Mortality ratio	Mortality (%)	LD 50
Control PBS 0.1ml	0	0	6	0	14	14	0/14	0	
CFU 10^8 2.1	0.322	2	4	2	8	10	2/10	20	10^8 35 cfu/ml
CFU10^7 3.4	0.531	3	3	3	4	7	3/7	42	
CFU$10^6$4.2	0.623	5	1	5	1	6	5/6	83	
CFU10^5 5.4	0.732	6	0	6	0	6	6/6	100	

gen. Nitrite results from feed can disrupt the oxygen transport in live fishes. Hardness of culture waters depends on levels of calcium and magnesium.

High total dissolved solids value directly indicates the presence of organic matter in culture waters. Culture water with high organic matter not only pollutes the tank, but also surrounding areas and support growth of different pathogens like causative agents of fulminant sepsis of *P. sutchi*. Some feed companies using animal meets, in place of soya while making feed pellets, it may be the indirect reason for entry of hospital landscape organisms in to aqua culture settings. Ponds treated with variety of chemotherapeutics to control different diseases, also affect the normal flora of pond bottom, it is also another reason to develop multi drug resistant bugs in to culture waters. After observing the gross symptoms of fish we postulated that emerging of new bacterial member's involvement in disease. Artificial challenge studies determined that *the* isolate can become pathogenic to *P. sutchi*. Out of 5 groups, control group fishes were injected with phosphate buffered saline, no mortality was observed, up to the end of the experiment. CFU 10^5 5.4 group shows 100% mortality of fishes within 48 h. As per Reed and Muench formula LD$_{50}$ $10^{8.35}$ was determined for *P. sutchi*. To the best of knowledge, there was no track record of *A. schindleri* eye infection in fishes till date around the globe.

REFERENCES

Andreas PJSA, Tawatchai K, Temdoung S, Anders D (2002). Impact of integrated fish farming on antimicrobial resistance in a Pond environment Appl. Enviro. Microbiol. 68(12): 6036–6042.

APHA (1998) Standard methods for the examination of water and waste-water analysis. 20th (Ed) Washington D.C. 12-13 pp.

Buisson Y, Tran VNG, Ginot L, Bouvet P, Shill H, Driot L, Meyran M (1990). Nosocomial outbreaks due to amikacin-resistant tobramycin sensitive Acinetobacter species: correlation with amikacin usage. J. Hosp. Infect. 15: 83–93.

Bouvet PJM, Grimont PAD (1986) Taxonomy of the genus Acinetobacter with the recognition of *Acinetobacter baumannii* Sp. nov., *Acinetobacter hemolyticus* Sp. nov., *Acinetobacter johnsonii* Sp. nov and *Acinetobacter junii* Sp. Nov., emended descriptions of *Acinetobacter calcoaceticus* and *Acinetobacter Iwoffii*. Intl. J. Syst. Bacteriol. 36: 228-240.

Bukhary Z, Mahmood W, Al-Khani A, Al-Abdel HM (2005) Treatment of nosocomial meningitis due to a multidrug resistant *Acinetobacter baumannii* with intra ventricular colistin. Saudi Med. J. 26: 656–658.

Fierobe L, Lucet JC, Decre D, Muller-Serieys C, Joly GML, Mantz J, Desmonts JM (2001). An outbreak of imipenem resistant

Acinetobacter baumannii in critically ill surgical patients. Infect Control Hosp. Epidemiol. 22: 35–40.

Heritier C, Poirel L, Nordmann P (2006). Cephalosporinase ove expression as a result of insertion of ISAba1 in Acinetobacte baumannii. Clin. Microbiol. Infect. 12: 123–130.

Joly GML, Decre D, Herrman JL, Bourdelier E, Bergogne B (1995 Bactericidal in-vitro activity of b-lactams and beta lactamase inhibitors, alone or associated, against clinical strains o' Acinetobacter baumannii: effect of combination with aminoglycosides J. Anti microb. Chemother 36: 619–629.

José Américo de Sousa, Ângela Teresa Silva-Souza (2001) Bacteria Community Associated with Fish and Water from Congonhas River Sertaneja, and Paraná, Brazil Braz. arch. biol. technol. 44: 4

Kalidas R, Rajdeep S (2012) Multidrug-resistant acinetobacter infectior and their susceptibility patterns in a tertiary care hospital, Niger Med J. 53: 126-128.

Levi I, Rubinstein E (1996). *Acinetobacter* infection and overview of clinical features In Bergogne-Be´re´zin, E., Joly-Guillou, M.L., and Towner, K.J. (eds.) Acinetobacter –Microbiology, Epidemiology, Infection, Management. CRC Press, New-York, Chap. 5: 101–115.

Mussi MA, Limansky AS, Viale AM (2005). Acquisition of resistance to carbapenems in multi-drug-resistant clinical strains of Acinetobacte baumannii: natural insertional inactivation of a gene encoding a member of a novel family of b-barrel outer membrane proteins Antimicrob. Agent Chemother. 49: 1432–1440.

Marian M (1990). Cahill Bacterial flora of fishes. Rev. Microbial. Ecol 19: 21-41.

Nemec A, De BT, Tjernberg I, Vaneechoutee M, Van DTJ, Dijkshoorn L (2001). *Acinetobacter ursingii* Sp. Nov., and *Acinetobacter schindler* Sp. Nov., isolated from human clinical specimens. Intl. J. Syst. Evol. Microbiol. 51: 1891-1899.

Ree LJ, Muench H (1938) A simple method of estimating fifty percent end points. Am. J. Hyg. 27: 493-497.

Rauta PR, Kuldeep K, Sahoo PK (2011) emerging new multi drug resistant bacterial pathogen, *Acinetobacter baumannii* associated with snake head *Channa striatus* eye infection. Curr. Sci. 101 (4): 548-553.

Trust TJ, Sparrow RAH (1974), the bacterial flora in the alimentary tract of freshwater salmonid fishes. Canadian J. Microbiol. 20(9): 1219-1228, 10.1139/m74-188.

Tamura K, Daniel P, Nicholas P, Glen S, Masatoshi N, Sudhir K (2011) MEGA5 Molecular Evolutionary Genetics Analysis Using Maximum Likelihood, Evolutionary Distance, and Maximum Parsimony Methods Mol. Biol. Evol. 28(10), 2731–2739.

Tu TD, Nguyen TNN, Nguyen QT, Dang TMT, Nguyen AT (2008). Common diseases of pangasius catfish farmed in Vietnam (Health Management) http://pdf.gaalliance.org/pdf/GAA-Dung-x pdf.

Trevors JT, Van EJD (1977). Microbial interactions in soil. In: Van Elsar, J. D., Trevors, J. T. Wellington, E. M. H. (Eds.), Modern soil microbiology. Marcel Dekker, Inc., New York, NY, 215- 243.

Vila J, Marcos A, Marco F, Abdalla S, Bergara Y, Reig R, Gomez LR, Jimenez AT (1993). In vitro antimicrobial production of ß-lactamases, amino glycoside modifying enzymes, and chloramphenicol acetyl transferase by and susceptibility of clinical isolates of Acinetobacter baumannii. Anti microb. Agents Chemother. 37:138–141.

Wolff M, Brun BC, Lode H (1997) the changing epidemiology of severe infections in ICU. Clin. Microb. Infect. (suppl) 3: S36–S47.

22

Determination of heavy metals and genotoxicity of water from an artesian well in the city of Vazante-MG, Brazil

Regildo Márcio Gonçalves da Silva[1] , **Eni Aparecida do Amaral**[2], **Vanessa Marques de Oliveira Moraes**[1] **and Luciana Pereira Silva**[1]

[1]Universidade Estadual Paulista (UNESP), Departamento de Ciências Biológicas - Laboratório de Fitoterápicos, Faculdade de Ciências e Letras de Assis, Avenida Dom Antônio 2100, CEP: 19806-900, Assis, São Paulo, Brasil.
[2]Centro Universitário de Patos de Minas (UNIPAM), Laboratório de Química Instrumental e Central Analítica.

The city of Vazante-MG is of great socioeconomic and environmental interest because it is the most important zinc producer district of Brazil. The mineral processing and geochemical processes may determine high concentrations of heavy metals in water intended for human consumption. Thus, the present study aimed to quantify and evaluate the heavy metal genotoxicity of artesian water in the city by Atomic absorption spectrophotometer analysis and testing with the *Allium cepa* test, respectively. This study reveals a chemical contamination in well water in the city, caused by the presence of heavy metals. Therefore, it can be considered that the high levels of heavy metals found in water samples are correlated with the genotoxic events observed in root cells of *A. cepa*.

Key words: *Allium cepa*, micronucleus, atomic absorption, chromosome aberration, mitotic index.

INTRODUCTION

The Vazante-MG region is of great socioeconomic and environmental interest, since it is the most important zinc producer district of Brazil (Hitzman et al., 2003). The inability of differentiating geogenic anomalies from those that result from processes of contamination related to human studies, suggests the need of assays to evaluate the presence of heavy metals in soil and water in order to contribute to a better evaluation of the occurrence of contamination by these metals (Borges Júnior et al., 2008). Preliminary evaluation of a suspected area of contamination is performed based on information available (CETESB, 1999, 2001). The area is considered contaminated if the concentration of elements or substances of interest are above the given threshold, which indicates the potential deleterious effect on human and animal health (Junior Borges et al., 2008a).

The effects of mineral processing together with the geochemical processes that naturally occur in reason of the soil characteristics and the entrainment of heavy particles to the aquatic system may provide high concentrations of heavy metals in water intended for domestic consumption (Yabe et al., 1998;, Guedes et al., 2005). The presence of heavy metals in groundwater may occur due to contact with rivers and lakes contaminated with sewage or by leaching by precipitation of contaminated soils (Di Natale et al., 2008).

Heavy metals are among the most common inorganic pollutants in water (Chandra et al., 2005). They are highly distributed over the earth's crust (Arambasic et al., 1995; Min et al., 2013) and represent one of the most toxic environmental pollutants (Ghosh, 2005; Sharma, 2009). Intoxication with heavy metals has been observed in many parts of the world, usually related to chronic exposure in environment through contamination of drinking water (Hang et al., 2009; Singh and Kalamdhad, 2013). Epidemiological evidence has shown that a long-term exposure is highly associated with increased risk of development of several diseases, including cancers (Zhuang et al., 2009). *In vivo* and *in vitro* assays have shown that heavy metals induce chromosomal aberrations and micronucleus in plant and animal (Rank et al., 1998; Majer et al., 2002; Rodriguez-Cea et al., 2003). Therefore it became important to carry out the environmental monitoring of water intended for human consumption.

Only with the chemical analyzes of water, it is not possible to determine the ecotoxicological risk that chemicals present in it can cause to the bodies, since such analysis alone does not indicate toxicity (Fuentes et al., 2006). Therefore, ecotoxicity tests among them those of environmental mutagenesis, have been proposed and applied to understand the genetic and physiological responses of exposed organisms (White et al., 2004; Chen et al., 2004). Bioassays with plants, such as the *Allium cepa* test, have some advantages over the tests in mammalian cells and microorganisms for environmental monitoring (Grant, 1994; Radic et al., 2014; Osakca & Silah, 2012). Plant assays are highly sensitive to many environmental pollutants, including heavy metals (Fiskesjo, 1985; Smaka-Kincl et al., 1996; Steinkellner et al., 1998; Fatima et al., 2005; Yi et al., 2007; Egito et al., 2007; Pesnya and Romanovsky, 2013). The use of the *A. cepa* bioassay is suggested because it is known that many plants are damaged by heavy metal contamination (Minissi et al., 1997; Amaral et al., 2007; Smith, 2001). This study aimed to quantify and evaluate the heavy metal genotoxicity of artesian water in the city of Vazante-MG/Brazil by atomic absorption spectrophotometer analysis and testing of the *A. cepa,* respectively.

MATERIALS AND METHODS

Samples collection

Water samples were collected at Vazante - MG/Brazil (S 17° 59'27 "W and 46° 54'04", altitude, 638 m) in May 2007, directly in wells registered by the Companhia de Saneamento de Minas Gerais (COPASA). Two collection points were determined: Sample 1 (S1): Water from the borehole, without treatment; Sample 2 (S2): water from an artesian well with simplified treatment (disinfection and fluoridation) in accordance with the rules of COPASA.

Three samples of 5 liters were collected from each point considered, being collected at intervals of 10 min and stored in sterile flasks, totaling 15 l per point. The pH of the samples was measured in the field, using a manual pH meter (Lutron pH-208).

The samples were transported in an isothermal box to th Laboratory of Chemistry and Instrumental Analytical Center of th University Center of Patos de Minas - UNIPAM.

Determination of heavy metals

For the heavy metals analyses, from each sample 100 mL wa taken, 20 mL of nitric acid PA was added and then heated to eva porate until it remained 60 mL solution. After reaching room tempe rature, 40 mL of ultrapure water was added to obtain a final solutio of 100 ml of sample for testing. The reading of the heavy metals i water was performed in triplicate and measured by the atomi absorption spectrophotometer Perkin Elmer 3300. In the preser study, we analyzed the following metals: Cadmium, lead, coppe and zinc. The gases used for reading were acetylene and com pressed air in the flame analysis with hollow cathode lamp, procedure performed in accordance with Santos et al. (2006).

Allium cepa test

Treatment

The experiment was conducted at the Laboratory of Plar Physiology, Centro Universitário de Patos de Minas, Minas Gerai: and analyzed at the Laboratory of Herbal Medicines, Universidad Estadual Paulista, Assis-SP/ Brazil. The experimental protocol wa: essentially performed as described by Ma et al. (1995). Twelve (12 bulbs were exposed to water samples collected, six for sample collected in S1 and six for those collected in S2. Twelve (12) othe bulbs were destined to negative control groups (NC) and positiv control (PC), which were prepared using mineral water an methylmetanosulfonate (MMS) to 10 mg / L (MMS, Sigma-Aldrich® CAS 66-27 - 3), respectively.

Exposures were performed for a fixed period of 48 h for all treat ments, except for the PC group that was exposed for 6 h accordin to the methodology described by Fiskesjö (1988) and Majer (2003) After the exposure period of the roots, they were fixed in acetic aci and ethanol solution (1:3) for 24 h. After fixation, the roots wer transferred to a solution of 70% ethanol and kept in refrigerator a an average temperature of 4°C.

Determination of mitotic index, chromosome aberration an micronucleus

For preparation of the slides, roots were hydrolyzed in 1 N HCl a 60°C for 8 min and then stained with 2% solution of carmine in 45% acetic acid. The roots were then placed on a slide and the firs millimeter removed from the apex of the root, so that the meri stematic region corresponding to 2 mm and F1 cells were isolatec for analysis by optical microscope. 1000 cells were counted pe slide in an increase of 400 times, with 5 slides per treatment anc control, and mitotic division stages, aberrant anaphases anc telophases, and the frequency of micronucleus were quantified. The mitotic index was calculated according to the equation:

IM [%] = the number of dividing cells (1000 per slide) / number o cells analyzed x 100

For analysis of chromosomal aberrations (aberrant anaphases anc telophases) and micronucleus frequency were performed according to methods previously described by Grant (1982) and in accordance with adjustments made by Yildiz et al. (2009).

Determining the length of the root

After the period of exposure and collection of the roots, the

Table 1. Results of MI, CA (anaphase and telophase) and MN in *Allium cepa* meristematic cells and root end length after treatment with water samples.

Sampling	Mitotic index	% Aberration chromosome		Micronucleated cells (%)	Length root (mm)
		Anaphase	Telophase		
NC	12.68±0.79	0.86±0.02	0.53±0.07	0.112±0.003	48.24±2.27
S1	5.11±0.13[ab]	11.47±0.52[ab]	9.11±0.16[ab]	3.374±0.123[ab]	19.31±2.09[ab]
S2	9.14±0.57[ab]	7.94±0.37[a]	5.36±0.37[a]	2.658±0.017[a]	23.17±1,93[ab]
PC	13.46±1.17	5.06±0.46	4.23±0.16	1.680±0.038	33.42±2.46

5000 cells analyzed per treatment. Mean±S.D. [a], Significantly different from negative control ($p < 0.05$), according to Kruskal–Wallis test. [b], significantly different from positive control ($p < 0.05$), according to Kruskal–Wallis test.

measurement of the length in millimeter of the roots was performed with the help of a digital caliper (DIGIMESS®), having a total of 25 roots per treatment.

Statistical analysis

The mitotic index, frequencies of chromosomal aberrations and micronucleus obtained for each treatment during the period between exposure and the samples were compared with the controls and analyzed statistically using the Kruskal-Wallis test ($p < 0.05$), as described by Grisolia et al. (2005) and Rudder et al. (2008).

RESULTS AND DISCUSSION

The pH of samples S1 and S2 varied from a minimum of 6.55 (S1) to a maximum of 6.65 (S2), with an average of 6.60. In its resolution of CONAMA (2005), permitted range is 6.5 to 7.5, so all values remained in that range (Guedes et al., 2005).

Figure 1 shows the concentrations of cadmium, copper, lead and zinc found in the different water samples (S1 and S2) and the maximum tolerable in the environment of each metal recommended by the WHO (1998) and according to CONAMA (2005). Both samples showed high levels of all analyzed metals which exceed the maxi-mum amount indicated and recommended by the rele-vant authorities. The sample S2 showed lower values when compared to sample S1, but all metals exceeded the maximum tolerated. The cadmium concentration in the sample S2 was 0.045 mg/L, or 4400% above recom-mended levels, the copper concentration was 0.086 mg/L, 855% above the indicated concentration, zinc showed 0.195 mg/L, exceeding 8% maximum tolerable concentration and lead showed 1341mg/L, being the metal with the highest value of the S2 sample, exceeding 13310% the recommended maximum. In relation to the S1 sample, it showed values for cadmium in excess of 4900% than the recommended maximum value, the value of copper exceeded 1044%, zinc exceeded 15% and lead exceeded 29410%, being the last one the highest of all metals analyzed according to the indicated maximum values (Figure 1).

According to Raskin and Ensley (2001) and Andrade et al. (2009) increased levels of heavy metals may be associated with destruction of vegetation cover in mining areas which exacerbates soil degradation, promoting water and wind erosion and leaching of contaminants into groundwater, leading to progressive degree of contami-nation in other areas. As reported by Rigobello et al. (1988) and Borges Júnior et al. (2008b) the region of Vazante has high levels of zinc and lead, being the lar-gest zinc producer district of Brazil.

In recent decades, environmental contamination with heavy metals has risen dramatically. It is known that certain heavy metals can cause DNA damage and carcinogenic effects in animals and humans, and are probably, related to its mutagenic activity (Ernst, 2002; Arora et al., 2008; Megateli et al., 2009). According to Knasmüller et al. (2009), the standard tests used to detect heavy metals is currently problematic because several carcinogenic metals result in negative information on bacterial gene mutation assays and genotoxicity assays with mammalian cells, but tests performed in plant cells have become known for being a quick and useful test system in biomonitoring (Majer et al., 2005). The advantages of these tests include the similarity of the plants chromosomes organization with the human, its sensitivity to changes in environment (Grant, 1994) and the possibility of studying the effects on a wide range of environmental conditions. Repeated application of these tests to assess the genotoxic risks present in natural waters (rivers and lakes), wastewater and drinking water has affirmed its utility (Blagojevic et al., 2009).

Thus, this assay evaluated the genotoxic activity of different water samples by mean of the *A. cepa* test. Table 1 shows the results of the mitotic index (MI), chro-mosome aberrations (CA), micronucleus (MN) and root length of *A. cepa* after treatment with water samples. The sample S1 showed significant differences in relation to positive and negative controls in all parameters analyzed, and its average MI was the lowest (5.11) than controls, NC (12.68), PC (13.45) and the length of the roots of the sample (19.31 mm) also had decreased compared to controls: NC (48.24 mm), PC (33.42 mm). Since the percentage of MN and CA increased compared to con-trols, the frequency of MN of S1 was 3.374%, while the frequencies of the NC was 0.112 and 1.680% of the PC. The percentage of AC of S1 also increased, both for anaphase (11.47%) and telophase aberrant (9.11%),

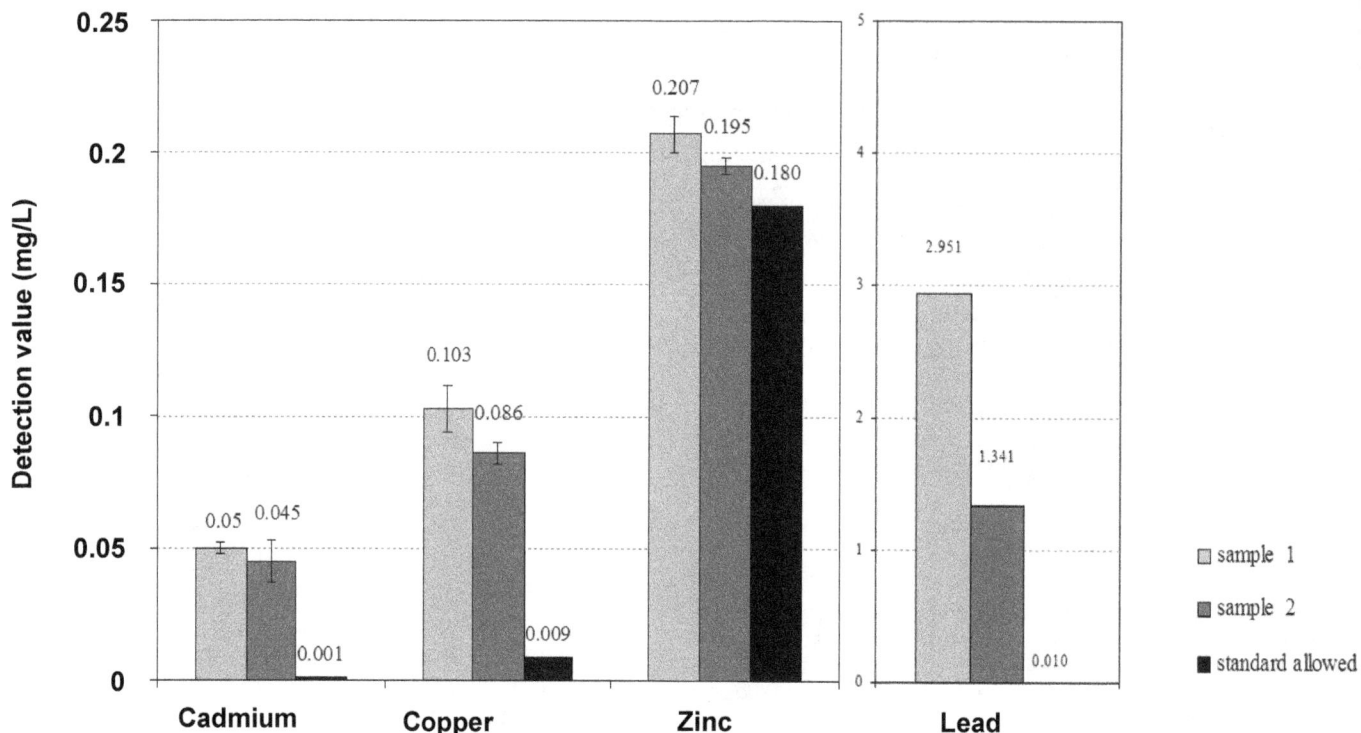

Figure 1. Concentration of heavy metals (cadmium, copper, lead and zinc) found in the water samples (S1 and S2) and maximum permitted under the Regulatory Determination CONAMA (2005) and recommended by WHO (1998).

while the negative and positive controls showed anaphase to 0.86 and 5.06% 0 and telophase, 53 and 4.23%, respectively.

The S2 differed from the positive and negative controls in relation to MI and length of roots. Both sample showed a decrease parameter, and the MI of the medium S2 is 9.14, while for the PC was 13.46 and the NC was 12.68. Values for the average length of the roots were 23.17 mm and S2 to the positive and negative controls were 33.42 and 48.24 mm, respectively. As for the parameters CA and MN percentage of the sample S2 showed no statistical difference in relation to the PC for both anaphase and telophase, but there were differences when compared to NC, and the percentages of both MN (2.658%) and AC (anaphase = 7.94% and telophase = 5.36%) increa-sed over the rate of MN, NC (0.112%) and the rate of CA of the same control (anaphase = 0.86% and telophase = 0.53%) (Table 1).

According to the results observed in this study, we consider that the high levels of heavy metals found in water samples are directly related to the genotoxic events observed in root cells of *A. cepa*. According to studies carried out by Seth et al. (2008), the high content of cadmium is associated with occurrence of chromosomal aberrations and increased frequency of micronuclei in the root of *A. cepa*. As shown by Glinska et al. (2007) and Ferraz et al. (2009), copper and its connections with macromolecules proved to be an effective cytotoxic agent and genotoxic in cell cultures and *in vivo* assays. Liu et

al. (1994 and 2003) and Seregin et al. (2004) showed that the lead and cadmium inhibited root growth as a result of disruption of cell cycle and Wierzbicka (1988, 1989 and 1999), and Samardakiewicz Woz'ny (2005) Fusconi et al. (2006) showed a decrease in the mitotic cells of the root, where this value was accompanied by reduction in the number of cells in metaphase and ana-phase. Furthermore, heavy metals, lead and cadmium induced c-mitosis, chromosomal adhesion and bridges, and besides that, lead also caused chromosome delay, nucleus with more condensed chromatin and inhibited cytokinesis.

Conclusion

The present study reveals a chemical contamination in artesian well water in the city of Vazante-MG caused by the presence of heavy metals. The decrease in mitotic index, reducing the average length of roots and increased frequency of chromosomal aberrations and micronucleus in root meristematic cells of *A. cepa* exposed to treatment may be correlated with the presence of certain heavy metals determined in our assay, as well as the interaction with other classes of environmental contaminants, which are probably the agents that together induced the genotoxicity observed in this assay. Taken together, these results show the importance of evaluating the genotoxicity of water wells in areas with mineral richness, especially in areas near active mining sites.

ACKNOWLEDGEMENTS

The authors acknowledge the technical and scientific support of Professor. Dr. Antônio Taranto Goulart Laboratory of Chemistry and Instrumental Analytical Center of the UNIPAM, Patos de Minas-MG/Brasil in analyzes of heavy metals.

REFERENCES

Amaral AM, Barbério A, Voltolini JC, Barros L (2007). Avaliação preliminar da citotoxicidade e genotoxicidade, da água da bacia do rio Tapanhon (SP- Brasil) através do teste *Allium (Allium cepa)*. Revista Brasileira de Toxicologia 20: 1-2, 65-72.

Andrade MG, Melo VFM, Gabardo J, Souza LCP, Reissmann CB (2009a). Metais pesados em solos de área de mineração e metalurgia de chumbo I – Fitoextração. Revista Brasileira Ciência do Solo 33: 1879-1888.

Andrade MG, Melo VFM, Gabardo J, Souza LCP, Reissmann CB (2009b). Metais pesados em solos de área de mineração e metalurgia de chumbo II - Formas e disponibilidade para plantas. Revista Brasileira Ciência do Solo 33:1889-1897.

Arambasic MB, Bjelic S, Subakov G (1995). Acute toxicity of heavy metals (copper, lead, Zinc), phenol and sodium on *Allium cepa* I., *Lepidium sativum* I. and *Daphnia magna* st.: Comparative investigations and the practical Applications. Water Res. 29:2, 497-503.

Arora M, Kiran B, Rani S, Rani A, Kaur B, Mittal N (2008). Heavy metal accumulation in vegetables irrigated with water from different sources. Food Chem. 111:811-815.

Blagojevic J, Stamenkovic G, Vujoševic Mladen (2009). Potential genotoxic effects of melted snow from an urban area revealed by the *Allium cepa* test. Chemosphere 76:1344-1347.

Borges Júnior M, Mello JWV, Ernesto C, Schaefer GR, Dussin TM, Abrahão WAP (2008a). Valores de Referência Local e Avaliação da Contaminação por Zinco em Solos adjacentes a Áreas mneradas no Município de Vazante-MG. Revista Brasileira de Ciência do Solo 32:2883-2893.

Borges Júnior M, Mello JWV, Ernesto C, Schaefer GR, Dussin TM, Abrahão WAP (2008b). Distribuição e Formas de Ocorrência de Zinco em Solos no Município de Vazante – MG. Revista Brasileira de Ciência do Solo 32:2183-2194.

Chandra S, Chauhan LKS, Murthy RC, Saxena PN, Pande PN, Gupta SK (2005). Comparative biomonitoring of leachates from hazardous solid waste of two industries using *Allium* test. Science of the Total Environment 347:46- 52.

Chen G, White PA (2004). The mutagenic hazards of aquatic sediments: A review. Mutation Res. 567:151–225.

Companhia de Tecnologia de Saneamento Ambiental - CETESB (1999) Manual de gerenciamento de áreas contaminadas. Projeto CETESB-GTZ. Cooperação Técnica Brasil-Alemanha. 2ªed. São Paulo, 389p.

Companhia de Tecnologia de Saneamento Ambiental -CETESB. (2001) Relatório de estabelecimento de valores orientadores para solos e águas subterrâneas no Estado de São Paulo. São Paulo, p. 247.

Di Natale F, Di Natale M, Greco R, Lancia A, Laudante C, Musmarra D (2008). Groundwater protection from cadmium contamination by permeable reactive barriers. J. Haz. Mat. 160:428-434.

Egito LCM, Medeiros MG, Medeiros SRB, Agnez-Lima LF (2007). Cytotoxic and genotoxic potential of surface water from the Pitimbu river, northeastern/RN Brazil. Genet. Mol. Biol. 30 (2):435-441.

Ernst E (2002). Toxic heavy metals and undeclared drugs in Asian herbal medicines. Trends Pharmacol. Sci. 23 (3).

Fatima RA, Ahmad M (2005). Certain antioxidant enzymes of *Allium cepa* as biomarkers for the detection of toxic heavy metals in wastewater. Sci. Total Environ. 346:256-273.

Ferraz KO, Wardell SMSV, Wardell JL, Louro SRW, Beraldo H (2009). Copper (II) complexes with 2-pyridineformamide-derived thiosemi-carbazones: Spectral studies and toxicity against *Artemia salina*. Spectrochimica Acta Part A. 73:140-145.

Fiskesjo G (1985). The *Allium* test as a standard in environmental monitoring. Hereditas 102:99-112.

Fiskesjö G (1988). The *Allium* test - an alternative in environmental studies: the relative toxicity of metal ions. Mutation Res. 197:243-260.

Fuentes A, Llorén M, Sáez J, Aguilar MI, Pérez-Marín AB, Ortuño JF, Meseguer VF (2006). Ecotoxicity, phytotoxicity and extractability of heavy metals from different stabilised sewage sludges. Environ. Pollut. 143:355-360.

Fusconi A, Repetto O, Bona E, Massa N, Gallo C, Dumas-Gaudot E, Berta G (2006). Effects of cadmium on meristem activity and nucleus ploidy in roots of *Pisum sativum* L. cv. Frisson seedlings. Environ. Exp. Bot. 58:253-260.

Ghosh M, Singh SP (2005). A Review on Phytoremediation of Heavy Metals and Utilization of its Byproducts. Appl. Ecol. environ. Res. 3:11-18.

Glinska S, Bartczak M, Oleksiak S, Wolska A, Gabara B, Posmyk M, Janas K (2007). Effects of anthocyanin-rich extract from red cabbage leaves on meristematic cells of *Allium cepa* L. roots treated with heavy metals. Ecotoxicol. Environ. Safety 68:343-350.

Grant WF (1982). Chromosome aberration assays in *Allium*. A report of the U.S. environmental protection agency gene-tox program. Mutation Res. 99:273-91.

Grant WF (1994). The present status of higher plant biossays for detection of environmental mutagens. Mutation Res. 310:175-185.

Grant WF (1994). The present status of higher plant biossays for detection of environmental mutagens. Mutation Res. 310:175-185.

Grisolia CK, Oliveira ABB, Bonfim H, Klautau-Guimarães MN (2005). Genotoxicity evaluation of domestic sewage in a municipal wastewater treatment plant. Genet. Mol. Biol. 28 (2):334-338.

Guedes JA, Lima RFS, Souza LC (2005). Metais pesados em água do rio Jundiaí - Macaíba/RN. Revista de Geologia 18(2):131-142.

Hang X, Wanga H, Zhou J, Du C, Chen X (2009). Characteristics and accumulation of heavy metals in sediments originated from an electroplating plant. J. Haz. Materials. 163:922-930.

Hitzman MW, Reynolds NA, Sangster DF, Allen CR, Carman CE (2003). Classification, genesis, and exploration guides for nonsulfides zinc deposits. Econ. Geol. 98:685-714.

Knasmüller S, Gottmann E, Steinkellner H, Fomin A, Pickl C, Paschke A, Göd R, Kundi M (2009). Detection of genotoxic effects of heavy metal contaminated soils with plant bioassays. Mutation Res. 420:37-48.

Leme DM, Marin-Morales MA (2008). Chromosome aberration and micronucleus frequencies in *Allium cepa* cells exposed to petroleum polluted water - A case study. Mutation Res. 650:80-86.

Liu D, Jiang W, Wang W, Zhao F, Lu C (1994). Effects of lead on root growth, cell division, and nucleolus of *Allium cepa*. Environ. Pollut. 86(1):1-4.

Liu W, Li PJ, Qi XM, Zhou QX, Zheng L, Sun TH, Yang YS (2005). DNA changes in barley (*Hordeum vulgare*) seedlings induced by cadmium pollution using RAPD analysis. Chemosphere 61:158-167.

Ma TH, Xu Z, Xu C, Mcconnel H, Rabago EV, Arreola GA, Zhang H (1995). The Improved *Allium/Vicia* root tip micronucleus assay for clastogenicity of environmental pollutants. Mutation Res. 334:185-195.

Majer BJ, Gottman E, Knasmüller S (2003). The micronucleus test with *Vicia faba* and *Allium cepa*. In: J. Maluszynska, M. Plewa (Eds): Bioassays in plant cells for improvement of ecosystem and human health. Wydawnictvo Uniwersytetu Slaskiego. Katowice, pp. 150.

Majer BJ, Grummt T, Uhl M, Knasmuler S (2005). Use of plant bioassays for the detection of genotoxins in the aquatic environment. Acta Hydroch. Hydrob. 33:45-55.

Majer BJ, Tscherko D, Paschke A, Wennrich R, Kundi M, Kandeler E, Knasmüller S (2002). Effects of heavy metal contamination of soils on micronucleus induction in *Tradescantia* and on microbial enzyme activities: a comparative investigation. Mutation Res. 515:111-124.

Megateli S, Semsari S, Couderchet M (2009).Toxicity and removal of heavy metals (cadmium, copper, and zinc) by *Lemna gibba* Smain. Ecotoxicol. Environ. Safety. 72:1774-1780.

Min X, Xie, X, Chai L, Liang Y, Li M, Ke Y. (2013). Environmental availability and ecological risk assessment of heavy metals in zinc leaching residue. Transactions Nonferrous Metal Society of China 23:208−218.

Minissi S, Lombi E (1997). Heavy metal content and mutagenic activity, evaluated by *Vicia faba* micronucleus test, of Tiber river sediments.

Mutation Res. 393:17-21.

Ministério Do Meio Ambiente Conselho Nacional Do Meio Ambiente-CONAMA. Resolução Nº 357, De 17 De Março De 2005.

Office of World Health Reporting. World Health Organization, (1998). World Health Organization Geneva.

Ozakca DU, Silah H (2012). Genotoxicity effects of Flusilazole on the somatic cells of Allium cepa. Pesticide Biochem. Physiol. 107:38–43.

Pesnya DS, Romanovsky AV (2013). Comparison of cytotoxic and genotoxic effects of plutonium-239 alpha particles and mobile phone GSM 900 radiation in the Allium cepa test. Mutation Res. 750:27– 33.

Radić S, Vujčić V, Cvetković Ž, Cvjetko P, Oreščanin V (2014). The efficiency of combined CaO/electrochemical treatment in removal of acid mine drainage induced toxicity and genotoxicity. Sci. Total Environ. 466-467, 84-89.

Rank J, Nielsen MH (1998). Genotoxicity testing of wastewater sludge using the Allium cepa anaphase-telophase chromosome aberration assay. Mutation Res. 418:113-119.

Raskin I, Ensley B (2000). Phytoremediation of toxic metals - using plants to clean up the environment. Plant Sci, Amsterdam 160:1073-1075.

Rigobello AE, Branquinho JA, Dantas MGS, Oliveira TF, Nieves Filho W (1988). Mina de zinco de Vazante, Minas Gerais. In: BRASIL. Ministério das Minas e Energia, DNPM/CVRD. Principais dépósitos minerais do Brasil (2), 670.

Rodriguez-Cea A, Ayllon F, Garcia-Vazquez E (2003). Micronucleus test in freshwater fish species: an evaluation of its sensitivity for application in field surveys. Ecotoxicol. Environ. Safety. 56:442-448.

Samardakiewicz S, Woz´ny A (2005). Cell division in Lemna minor roots treated with lead. Aquat. Bot. 83:289-295.

Santos DM, Bossini JAT, Preussler KH, Vasconselos EC, Carvalho-Neto FS, Carvalho-Filho MAS (2006). Avaliação de Metais Pesados na Baía de Paranaguá, PR, Brasil, sob Influência das Atividades Antrópicas. J. Braz. Soc. Ecotoxicol. 1:157-160.

Seregin IV, Shpigun LK, Ivanov VB (2004). Distribution and toxic effects of cadmium and lead on maize roots. Russ. J. Plant Physiol. 51:525-533.

Seth CS, Misra V, Chauhan LKS, Singh RR (2008). Genotoxicity of cadmium on root meristem cells of Allium cepa: cytogenetic and Comet assay approach. Ecotoxicol. Environ. Safety 71:711-716.

Sharma S (2009). Study on impact of heavy metal accumulation i Brachythecium populeum (Hedw.) B.S.G. Ecological Indicator 9:807-811.

Smaka-Kincl V, Stegnar P, Lovka M, Toman MJ (1996). The evaluatio of waste, surface and ground water quality using the Allium te: procedure. Mutation Res 368:171-179.

Soares CRFS, Accioly AMA, Marques TCLLSM, Siqueira JO, Moreir FMS (2001). Acúmulo e Distribuição de Metais Pesados nas Raízes Caule e Folhas de Mudas de Árvores em Solo Contaminado po Rejeitos de Indústria de Zinco. Revista Brasileira de Fisiologi Vegetal 13(3):302-315.

Steinkellner H, Mun-Sik K, Helma C, Ecker S, Ma Te-Hsiu, Horak C Kundi M, KnasmuÊller S (1998). Genotoxic Effects of Heavy Metal: Comparative Investigation with Plant Bioassays. Environ. Mo Mutagenesis. 31:183-191.

White PA, Claxton LD (2004). Mutagens in contaminated soil: a review Mutation Res. 567:227-345.

Wierzbicka M (1988). Mitotic disturbances induced by low doses c inorganic lead. Caryologia 41:143-160.

Wierzbicka M (1989). Disturbances in cytokinesis caused by inorgani lead. Environ. Exp. Bot. 29:123-133.

Wierzbicka M (1999). The effect of lead on the cell cycle in the roc meristem of Allium cepa L. Protoplasma 207:186-194.

Yabe MJS, Oliveira E (1998). Metais pesados em águas superficiai como estratégia de caracterização de bacias Hidrográficas. Químic Nova 21(5):551-556.

Yi H, Wu L, Jiang L (2007). Genotoxicity of arsenic evaluated by Alliun root micronucleus assay. Sci. Total Environ. 383:232-236.

Yildiz M, CigErci IH, Konuk M, Fidan AF, Terzi H (2009). Determinatio of genotoxic effects of copper sulphate and cobalt chloride in Alliur cepa root cells by chromosome aberration and comet assays Chemosphere. 75:934-938.

Zhuang P, McBride MB, Xia H, Li N, Li Z (2009). Health risk from heav metals via consumption of food crops in the vicinity of Dabaosha mine, South China. Sci. Total Environ. 407:1551-1561.

23

Identification of phytochemical components of aloe plantlets by gas chromatography-mass spectrometry

Mansoor Saljooghianpour and Taiebeh Askari Javaran

Islamic Azad University, Iranshahr Branch, Iranshahr, Iran.

Aloe vera plants were collected from Blochestan, Iran and were transferred to tissue culture laboratory. Shoot tip explants were inoculated on solid MS medium supplemented with 0.5 mgl^{-1} benzyl adenine + 0.5 mgl^{-1} α-naphthalene acetic acid and sub-cultured on the same medium for plantlet production and propagation once every four weeks. After plantlets production, extracts of *A. vera* plantlet were analyzed by gas chromatography-mass spectrometry (GC-MS). According to the results, 26 phytochemical compounds were identified. Results indicate that these compounds of micropropagated plantlets are similar to the phytochemical compounds identified by other researchers in aloe plants. With attention on the obtained results of GC-MS analysis, the obtained compounds of micropropagated plantlets did not vary in relation to aloe plants. These results also indicate that the use of propagated plantlets by tissue culture to produce and extract phytochemical compounds is useful and efficient, as was observed and expected. So, we can use this method (tissue culture) instead of aloe cultivation which is limited in some regions of the world.

Key words: Aloe medicinal plant, phytochemical components, micropropagation, tissue culture, gas chromatography-mass spectrometry (GC-MS) analysis.

INTRODUCTION

Aloe vera is a medicinal, cosmetic and ornamental plant. The genus Aloe is a perennial succulent herb growing in tropical and subtropical parts of the world. Therefore, aloe cultivation is limited in these regions. There are over 300 species of aloe; most of them are native to South Africa, Madagascar and Arabia.

The different species have somewhat different concentrations of active ingredients (Yagi et al., 1998; Van Wyk et al., 1995). At least, a quarter of Aloe genera is valued for traditional medicine (Grace et al., 2009), while a small number is wild harvested or cultivated for natural products prepared from the bitter leaf exudate or gel-like leaf mesophyll; *A. vera* is commonly cultivated and supports a global natural products industry. Today, *A. vera* gel is an active ingredient in hundreds of skin lotions, sun blocks and cosmetics (Grindlay et al., 1986).

Aloe gel is 99% water with a pH of 4.5 and is a common ingredient in many non-prescription skin salves. Aloe extracts have been used to treat canker sores, stomach ulcers and even AIDS. The gel contains an emollient polysaccharide, glucomannan, which is a good moisturizer utilized in many cosmetics (Henry, 1979). Acemannan, the major carbohydrate fraction in the gel demonstrates antineoplastic and antiviral effects (Mc Daniel et al., 1990). The gel also contains bradykininase, an anti-inflammatory agent, which prevents itching, and salicylic acid as well as other antiprostaglandin compounds that relieve inflammation (Yagi et al., 1982). Other important pharmacological activities of *A. vera* are anti-diabetic (Rajasekaran et al., 2006), antiseptic (Capasso et al., 1998), anti-tumor (Winter et al., 1981), and wound and burn healing effect (Heggers et al.,

Figure 1. *Aloe vera* plant.

1993). The sticky liquid latex is derived from the yellow-ish-green pericyclic tubules that line the leaf (rind); this is the part that yields laxative anthraquinones. The leaf lining (latex, resin or sap) contains anthraquinone glycol-sides (aloe-emodin and barbaloin) which are potent stimulant laxatives.

Sexual reproduction by seeds due to male sterility in aloe plants is almost not effective and vegetative propagation through lateral shoots or lateral buds is only possible during growing seasons (Nayanakantha et al., 2010), and is slow and very expensive for commercial plant production (Meyer and Staden, 1991). To overcome slow propagation rate, micro propagation will be a very useful technique for mass production of aloe.

A. vera has been cultured *in vitro* by various researchers (Natali et al., 1990; Roy and Sarkar, 1991; Abrie and Staden, 2001). The technique of tissue and organ culture is used for rapid multiplication of plants, for genetic improvement of crops, for obtaining disease-free clones and for preserving valuable germplasm. One of the major applications of plant tissue culture is micropropagation or rapid multiplication. As compared to conventional propagation, micropropagation has the advantage of allowing rapid propagation in limited time and space.

Gas chromatography-mass spectrometry (GC-MS) is a method that combines the features of gas liquid chromatography and mass spectrometry to identify

different substances within a test sample. GC-MS car provide meaningful information for components that are volatile, non-ionic, thermally stable and have relatively low molecular weight.

In this present study, we used micropropagated plant-lets of *A. vera* for evaluation of phytochemical com-ponents by GC-MS analysis.

MATERIALS AND METHODS

A. vera plants (Figure 1) were collected from Blochestan farmland in Iran and were transferred to Tissue Culture Laboratory of Kara Agricultural Biotechnology Research Institute in September 2010. Shoot tip explants containing one to two buds were cut and washed with tap water for 10 min, and after surface sterilization using 2% (w/v) NaOCl for 20 min, they were thoroughly rinsed with sterile water. The explants were thoroughly washed with sterile double distilled water for four to five times to remove any trace of the sterilant. Then, the explants were inoculated on solid Murashige and Skoog (MS) (Murashige and Skoog, 1962) medium supple-mented with 0.5 mgl^{-1} benzyl adenine + 0.5 mgl^{-1} α-naphthalene acetic acids into jars (250 ml capacity) containing 40 ml of the above-mentioned medium. Samples were sub-cultured every four weeks once on the same medium for plantlet production and propagation (Figure 2).

Preparation of plant extract

The micropropagated *A. vera* plantlets were washed with distilled water and were kept in-room temperature to be dried by air. Dried

Figure 2. Micropropagated plantlets of *A. vera*.

plantlets were crushed to the small pieces and were powdered and kept in polythene bags for further use. Aqueous extract of the studied samples were used to carry out the qualitative and quantitative analysis using standard procedures to identify the phytochemical components as described by Sofowara (1993) and Trease and Evans (1989).

Extracts of *A. vera* plantlets were analyzed by GC-MS. GC analysis was performed using a Hewlett-Packard 6890 chromatograph equipped with a flame ionization detector and injector MS transfer line with temperature of 280°C, respectively. A fused silica capillary column Hp- 5ms (5% phenyl : 95% dimethyl siloxane 30 M × 0.25 mm film thickness 0. 32 Lm) was used. The oven temperature was programmed from 110°C (isothermal for 2 min), with an increase of 10°C/min, to 200°C, then 5°C/min to 280°C, ending with a 9 min isothermal at 280°C. The carrier gas helium was at a flow rate of 1 ml/min. GC-MS analyses were carried out on an Agilent Technologies Network mass spectrometer (model 5973) coupled to H.P. gas chromatograph (model 6890) equipped with NBS 75K Library Software database. The capillary column and GC conditions were as described above. Mass spectra were taken at 70 eV; the scanning rate was 1 scan/s and the run time was 90 min. Compound identification was accomplished by comparing the GC relative retention times and mass spectra to those of authentic substances analyzed under the same conditions, by their retention indices (RI) and by comparison with reference components.

RESULTS AND DISCUSSION

The utilization of GC-MS was effective and useful for the identification of the bioactive compounds in *A. vera*. According to the results, 26 bio-active phytochemical compounds were identified in the GC-MS analysis of *A. vera* plantlets. The identification of phytochemical com-pounds is based on the peak area, molecular weight and molecular formula (Table 1).

Results indicate that these compounds of micro-propagated plantlets are similar to the phytochemical compounds identified by other researchers in aloe plants (Sathyaprabha et al., 2010; Lakshmi et al., 2011). 10 compounds with biological activities were found in *Aloe vera*. The main compounds include oleic acid (14.49), 11,14-eicosadienoic acid, methyl ester (2.71), n-hexadecanoic acid (20.41), 1,2-benzenedicarboxylic acid, butyloctyl ester (2.28), hexadecanoic acid, methyl ester (1.45), tetradecanoic acid (1.03), (4,7-dinitronaphthalen-1-yl)-(4-methoxyphenyl)diazene (0.09), 1-heptanol, 2-propyl- (3.77), 1,2-benzenedicarboxylic acid, diisooctyl ester (13.56) and squalene (6.57). These compounds of *A. vera* were shown to have the activity of anticancer, antimicrobial, etc. These results also indicate that the use of propagated plantlets by tissue culture to produce and extract phytochemical compounds is useful and efficient, as was observed and expected. So, we can use this method (tissue culture) instead of aloe cultivation which is limited in some regions of the world for production and extraction of phytochemical compounds.

The composition of identified active compounds in *A. vera* is the subject of future research studies. With attention on the obtained results of GC-MS analysis, the obtained compounds of micropropagated plantlets do not vary in relation to aloe plants, and the bioactive phytochemical compounds have not changed within micropropagated plantlets in relation to wild plant (Figures 3 and 4). With regards to the variations of environmental and growth conditions, the phytochemical profiles of individual plants changes. Wild plants may produce secondary metabolites, which have no apparent

Table 1. Identified components of *A. vera* plantlet by GC-MS.

RT	Name of the compound	Molecular Formula	Molecular Weight	Peak area (%)
3.06	p-Xylene	C_8H_{10}	106	3.13
3.78	1,5-Heptadien4-one,3,3,6-trimethyle-	$C_{10}H_{16}O$	152	1.69
7.03	1- Heptanol, 2- propyl-	$C_{10}H_{22}O$	158	3.77
8.25	Tridecane	$C_{13}H_{28}$	184	0.1
9.59	7- Tetradecane, (z)-	C14H28	196	0.17
10.87	Tetradecane	$C_{14}H_{30}$	198	0.32
12.14	Hexadecane	$C_{16}H_{34}$	226	0.38
13.64	12,15-Octadecadiynoic acid,methyle ester	$C_{19}H_{30}O_2$	290	0.18
13.96	(4,7-Dinitronaphthalen-1-yl)-(4-methoxyphenyl)diazene	$C_{17}H_{12}N_4O_5$	352	0.09
14.44	Tetradecanoic acid	$C_{14}H_{28}O_2$	228	1.03
15.87	Octadecane, 3- ethyl-5-(2-ethylbutyl)-	$C_{26}H_{54}$	366	0.21
17.28	Undecane	$C_{11}H_{24}$	156	0.45
17.42	1,2-Benzenedicarboxylic acid, diisooctyl ester	$C_{24}H_{38}O_4$	390	13.56
19.38	9-Octadecenoic acid, (2-phenyl-1,3dioxolan-4-yl)methyle ester.cis-	$C_{28}H_{44}O_4$	444	3.04
20.74	9,12,15- Octadecatrienoic acid, 2-((trimethylsilyl)oxy)-1-(((trimethylsilyl)oxy)methyl) ethyl ester, (ZZZ)-	$C_{27}H_{52}O_4Si_2$	496	2.08
20.40	Oleic acid	$C_{18}H_{34}O_2$	282	14.49
24.68	Hexadecanoic acid, methyl ester	$C_{17}H_{34}O_2$	270	1.45
25.94	1,2-Benzenedicarboxylic acid, butyloctyl ester	$C_{20}H_{30}O_4$	334	2.28
27.26	n- Hexadecanoic acid	$C_{16}H_{32}O_2$	256	20.41
28.71	11,14-Eicosadienoic acid, methyl ester	$C_{21}H_{38}O_2$	322	2.71
29.11	1-Monolinoleoylglycerol trimethylsilyl ether	$C_{27}H_{54}O_4Si_2$	498	2.63
30.48	Eicosane	$C_{20}H_{42}$	282	3.36
31.20	Heptacosane	$C_{27}H_{56}$	380	6.08
32.77	Octacosane	$C_{28}H_{58}$	394	9.52
34	Squalene	$C_{30}H_{50}$	410	6.57
35.43	Hentriacontane	$C_{31}H_{64}$	436	8.14

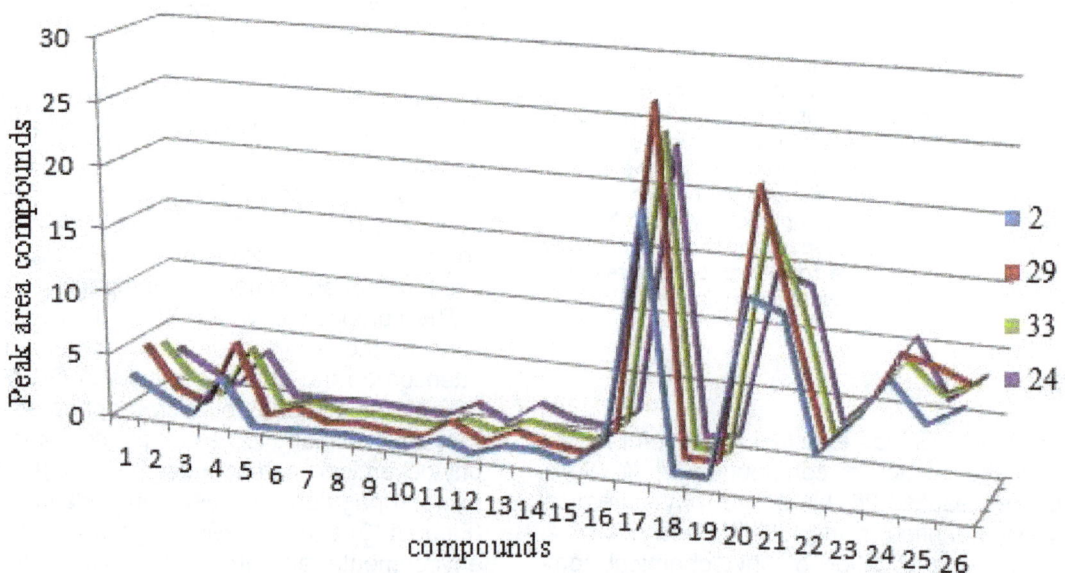

Figure 3. GC-MS graph of *A. vera* plantlets (2, 29, 33 and 24 accessions).

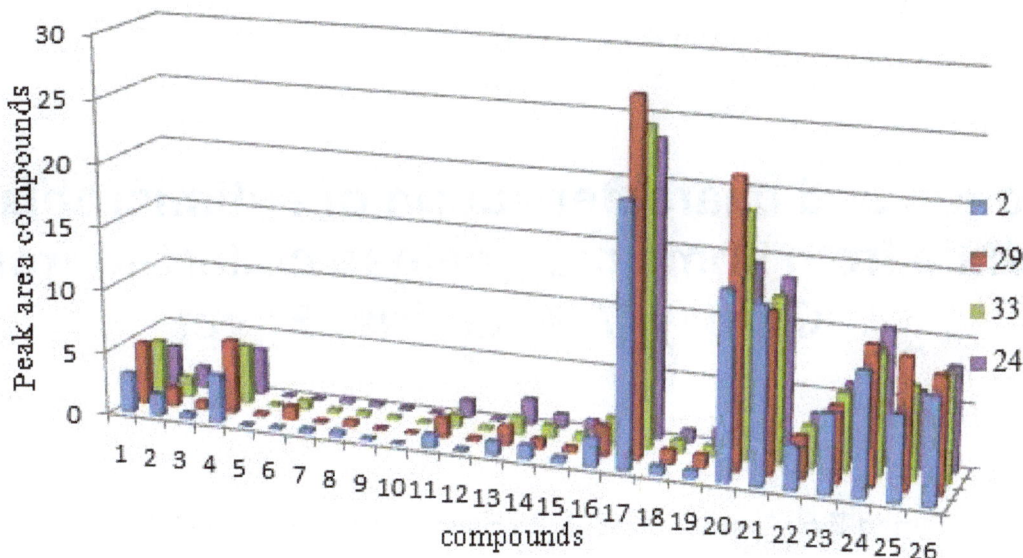

Figure 4. GC-MS graph of *A. vera* plantlet (2, 29, 33 and 24 accessions).

role in primary plant growth or development processes. These molecules are often unique in plants of a single species and are increased during times of high stress such as drought, fire and bacterial infection stresses in micropropagated plantlets.

REFERENCES

Abrie A, Staden JV (2001). Micropropagation of endangered Aloe Polyphylla. Plant G. Regu. 33(1):19-23.

Capasso F, Borrelli F, Capasso R (1998). Aloe and its therapeutics use. Phytoth. Res. 12:121-127.

Cock IE (2011). Problems of Reproducibility and Efficacy of Bioassays Using Crude Extracts, with reference to *Aloe vera*. Pharmacog. Comm. 1(1):52-62

Grace OM, Simmonds MJS, Smith GF, van Wyk AE (2009). Chemosystematic evaluation of Aloe section Pictae (Asphodelaceae). Bioch. Sys. Eco. 1:1-6.

Grindlay D, Reynolds T (1986). The *Aloe vera* phenomenon: a review of the properties and modern uses of the leaf parenchyma gel. J. Ethnopharmacol. 16:117-51.

Heggers J, Pelley R, Robson M (1993). Beneficial effects of Aloe in wound healing. Phytoth. Res. 7:S48-S52.

Henry R (1979). An updated review of *Aloe vera*. Cosmetics and Toiletries. 94:42-50.

Lakshmi PTV, Rajalakshmi P (2011). Identification of Phyto-Components and Biological Activities of *Aloe vera* Through the Gas Chromatography-Mass Spectrometry. Inter. Res. J. Pharm. 2(5):247-249.

McDaniel H, Carpenter R, Kemp M, Kahlon J, Mc Analley B (1990). Extended survival and prognostic criteria for Acemannan (ACE-M) treated HIV Patients. Antiviral Res. 1:117-125.

Meyer HJ, Staden JV (1991). Rapid in vitro propagation of *Aloe barbadensis* Mill. Plant cell Tis. and Org. Cul. 26:167-171.

Murashige T, Skoog F (1962). A revised medium for rapid growth and bioassays of tobacco tissue cultures. Physiol. Plantarum. 15:473-497.

Natali L, Sanchez IC, Cavallini A (1990). In vitro culture of Aloe barbadensis Mill: Micropropagation from vegetative meristems. Plant Cell Tis. Org. Cul. 20:71-74.

Nayanakantha NMC, Singh BR, Gupta AK (2010). Assessment of Genetic Diversity in Aloe Germplasm Accession from India Using RAPD and Morphological Markers. Ceylon J. Sci. Biol. Sci. 39(1):1-9.

Rajasekaran S, Sivagnanam K, Subramanian S (2006). Modulatory effects of *Aloe vera* leaf gel extract on oxidative stress in rats treated with steptozotocin. J. Pharm. Pharmacol. 57(2):241-246.

Roy SC, Sarkar A (1991). In vitro regeneration and micro propagation of *Aloe vera*. Scientia Hort. 47(1-2):107-114.

Sathyaprabha G, Kumaravel S, Ruffina D, Praveenkumar P (2010). A Comparative study on Antioxidant, Proximate analysis, Antimicrobial activity and phytochemical analysis of *Aloe vera* and *Cissus quadrangularis* by GC-MS. J. Pharm. Res. 3(12):2970-2973.

Sofowara A (1993). Medicinal plants and Traditional medicine is Africa spectrum Books LTD. Ibadan, Nigeria p.289.

Trease GE, Evans WC (1989). Pharmacognosy. 11th edn Brailliar Tiridel can Macmillian publishers.

Van Wyk BE, Yenesew A, Dagne E (1995). Chemotaxonomic survey of anthraquinones and pre-anthraquinones in roots of Aloe species. Bioch. Sys. Eco. 23:267–275.

Winter WD, Benavides R, Clouse WJ (1981). Effects of Aloe extracts on human normal and tumor cells in vitro. Econ. Bot. 35:89-95.

Yagi A, Harada N, Yamada H, Iwadare SIN (1982). Antibradykinin active material in Aloe saponaria. J. Pharm. Sci. 71:1172-1174.

Yagi A, Tsunoda M, Egusa T, Akasaki K, Tsuji H (1998). Immunochemical distinction of *Aloe vera*, *A. arborescens*, and *A. chinensis* gels [letter]. Planta Medica. 64:277-278.

24

Production and characterization of antimicrobial active substance from some macroalgae collected from Abu-Qir bay (Alexandria) Egypt

Mohamed E.H. Osman, Atef M. Aboshady and Mostafa E. Elshobary

Botany Department, Faculty of Science, Tanta University, Tanta, Egypt.

The antimicrobial activity of three different macroalgal species [*Jania rubens* (Linnaeus) Lamouroux; *Ulva fasciata* Delile and *Sargassum vulgare* C. Agardh] belonging to Rhodophyta, Chlorophyta and Phaeophyceae, respectively, were collected seasonally in 2007 to 2008 from Abu-Qir bay (Alexandria, Egypt). The different macroalgal species were tested against pathogenic microbes such as *Bacillus subtilis*, *Staphylococcus aureus* and *Streptococcus aureus* as gram-positive bacteria, *Escherichia coli*, *Salmonella typhi* and *Klebsiella pneumoniae* as gram-negative bacteria and one yeast strain, *Candida albicans*. The influence of sampling season on the antimicrobial activity of the collected seaweeds showed strong activity in spring followed by winter, summer and autumn, respectively. However, the strongest antimicrobial activity was recorded in 70% acetone extract of *U. fasciata* collected during winter against all the tested microorganisms. This extract was purified using column chromatography (CC) and thin layer chromatography (TLC). The nature of this purified antimicrobial material was detected using different chemical analysis (UV, IR, ^{1}H NMR and MS) which indicated that it is an aromatic compound and has different active groups (-NH$_2$, -C=O, -NO$_2$, phenyl ring and -CH$_3$). The molecular weight of the compound was determined (662) and its structure was characterized as a derivative of phthalate ester [(E)-1-(10-acetamido-2-nitrodec-9-enyl) 2-(10-acetamido-2-nitrodecyl) 4-methylphthalate]. This is the first evidence of the isolation of phthalate esters derivative from green seaweeds (*U. fasciata*) that has broad antimicrobial activity.

Key words: Antimicrobial, pathogenic microbes, season, seaweeds.

INTRODUCTION

Infectious diseases are a major cause of morbidity and mortality worldwide (WHO, 2004). The increase in failure of chemotherapeutics and antibiotic resistance exhibited by pathogenic microbial infectious agents has led to the screening of several medicinal plants for their potential antimicrobial activity (Colombo and Bosisio 1996; Cordell 2000; Scazzocchio et al., 2001). Synthetic drugs are not only expensive and inadequate for the treatment of diseases, but are also often with adulterations and side effects. Therefore, there is a need to search for new infection-fighting strategies to control microbial infections (Sieradzki and Tomasz, 1999). There are numerous reports of compounds derived from macroalgae with a broad range of biological activities, such as the anti-microbial activities (Reichelt and Borowitzka, 1984; Ballantine et al., 1987; Ballesteros et al., 1992; Vlachos et al., 1996), antiviral diseases (Trono, 1999), antitumors and anti-inflammatories (Scheuer, 1990) as well as neurotoxins (Kobashi, 1989). Subsequent chemical investigations of bioactive extracts led to the discovery of many struc-

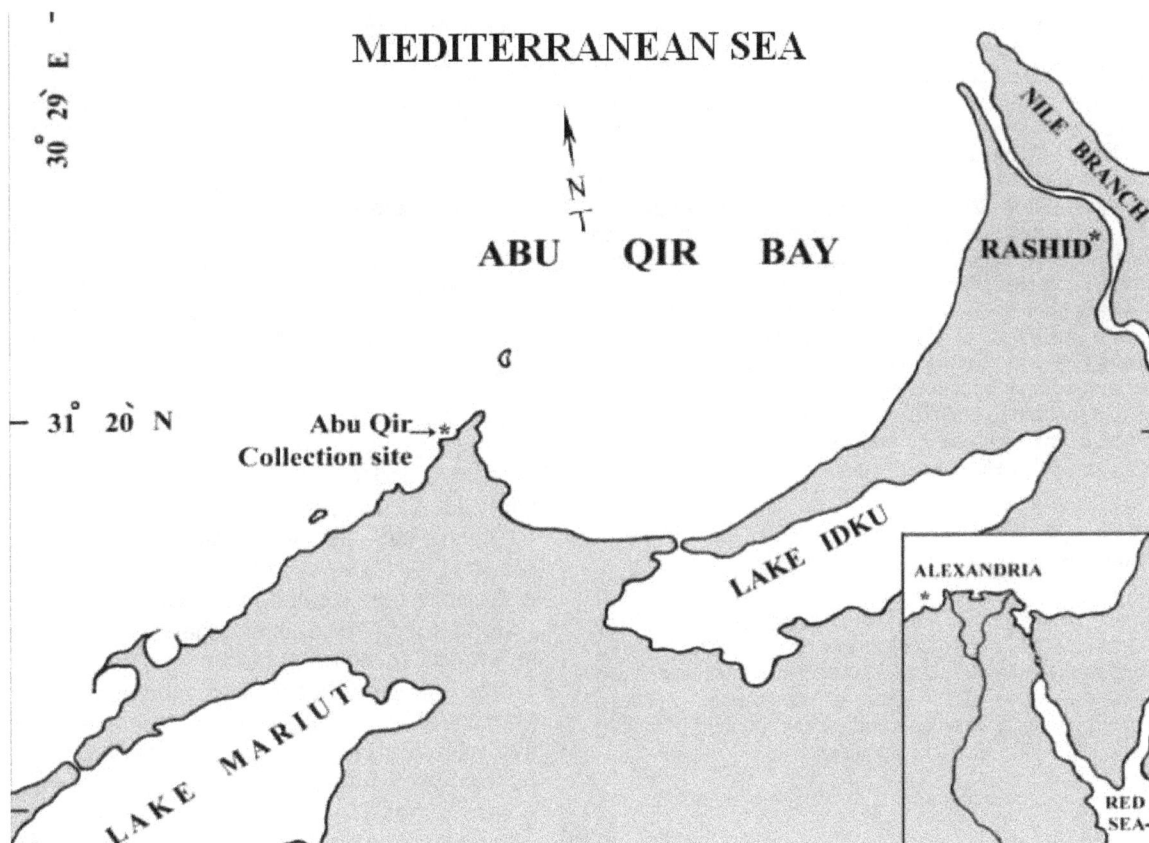

Figure 1. Map of Abu Qir showing collection site.

turally diverse antimicrobial metabolites from marine plants (Blunt et al., 2003, 2004, 2005; Faulkner, 2002). While, marine plant's metabolites have been studied extensively for their biomedical potential, their activities against human pathogens provide little information about their ecological role in antimicrobial activities and chemical structure of these antimicrobial compounds.

Alexandria has an extensive coast where, seaweeds from virtually all groups are present. The aims of this work were the search of novel compounds of potential antimic-robial value extracted from seaweeds, the study of the effect of seasonal variation on antimicrobial production, purification and elucidation of the structure of the anti-microbial compounds.

MATERIALS AND METHODS

Collection of algae

Three species of seaweeds from different divisions (*Jania rubens,* from Rhodophyta, *Ulva fasciata* from Chlorophyta and *Sargassum vulgare* from Phaeophyceae) were collected seasonally by hand in 2007 to 2008 from Rocky Bay of Abu Qir (N 31°19` E030°03`) (Figure 1). All samples were brought to the laboratory in plastic bags containing sea water to prevent evaporation. The algae were cleaned from epiphytes and rock debris and were given a quick fresh water rinse to remove surface salts. After collection, the samples were cleaned, air dried in the shade at room temperature

(25 to 30°C) in the dark on absorbent paper and grounded to fine powder in an electrical coffee mill. The specimen from the collected seaweeds was preserved for identification and all the seaweeds were identified following Abbott and Hollenberg (1976) and Taylor (1985) and Aleem (1993).

Tested micro-organisms

Seven bacterial strains were obtained from the Culture Collection of Botany Department, Faculty of Science, Tanta University. They included *Bacillus subtilis, Staphylococcus aureus* and *Streptococcus aureus* as gram-positive bacteria, *Escherichia coli, Salmonella typhi* and *Klebsiella pneumoniae* as gram-negative bacteria and one yeast strain (*Candida albicans*) as yeast.

Preparation of the extracts

The extraction was carried out with70% acetone. The extraction was carried out by soaking the dried materials in 70% acetone (1:15 v/v) on a rotary shaker at 150 rpm at room temperature (25 to 30°C) for 72 h. The extracts from three consecutive soakings were pooled and filtered using filter paper (Whatman no. 4). The obtained filtrate was freed from solvent by evaporation under reduced pres-sure. The residues (crude extracts) obtained were resuspended in 70% acetone to a final concentration of 100 mg/ml and then stored at -20°C in airtight bottle.

Antimicrobial activity test

15 ml of the sterilized media (nutrient agar (Oxoid) for bacteria and Sabouraud dextrose agar for yeast) were poured into sterile caped

test tubes. Test tubes were allowed to cool to 50°C in a water bath and 0.5 ml of uniform mixture of inocula (10⁸ CFU for bacteria and yeast) were added. The tubes were mixed using a vortex mixer vibrating at 1500 to 2000 rounds min⁻¹ for 15 to 30 s. Each test tube's contents were poured onto a sterile 100 mm diameter Petri dish for solidification (Mtolera and Semesi, 1996).

The antimicrobial activity was evaluated using well-cut diffusion technique (El-Masry et al., 2000). Wells were punched out using a sterile 0.7 cm cork borer in nutrient agar plates inoculated with the test microorganisms. About 50 µl of the different algal extracts were transferred into each well. For each microorganism, controls were maintained where pure solvent was used instead of the extract. All the plates were incubated at 4°C for 2 h to slow the growth of microorganisms and give suitable time for the antimicrobial agent to diffuse. To prevent drying, the plates were covered with sterile plastic bags and incubated at 37°C for 24 h. (Mtolera and Semesi, 1996). The result was obtained by measuring the diameter of the inhibition zone for each well, and expressed as millimeter.

Statistical analysis

The results are presented as mean ± standard deviation of the mean (n = 3). The statistical analyses were carried out using SAS program (1989 to 1996) version 6.12. Data obtained were analyzed statistically to determine the degree of significance between treatments using one and three way analysis of variance (ANOVA) at P ≤ 0.01 and P ≤ 0. 001 levels of significance.

Column chromatography

Selected active crude extracts (2 g) were fractionated by column chromatography on silica gel (EDWC, 60-120 mesh). Column (2 cm × 40 cm) was set up in benzene with silica gel (30 to 40 g) and eluted with gradients of solvents from 10:1% of benzene: acetone to 1:10% benzene: acetone (Solomon and Santhi, 2008). The collected fractions were evaporated to dryness with a rotary eva-porator and then, the dried samples were dissolved in pure acetone and assayed for their antimicrobial activity. The maximum absorp-tion of the active fractions was measured by spectrophotometer (UV 2101/ pc) using quartz cuvette containing the different fractions. Different active fractions with same absorption maximum were pooled together (Solomon and Santhi, 2008). The active fractions were tested for purity using TLC (thin layer chromatography). The purified fraction was lyophilized and subjected to the following analyses in order to reveal its structure as far as possible:

UV-spectra

The UV-spectra of the tested material were determined using UV2101/pc spectrophotometer. The wavelength ranged from 200 to 800 nm.

The infrared spectra (IR)

Using Perkin-Elmer 1430 infrared spectrophotometer, the molecular structure of the antimicrobial material was partially identified. Since the antimicrobial material is liquid at room temperature, so it can be examined directly as a thin film, "neat", between two clean and transparent NaCl plates. The measurements were carried out at infra red spectra between 400 to 4000 nm.

Nuclear magnetic resonance (H¹NMR) spectra

The sample was dissolved in Dimethyl-d^6 sulfoxide (d^6 DMSO). The different functional groups were identified using NMR (JNMPMX60SI).

Mass spectra (MS)

A mass spectrophotometer (MS-5988) was used. The product was subjected to a steam of high energy of electrons at elevated temperature up to 100°C. The cleavage fragments were yielded which were characterized by mass/charge from mass spectra data.

RESULTS

J. rubens and U. fasciata were present in all seasons whereas, S. vulgare was present in spring and summer only. This indicated that a certain level of temperature is required for these species to grow in a massive quantity to facilitate the collection procedures.

Concerning the antimicrobial activities of the different seaweeds collected in the various seasons, the results in Table 1 showed that U. fasciata showed stronger activity than J. rubens in autumn. S. aureus was the most sensitive microorganism to U. fasciata extract. However, K. pneumoniae was the most sensitive microorganism to J. rubens extract.

In winter, the extract of U. fasciata also was more active than J. rubens, where K. pneumoniae was the most sensitive tested microorganisms for U. fasciata and J. rubens extracts. In spring, the extract of J. rubens was more active than U. fasciata and S. vulgare. With regard to J. rubens, it showed high antimicrobial activity against B. subtilis whereas, K. pneumoniae and S. aureus were the most sensitive to U. fasciata and S. vulgare extracts respectively.

In summer, the extract of U. fasciata exhibited stronger antimicrobial activity than J. rubens and S. vulgare respectively. The results showed that the extracts of U. fasciata inhibited all the tested microorganisms and S. aureus was the most sensitive microorganism to U. fasciata and J. rubens extracts. However, B. subtilis and K. pneumoniae exhibited higher activity for S. vulgare.

The obtained results show that the highest activity of the different seaweeds extracts were those collected in spring followed by winter, summer and autumn, respectively (Figure 2). The antimicrobial activity of the different species with respect to the different seasons could be arranged in the following order, U. fasciata in winter > spring > autumn > summer followed by J. rubens in spring > winter > autumn > summer and S. vulgare in spring > summer. The stated results indicated that the promising seaweeds for the production of the antimicrobial material was U. fasciata (Chlorophyta) that was collected in winter season against all the tested microorganisms. Therefore we selected this species for further investigation.

The statistical analysis using three-way ANOVA con-firmed that the variation in antimicrobial activities in rela-

Table 1. The antimicrobial activity of 70% acetone extract of seaweeds from different seasons against different tested microorganisms, measured as diameter of inhibition zone (mm).

Season	Seaweed	Diameter of inhibition zone (mm)						
	Microorganism	B. subtilis	S. aureus	S. aureus	E. coli	S. typhi	K. pneumoniae	C. albicans
Autumn	J. rubens	13±1	9±0	14.5±0.5	9.5±0.5	9.5±0.5	15.2±1.2	10±0
	U. fasciata	11±0	11±1	15±0	12±1	12±0.5	13.5±0.5	12±0
	S. vulgare	N.P.	N.P.	N.P.	N.P	N.P.	N.P	N.P
Winter	J. rubens	14±0.1	12.1±0.1	13±0	16±0	12.2±0.7	17±0	12.7±0.2
	U. fasciata	22.2±0.2	19.3±1.2	20±1	19.5±0.5	21.8±1.0	24.6±0.5	19.5±0.7
	S. vulgare	N.P.	N.P.	N.P.	N.P	N.P.	N.P	N.P
Spring	J. rubens	18±1	12.7±0.7	14±0	14.3±0.5	15±1	15.2±1.2	14.7±0.2
	U. fasciata	15.7±0.2	12.8±2.0	14.3±1.5	12.2±0.2	12.6±0.5	17.25±2.25	16.5±0.5
	S. vulgare	12±1	15.6±0.3	11.5±0.5	12±1	11±0	11.5±0.5	11.8±0.3
Summer	J. rubens	9.1±1.0	10.5±0.5	12.5±0	9±1	10±0	8.5 ± 0.5	8.2±0.2
	U. fasciata	10.6±0.5	12±1	14.5±1.5	11±1	14±1	12 ± 0.2	12±0
	S. vulgare	8.3±0.7	0±0	8±0	7.5±0	7.5±0.5	8.2 ± 0.7	0±0

N.P. = Not present; (±) standard deviation of the mean (n=3).

tion to seasons, seaweeds and microorganisms, and their interaction on antimicrobial activity was significant at P ≤ 0.001 for all the treatments.

Purification and characterization of antimicrobial crude extract

Column chromatography

The results showed that 27 fractions were collected and only first four fractions had antimicrobial activity (Table 2). The UV absorption spectrums of these fractions were determined using spectrophotometer (UV 2101/ pc) at range of 200 to 800 nm. The obtained results are shown in Figure 3. The results indicated that the four fractions had the same absorptions peaks (three absorptions peaks at 333, 405 and 664 nm). Therefore, they pooled together. Thereafter, the different fractions were exposed to purity test using TLC technique (thin layer chromatography).

Thin layer chromatography (TLC)

The result showed that the examined four active fractions had approximately the same R_f value (Retention factor) and the statistical analysis using one-way ANOVA confirmed that the difference of R_f value of the four fractions compared with all the fractions sample was non-significant at P ≥ 0.01 (Table 3). Therefore, from UV analysis and TLC technique, the four active fractions were pooled together.

Elucidation of the chemical structure of the purified material isolated from 70% acetone of winter collected U. fasciata extract

UV spectra of the antimicrobial material

Before measuring the UV spectrum, the different pigments and impurities were removed by filtration using charcoal. Then, the UV spectrum of the purified antimicrobial material isolated from U. fasciata was carried out in pure acetone. This spectrum showed one absorption peak at 333 nm, indicating the presence of an aromatic compound (Figure 4).

The obtained compound was examined for antimicrobial activity. The results showed that the compound still had antimicrobial activity indicating that the compounds which had peak at 405 and

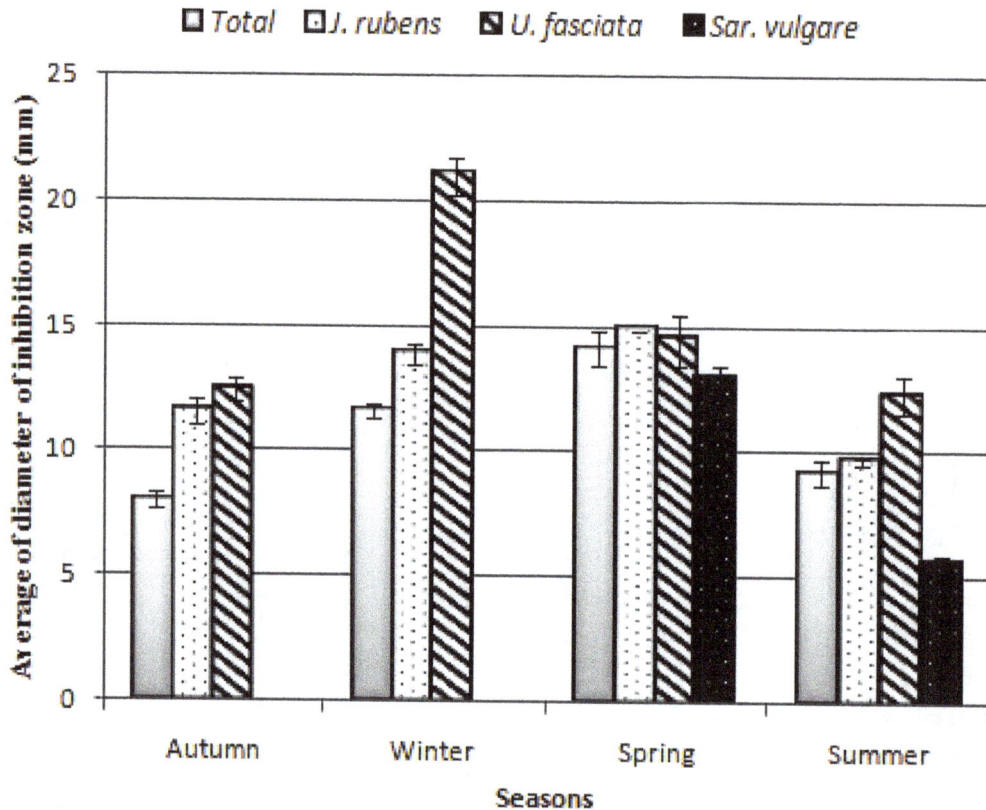

Figure 2. Antimicrobial activity of the different collected seaweeds that were collected in the different seasons.

Table 2. The antimicrobial activities of the different fractions obtained from the silica gel column chromatography against *K. pneumoniae*.

Number of fraction	Diameter of inhibition zone (mm)	Number of fractions	Diameter of inhibition zone (mm)
Fr.1	10 ± 0.023	Fr.15	-
Fr.2	13 ± 0.052	Fr.16	-
Fr.3	15 ± 0.034	Fr.17	-
Fr.4	17 ± 0.029	Fr.18	-
Fr.5	-	Fr.19	-
Fr.6	-	Fr.20	-
Fr.7	-	Fr.21	-
Fr.8	-	Fr.22	-
Fr.9	-	Fr.23	-
Fr.10	-	Fr.24	-
Fr.11	-	Fr.25	-
Fr.12	-	Fr.26	-
Fr.13	-	Fr.27	-
Fr.14	-		

(±)Standard error of the mean (n=3).

664 nm had not any antimicrobial activity. Therefore, we completed our investigation on the purified compound which had peak at 333 nm.

The infrared spectra (IR) of the antimicrobial material

The spectrum was subdivided into different regions, namely,

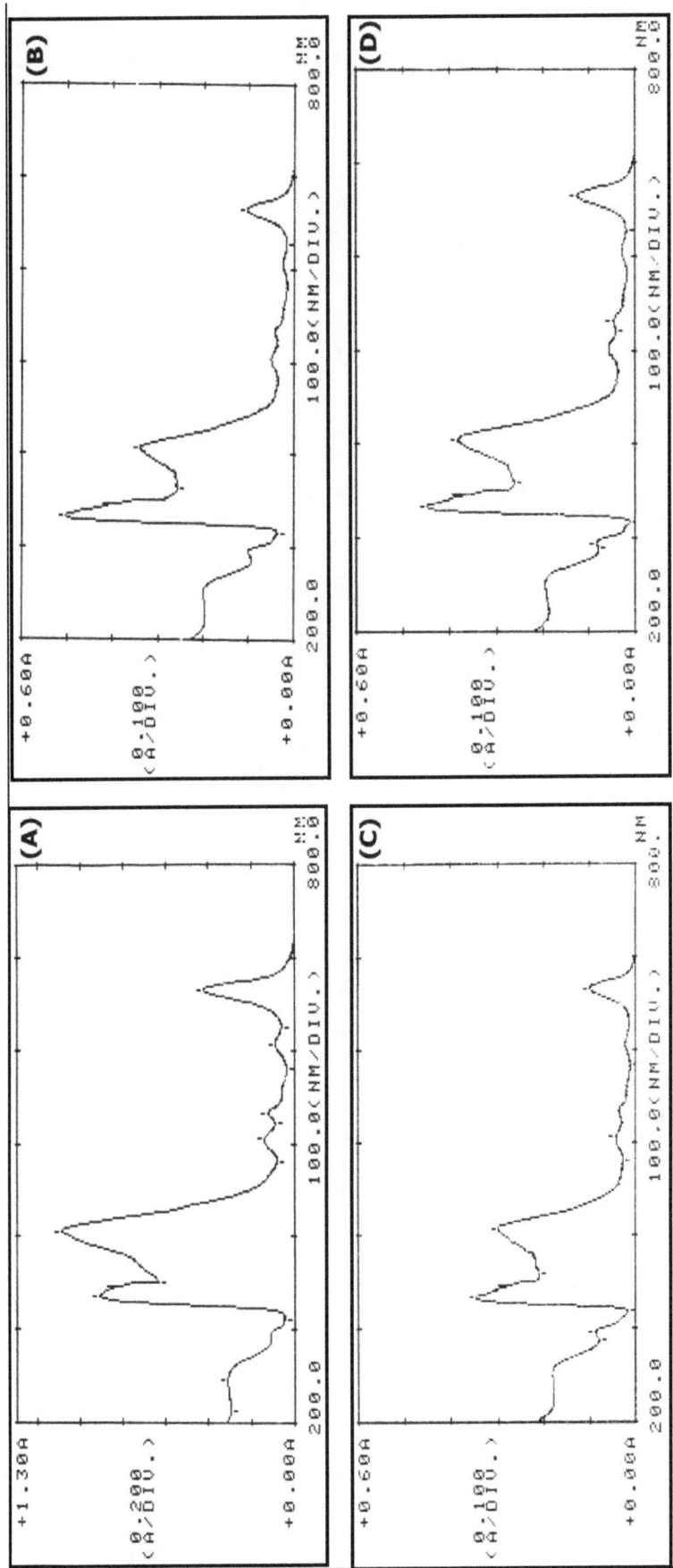

Figure 3. UV spectrophotometer scanning for the different active fractions separated by column chromatography.

Table 3. The retention factor of the spots from different fractions.

Sample	R_f
All fraction	0.38
Fraction1	0.377[(n.s.)]
Fraction 2	0.4[(n.s.)]
Fraction 3	0.372[(n.s.)]
Fraction 4	0.39[(n.s.)]
F-value	5.67[(n.s.)]

[(ns)] Non significant at P ≥ 0.01 using one way analysis of variance (ANOVA).

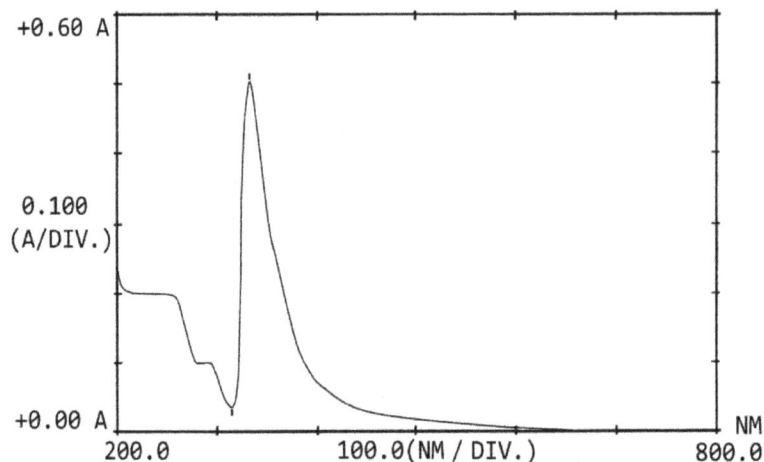

Figure 4. UV spectrum of the antimicrobial material produced by winter collected *U. fasciata* after purification by charcoal.

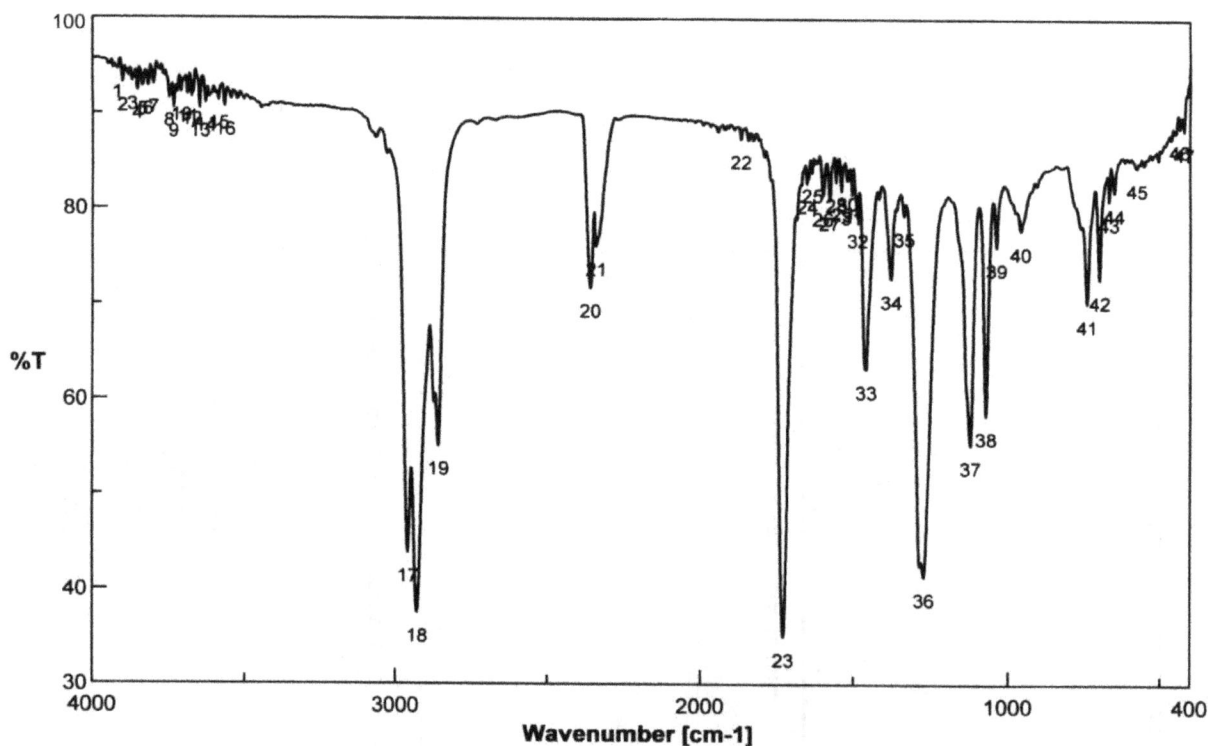

Figure 5. IR spectra of the purified antimicrobial material produced by winter *collected U. fasciata.*

the 2850 to 3050, 1710 to 1780, 1350 to 1470, 1020 to 1390 and 675 to 870 regions. The representative curve is shown in Figure 5.

Absorption in the 2850 to 3050 region: In this region, the υ CH aliphatic and υ CH aromatic strong stretch bands may appear. The IR revealed a strong band at 2958, 2927 and 2857 cm^{-1}, that can be attributed to the stretching vibrations of υ CH aliphatic and at 3000 to 3050 for

υ CH aromatic group.

Absorption in the1710 to 1780 region: This region comprised one band due to the stretching vibration of the υ C=O of COOR group at 1729 cm^{-1}.

Absorption in the 1020 to 1390 region: This region comprised two bands due to the stretching vibration of the υ C-N at 1122 and 1072 cm^{-1} and in addition, one bands

bands due to the stretching vibration of the υ N-O of nitro at 1272.79 cm^{-1}.

Absorption in the 870 to 675 region: This region comprised 3 bands due to the stretching vibration of the υ phenyl ring substitution band at 742,700 and 6698 cm^{-1}.

Proton magnetic resonance spectra

The ^{1}H NMR spectrum of the compound investigated was measured using in dimethyl-d^6 sulfoxide (d^6 DMSO) as solvent. The characteristic signals in ^{1}H NMR spectrum are represented graphically in Figure 6.

The signals were at δ 7.68 ppm (s,2H,2NH), at δ 6.89 to 7.47 ppm (m,3H of three aromatic protons), δ 5.320 ppm (s, 2H, CH=CH), δ 3.998 and 4.126 ppm (d, 2H, CH-NO$_2$), δ 3.342 ppm (s, 4H, 2 O-CH$_2$), δ 1.26, 1.6, 2.5, 2.73 ppm(s, 28H, 14 CH2) and at δ 1.9 and 2.1 (9H,3-CH$_3$).

Mass spectra of the antimicrobial material

The mass spectrum fragmentation pattern of the compound investigated is shown in Figure 7. It reveals the presence of peak at m/z 662 of relative abundance characteristic of the parent compound. According to the earlier mentioned chemical analysis, the chemical structure of the purified antimicrobial material isolated from *U. fasciata* was:

m/z: 57.03

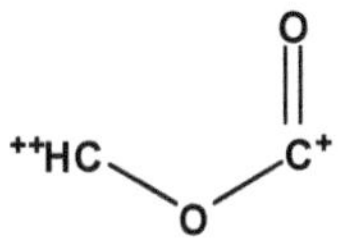

m/z: 85.03

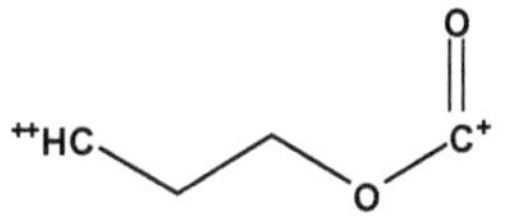

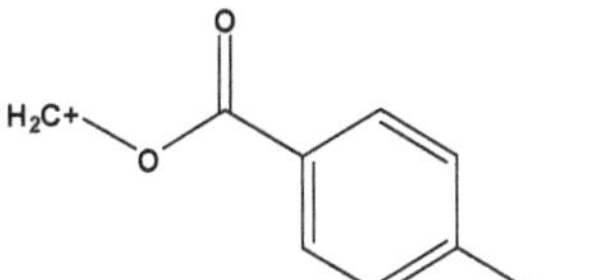

m/z: 149.17
m/z: 167.25

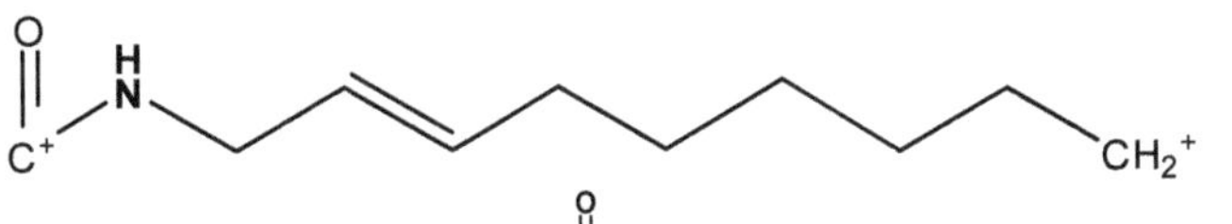

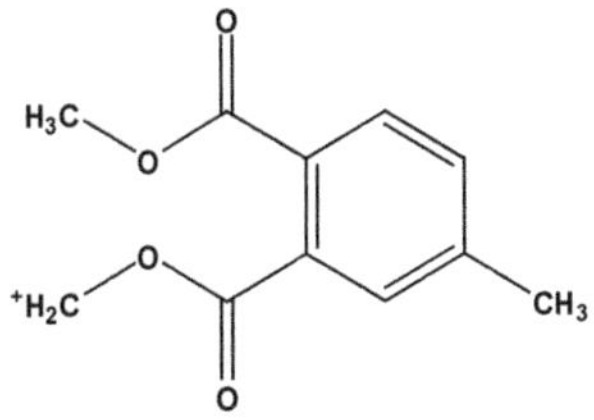

m/z: 207.20

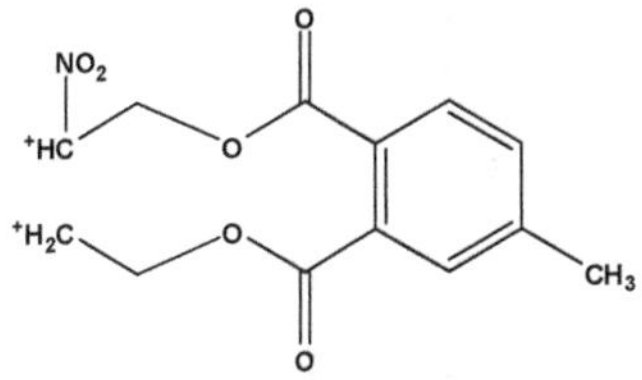

m/z: 279.24

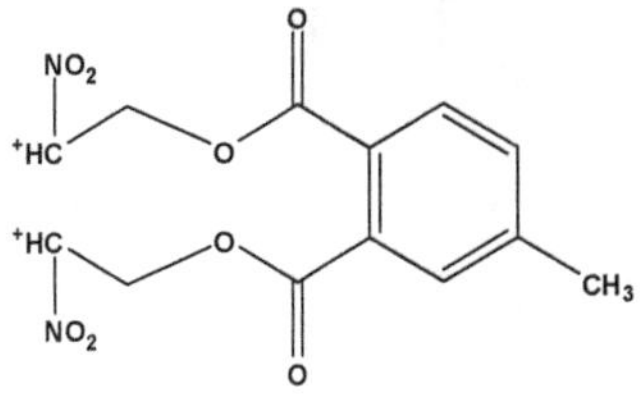

m/z: 324.24

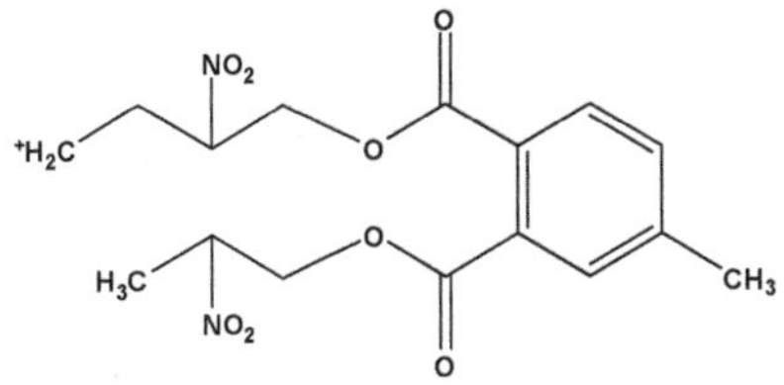

m/z: 367.33
m/z: 423.44

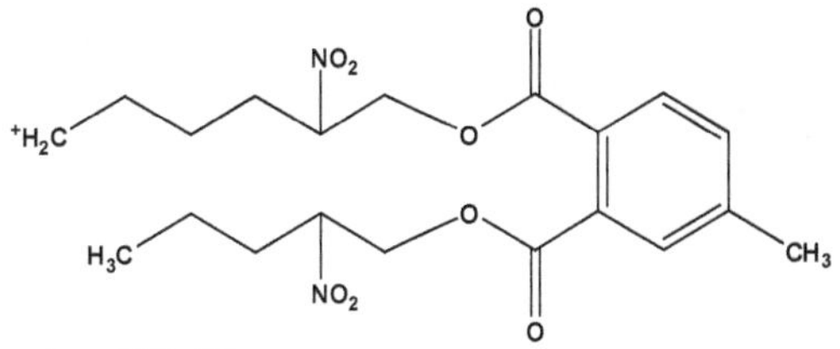

m/z: 479.54

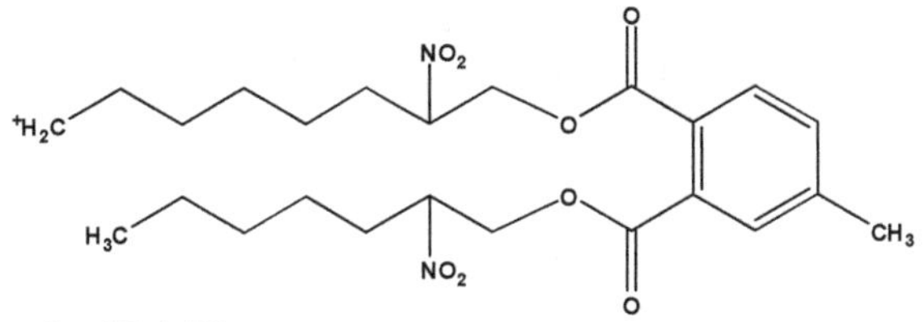

m/z: 591.72

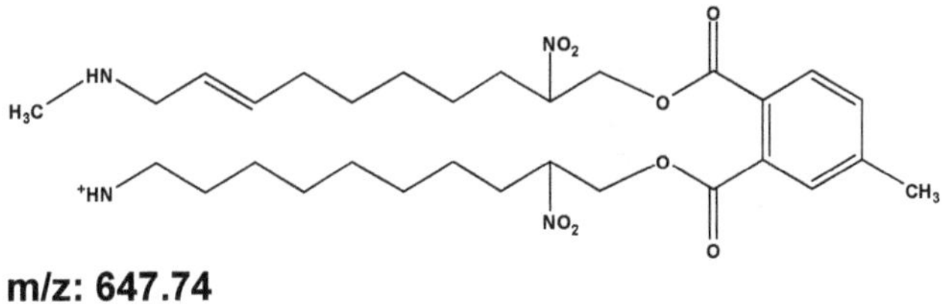

m/z: 647.74

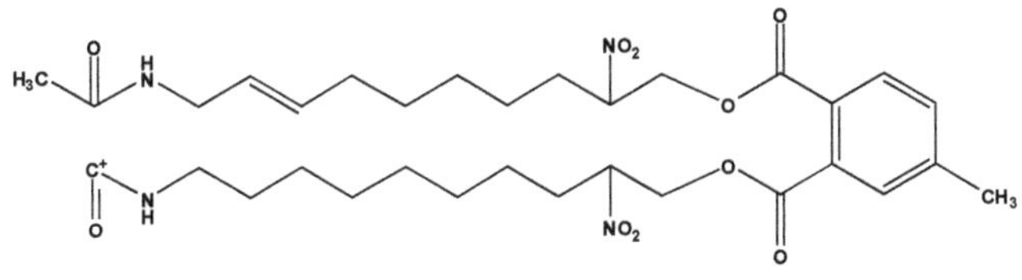

According to the stated data, the suggested structure of the isolated antimicrobial substance should be as the following:

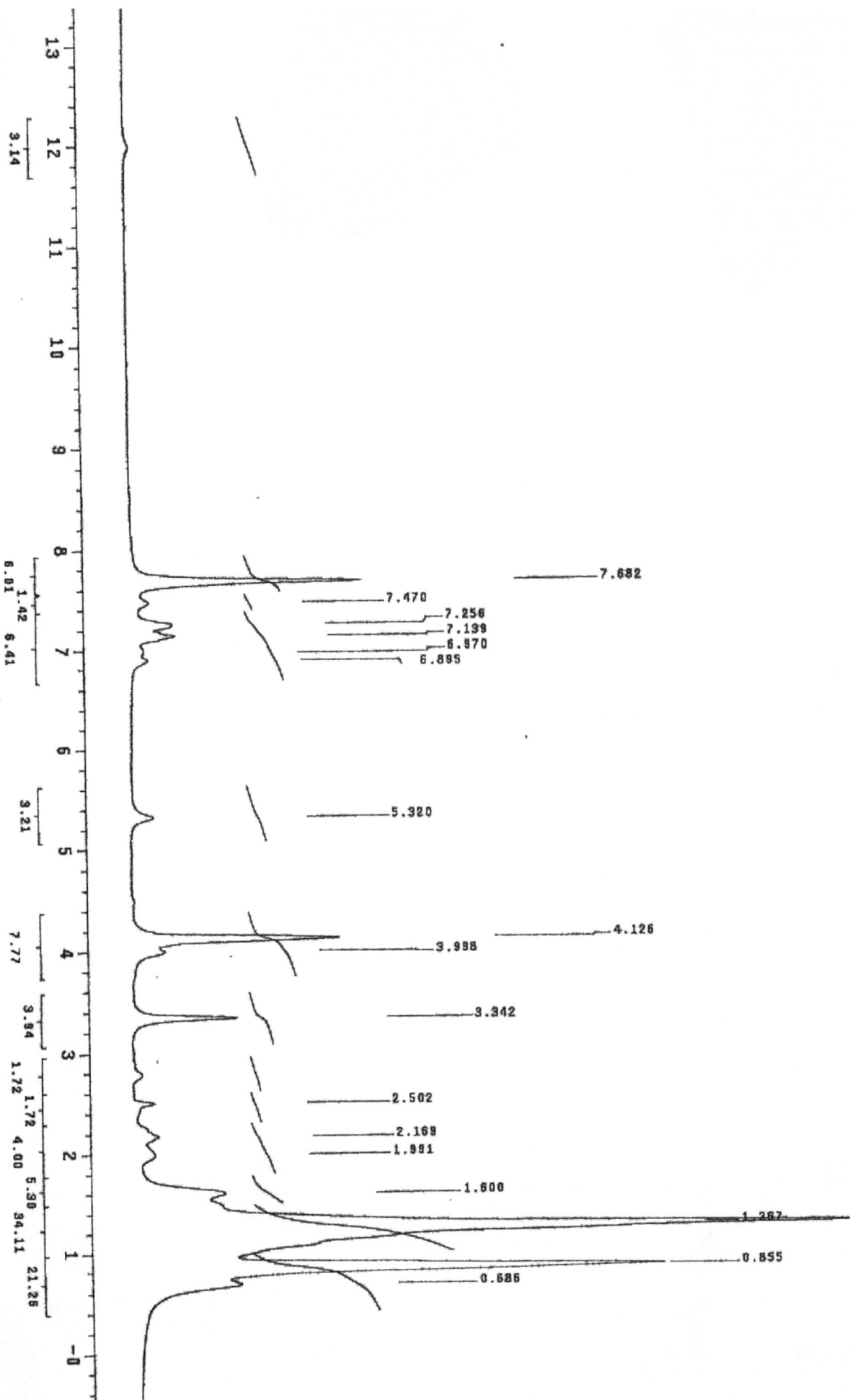

Figure 6. Proton magnetic resonance of the antimicrobial material produced by winter collected *U. fasciata*.

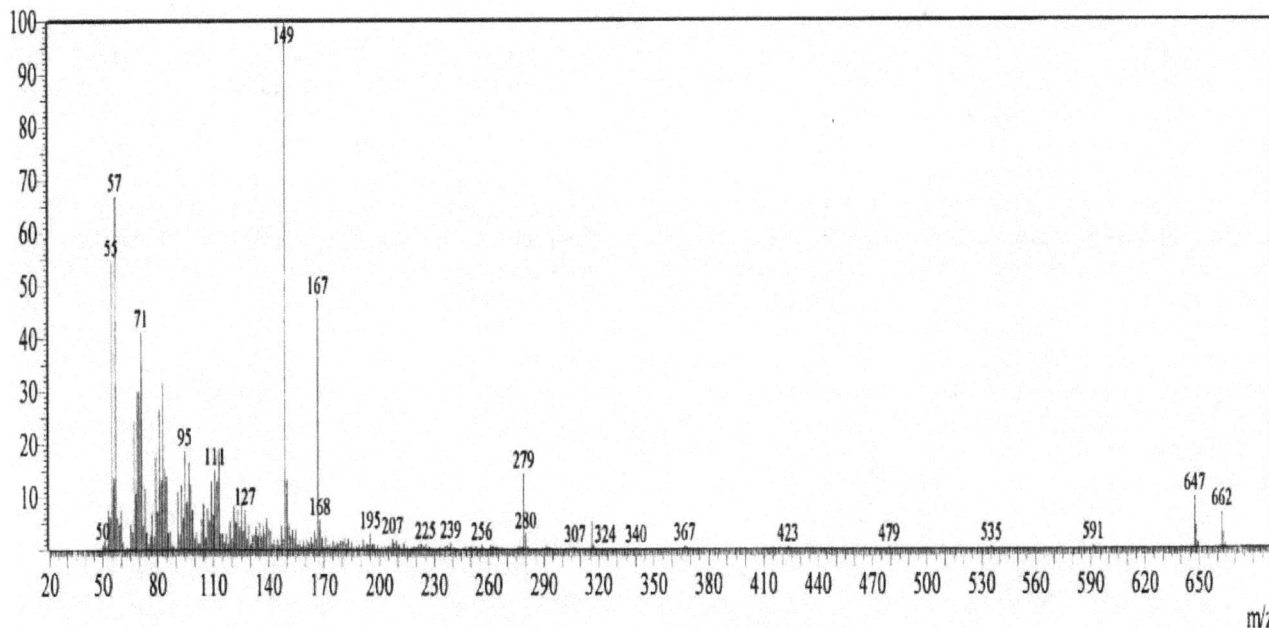

Figure 7. Mass spectra of the antimicrobial material produced by *U. fasciata*.

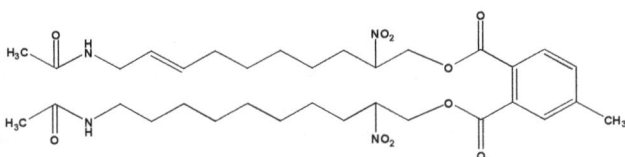

(*E*)-1-(10-acetamido-2-nitrodec-9-enyl), 2-(10-acetamido-2-nitrodecyl) 4- methyl phthalate. Chemical formula was $C_{33}H_{50}N_4O_{10}$; exact mass was 662.35; molecular weight was 662.77; elemental analysis was C: 59.80, H: 7.60, N: 8.45 and O: 24.14. This was the suggested chemical structure of the purified antimicrobial material isolated from *U. fasciata*.

DISCUSSION

This study is an endeavor for the study of the production, purification and structure elucidation of the antimicrobial compounds isolated from some species of seaweeds. Three species of seaweeds were tested for their capacity to produce antimicrobial substance. These species were collected from Abu-Qir, Alexandria coast, Egypt.

Lipid-soluble extracts from marine macroalgae have been investigated as a source of substances with pharmacological properties. Moreover, several different organic solvents have been used to screen algae for antibacterial activity (Mahasneh et al., 1995; Sukatar et al., 2006).

In this study, 70% acetone was used for extracting the bioactive compounds from different tested seaweeds and gave antimicrobial activity against all selected pathogens. This results are in accordance with those obtained by Wefky and Ghobrial (2008) and Fareed and Khairy (2008).

It is well know that some species of macro-algae possess antibacterial activities against pathogenic bacteria (Kumar and Rengasamy, 2000; Selvin and Lipton, 2004; Tüney et al., 2006; Karabay-Yavasoglu et al., 2007; Salvador et al., 2007; Chiheb et al., 2009). The results reported by the earlier mentioned authors are in accordance with our data, which demonstrated that all the tested seaweeds had antimicrobial activity against the tested microorganisms. However, Salvador et al. (2007) and Gonzalez del Val (2001) detected that some seaweeds have not any antimicrobial activity in all the seasons. These differences in activity may be due to different developmental stages, locality, extraction methods, etc. Also, the antimicrobial activity depends on both algal species and efficiency on extraction of their active(s) principle(s)

In relation to the taxonomic groups, Reichelt and Borowitzka (1984) and Salvador et al. (2007) screened many species of algae for their antibacterial activity. They reported that the members of the red algae exhibited high antibacterial activity. In contrast, in our study, the green alga (*U. fasciata*) was the most active one. These results are in agreement with the results of Kandhasamy and Arunachalam (2008) who reported that green algae (Chlorophyta) were the most active taxa than others and Fareed and Khairy (2008) which showed that *U. lactua* (Chlorophyta) was more active when compared with *J. rubens* (Rhodophyta).

Some pure compounds isolated from seaweeds have been identified as natural antimicrobial compounds.

However, the relationship between their ecologic role and antimicrobial activities is not fully understood in many studies. Chemical defenses can be very specific or very broad, depending on the method of extraction, the organisms, season of algal collection, different growth stages of the plant, experimental methods, etc.

In our study, we focused on the possibility that the antimicrobial activity of the tested seaweeds will also fluctuate seasonally. As regards to seasonal variation of bioactivity, for all of the tested seaweeds, spring was the season with the highest activity against the test microorganisms followed by winter; these results are in accordance with those obtained from Atlantic samples by Hornsey and Hide (1974), from Mediterranean samples by Khaleafa et al. (1975) and Stirk and Reinecke (2007) who reported that seasonal variation in antibacterial activity was observed with the extracts which have antibacterial activity in late winter and early spring.

In this study, U. fasciata was the most effective seaweed species, having antibacterial activity throughout the year compared with other seaweeds screened for their antibacterial activity. U. fasciata inhibited the growth of all the tested microorganisms and this result is in agreement with Parekh (1978), who reported that the extract of U. fasciata was found to be inhibiting for both gram positive and gram negative bacteria. Also, Selvin and Lipton (2004) reported that the green alga U. fasciata exhibited broad-spectrum antibacterial activity.

The results showed that U. fasciata extracts of winter collection exhibited stronger antimicrobial effects followed by spring season than the other seasons and this agreed with Stirk and Reinecke (2007), who demonstrated that U. fasciata collected in winter and spring seasons were active against the tested organisms compared with other seasons. This may be influenced by the seasonal variation as extracts of U. fasciata from winter and spring collection were more potent as compared with summer and autumn collections. The former represents the peak growing and reproductive season, while the later is the stasis and senescence period of U. fasciata growth. The highest antimicrobial action of winter collection is possibly due to the elevated biochemical constituents during the growing and reproductive phase of the U. fasciata. This hypothesis is further strengthened by some worker (Hornsey and Hide, 1974; Daly and Prince, 1981; Moreau et al., 1984; Rao and Indusekhar, 1989; Muñoz, 1992).

In this study, the antimicrobial material extracted from U. fasciata, which was collected from column chromatography was purified with charcoal and subjected to UV analysis. It showed maximum absorption spectra at 333 nm. Accordingly, the composition of the active antimicrobial material was suggested to contain aromatic ring.

The infrared (IR) spectroscopy indicated the presence of many functional groups in the antimicrobial material which were υ CH aliphatic at 2958, 2927 and 2857 cm^{-1}, υ CH aromatic at 3000 to 3050, υ C=O of COOR group at1729cm^{-1}, υ C-N at 1122 and1072 cm^{-1}, υ N-O of nitro at 1272 cm^{-1} and υ phenyl ring substitution band at 742.46, 700.033 and 669.178 cm^{-1}.

The ^{1}H NMR spectrum showed two protons of two (NH), three protons of aromatic protons, two protons of (CH=CH), two protons of two (CH-NO2), two protons of two (O-CH2), 28 protons of 14 (CH2-) and nine protons of three (-CH3). The mass spectroscopy of our antimicrobia material indicated that the molecular weight was 662.

According to the obtained data, the compound is a aromatic compound having the following structur $C_{33}H_{50}N_4O_{10}$ [(E)-1-(10-acetamido-2-nitrodec-9-enyl) 2 (10-acetamido-2-nitrodecyl) 4-methylphthalate]. This i the first evidence for the isolation of phthalate esters der vative from green seaweeds (U. fasciata) and that ha broad antimicrobial activity. However, some worker isola ted it from brown algae. Cho et al. (2005) isolated di-n octylphthalate which has antifouling activities from th brown seaweed Ishige okamurae. Also, Ganti et a (2006) isolated phthalic acid from Sargassum confusum.

These previous reports led us to suggest that the origi of the phthalate esters was natural and can be synthe sized by living organisms not derived from artificial pro ducts or a contaminant from environment.

Finally, we conclude that macroalgae from Abu Qi coast in Alexandria could be considered as potentia sources of bioactive compounds and the production o these compounds are affected by seasons. Our stud indicated that the activities of different seaweeds for th production of antimicrobial substances in various sea sons could be arranged in the following seasons spring > winter > summer >autumn and we obtained the highes activity from the winter collected U. fasciata. Thus, the production of the antimicrobial substance by seaweeds i season and species dependant. The structure of this anti microbial material is $C_{33}H_{50}N_4O_{10}$ [(E)-1-(10-acetamido-2 nitrodec-9-enyl) 2-(10-acetamido-2-nitrodecyl) 4-methyl phthalate], which is a strong anti-microbial activity agains the tested human pathogenic microorganisms.

REFERENCES

Abbott IA, Hollenberg IG (1976). Marine algae of California Stanfor University press.

Aleem AA (1993). The marine Algae of Alexandria , Egypt . Aleem AA (Ed.), Faculty of science, University of Alexandria, Egypt.

Ballantine DL, Gerwick WH, Velez SM, Alexander E, Guevara P (1987) Antibiotic activity of lipid-soluble extracts from Caribbean marine algae. Hydrobiol., 151/152: 463–469.

Ballesteros E, Martin D, Uriz MJ (1992). Biological activity of extracts from some Mediterranean macrophytes. Bot. Mar., 35: 481–485.

Blunt JW, Copp BR, Munro MHG, Northcote PT, Prinsep MR (2003) Marine natural products. Nat. Prod. Rep., 20: 1–48.

Blunt JW, Copp BR, Munro MHG, Northcote PT, Prinsep MR (2004). Marine natural products. Nat. Prod. Rep., 21: 1–49.

Blunt JW, Copp BR, Munro MHG, Northcote PT, Prinsep MR (2005) Marine natural products. Nat. Prod. Rep., 22: 15–61.

Chiheb I, Riadi H, Martinez-Lopez J, Dominguez SJF, Gomez VJA Bouziane H, Kadiri M (2009). Screening of antibacterial activity in marine green and brown macroalgae from the coast of Morocco. Afr J. Biotechnol., 8(7): 1258-1262.

Cho JY, Choi JS, Kang SE, Kim JK, Shin HW, Hong YK (2005) Isolation of antifouling active pyroglutamic acid, triethyl citrate and di-n-octylphthalate from the brown seaweed Ishige okamurae. J. Appl Phycol., 17: 431–435.

Colombo ML, Bosisio E (1996). Pharmacological activities of Chelidonium majus L (Papaveraceae). Pharmacol. Res., 33: 127–134.

Cordell GA (2000). Biodiversity and drug discovery a symbiotic relationship. Phytochemistry, 55: 463-480.

Daly JEL, Prince JS (1981). The ecology of *Sargassum pteropleuron* Grunow (Phaeophyceae, Fucales) in the waters off South Florida. III, Seasonal variation in alginic acid concentration. Phycology, 20: 352–357.

El-Masry HA, Fahmy HH, Abdelwahed ASH (2000). Synthesis and antimicrobial activity of some new benzimidazole derivatives. Molecule, *S*: 1429-1438 .

Fareed MF, Khairy HM (2008). *In vitro* Antimicrobial Activities of Seaweeds Collected from Abu-Qir Bay Alexandria, Egypt. World Appl. Sci. J., 5(4): 389-396.

Gonzalez del Val A, Platas G, Basilio A (2001). Screening of antimicrobial activities in red, green and brown macroalgae from Gran Canaria (Canary Islands, Spain). Int. Microbiol., 4: 35-40.

Ganti VS, Kim KH, Bhattarai HD, Shin, HW (2006). Isolation and characterisation of some antifouling agents from the brown alga *Sargassum confusum*. J. Asian Nat. Prod. Res. 8: 309–315.

Hornsey IS, Hide D (1974). The production of antimicrobial substances by British marine algae I. Antibiotic producing marine algae. Br. Phycol. J., 9: 353–361.

Kandhasamy M, Arunachalam KD (2008). Evaluation of in vitro antibacterial property of seaweeds of southeast coast of India. Afr. J. Biotechnol., 7(12): 1958-1961.

Karabay-Yavasoglu NU, Sukatar A, Ozdemir G, Horzum Z (2007). Antimicrobial activity of volatile components and various extracts of the red alga *Jania rubens*. Phytother. Res., 21: 153-156.

Khaleafa AF, Kharboush AM, Metwalli AA, Mohsen F, Serwi A (1975). Antibiotic fungicidal action from extracts of some seaweeds. Bot. Mar., 18: 163-165.

Kumar KA, Rengasamy R (2000). Evaluation of Antibacterial Potential of Seaweeds Occurring along the Coast of Tamil Nadu, India against the Plant Pathogenic Bacterium *Xanthomonas oryzae* pv. *oryzae* (Ishiyama) Dye. Bot. Mar. 43(5): 409–415.

Mahasneh I, Jamal M, Kashashneh M, Zibdeh M (1995). Antibiotic activity of marine algae against multi-antibiotic resistant bacteria. : Microbios., 83(334): 23-26.

Moreau J, Pesando D, Caram B (1984). Antifungal and anti bacterial screening of Dictyotales from the French Mediterranean coast. Hydrobiol., 116/117: 521-524.

Mtolera MSP, Semesi AK (1996). Antimicrobial activity of extraxts from six green algae from Tanzania. Curr. Trends Mar. Bot. Res. East Afr. Regul., 211-217.

Muñoz A (1992). Drogas del mar. Sustancias biomédicas de algas marinas. Servicio de Publicaciones Intercambio Científico, Universidad de Santiago de Compostela, Santiago de Compostela.

Parekh KS (1978). Antimicrobial and Pharmaceutically Active Substances from Marine Sources. Ph.D. Thesis, Saurastra University, Rajkot, India.

Rao CK, Indusekhar VK (1989). Seasonal variation in chemical constituents of certain brown seaweeds and seawater from Saurashtra coast. I. Carbon, nitrogen and phosphorus. Mahasagar. 22: 63–72.

Salvador N, Gomez-Garreta A, Lavelli L, Ribera MA (2007). Antimicrobial activityof Iberian macroalgae. Sci. Mar., 71: 101-113.

SAS (1996). Copyright (c) 1989-1996 by SAS Institute Inc., Cary, NC, USA. SAS (r) Proprietary Software Release 6.12 TS020.

Scazzocchio F, Comets MF, Tomassini L, Palmery M (2001). Antibacterial activity of Hydrastis canadensis extract and its major isolated alkaloids. Planta Med., 67: 561-563.

Scheuer PJ (1990). Some marine ecological phenomena: chemical basis and biomedical potential. Science, 248: 173-177.

Selvin J, Lipton AP (2004). Biopotentials of *Ulva fasciata* and *Hypnea musciformis* collected from the peninsular coast of India. J. Marine Sci. Technol., 12(1): 1-6.

Sieradzki KSW, Tomasz A (1999). Inactivation of the methicillin resistance gene mecA in vancomycin-resistant *Staphylococcus aureus*. Micro. Drug. Resist. 5: 253-257.

Solomon RDJ, Santhi VS (2008). Purification of bioactive natural product against human microbial pathogens from marine seaweed *Dictyota acutiloba* J. Ag. DOI 10.1007/s11274-008-9668-8.

Stirk WA, Reinecke DL (2007). Seasonal variation in antifungal, antibacterial and acetylcholinesterase activity in seven South African seaweeds. J. Appl. Phycol., 19: 271–276.

Sukatar A, Karabay-Yavasoglu NU, Ozdemr G, Horzum Z (2006). Antimicrobial activity of volatile component and various extracts of *Enteromorpha linza* (Linnaeus) J. Agardh from the coast of Izmir, Turkey. Ann. Microbiol., 56(3): 275-279.

Taylor WS (1985). Marine algae of the easterntropical and subtrobical coasts of Americas.ANN Arbor the university of Michigan press.

Trono JGC (1999). Diversity of the seaweed flora of the Philippines and its utilization. Hydrobiol., 1–6: 398-399,

Tüney I, Cadirci BH, Unal D, Sukatar A (2006). Antimicrobial activities of the Extracts of marine algae from the coast of Urla (Izmir, Turkey). Turk. J. Biol., 30: 171-175.

Vlachos V, Critchley AT, Von HA (1996). Establishment of a protocol for testing antimicrobial activity in southern African macroalgae. Microbios., 88: 115–123.

Wefky S, Ghobrial M (2008). Studies on the bioactivity of different solvents extracts of selected marine macroalgae against fish pathogens. Res. J. Microbiol., 3: 673-682.

WHO (2004). The world health report. Changing history. Statistical annex. Death by cause, sex and mortality stratum in WHO regions, estimates for 2002. Geneva, Switzerland, pp. 120-121.

Plant regeneration through indirect organogenesis of chestnut (*Castanea sativa* Mill.)

Mehrcedeh Tafazoli[1]*, Seyed Mohammad Hosseini Nasr[1], Hamid Jalilvand[1] and Dariush Bayat[2]

[1]Department of Forestry, Faculty of Natural Resources, Sari Agricultural Science and Natural Resources University, Sari, Iran.
[2]Deputy of Humid and Semi-humid Areas, Forests Ranges and Watershed Organization, Chalus, Iran.

To establish an effective protocol for plant regeneration through indirect organogenesis, effects of explants type, culture media and plant growth regulators on callus induction and shoot regeneration of chestnut (Castanea sativa Mill.) were investigated. Three different explants (root, nodal and internodal segment), two different media [Murashige and Skoog medium (MS) and Gamborg's B5 (B5)] and different plant growth regulators (6-benzylaminopurine (BA), thidiazuron (TDZ), indole-3-butyric acid (IBA) and indole-3-acetic acid (IAA)) with different concentration (0.2, 0.5, 1 and 1.5 mgL⁻¹) for shoot and root induction were chosen. The results show that nodal segment was the best explant for callus induction (69.4%) when cultured on MS medium supplemented with 1 mgL⁻¹ TDZ and MS was the best medium to induce callus formation (74.6%). The highest shoot multiplication (66.9%) was observed on MS medium with 0.2 mgL⁻¹ TDZ. Regenerated shoots were rooted in vitro on MS containing 1.5 mgL⁻¹ IBA. Also, plantlets with well developed root and shoot systems were acclimatized inside the green house and 80% of the plantlets survived on transfer to garden soil. This protocol provides a basis for future studies on genetic improvement.

Key words: Chestnut, node, internode segment, indirect organogenesis, callus formation, shoot regeneration.

INTRODUCTION

For a long time chestnut has been an important economic resource in Europe and more recently in Asia, also playing an important environmental role in many agroforestry systems (Bounous, 2005). Chestnut is a hardwood forest species of considerable agro-economic important tree species for both timber and nut production. However, this tree species is threatened by pollution, social and economic changes, and two major fungal diseases; ink disease (*Phytophthora* sp.) and chestnut blight [*Cryphonectria parasitica* (Murr.) Barr.] (Sauer and Wilhelm, 2005). Chestnut is a woody species, which is difficult to propagate either generatively by seed or vegetatively by grafting or cuttings (Osterc et al., 2005). However, as an alternative to conventional vegetative propagation methods, efforts have been made to establish reliable *in vitro* regeneration systems that allow clonal propagation (Vietez and Merkle, 2005; Troch et al., 2010). *In vitro* tissue culture techniques have been applied to chestnut regeneration since the 1980's (Rodriguez, 1982; Vieitez et al., 1983). Also *in vitro* establishment in chestnut is possible from both juvenile and mature material (Sánchez et al., 1997) and explants such as cotyledonary node (San-José et al., 2001), bud (Vieitez and Vieitez, 1980), nodal segment (Osterc et al., 2005), etc have been utilized for *in vitro* propagation of chestnut.

Production of regenerated plant through indirect organogenesis is one of the possible ways to contribute to genetic improvement because there are some advantages

*Corresponding author. Email: mehr_tafazoli@yahoo.com.

Abbreviations: BA, 6-Benzylaminopurine; **IAA,** indole-3-acetic acid; **TDZ,** thidiazuron; **IBA,** indole-3-butyric acid; **MS,** Murashige and Skoog medium; **B₅,** Gamborg's **B₅.**

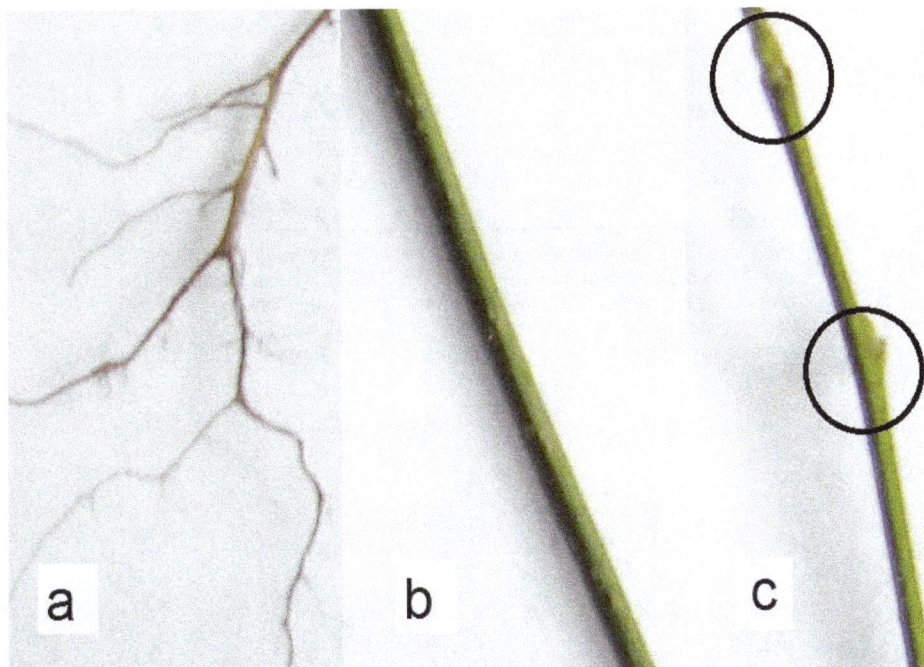

Figure 1. The explants used for experiments (a) root (b) internodal segments and (c) nodal segments.

of shoot regeneration from callus over direct shoot regeneration (Avilés et al., 2009). A callus phase is commonly included in tissue culture protocols with the objectives of generating variability to introduce new desirable traits and generating transgenic plants to introduce traits (Fereol et al., 2005; Zheng et al., 2005). Callogenic initiation implies an initial stage of differentiation from the parental tissue. Thus, to establish callus cultures, the determination of the initial tissue used is a fundamental factor in order to achieve the desired response (Bandyopadhyay et al., 1999). Earlier, zygotic embryos (Kvaalen et al., 2005), hypocotyls (Ahn et al., 2007), ovaries and ovules (Sauer and Wilhelm, 2005), internodes (Chandra and Bhanja, 2002), nodal segments (Arora et al., 2010) and cotyledonary node (Dhabhai and Batra, 2010) have been used in induction of callogenesis and indirect organogenesis.

The aim of this study was to investigate the effects of explant type, culture media and plant growth regulators on callus induction and shoot regeneration from chestnut, and also establish an efficient tissue culture protocol that would provide an efficient tool to be used in chestnut breeding programs.

MATERIALS AND METHODS

Plant materials and culture conditions

Seeds of *Castanea sativa* Mill. were collected from Visroud in Gilan, North of Iran. The area was about 350 ha with 200 to 600 m altitude. The seedlings were grown in a glasshouse under natural illumination (Hou et al., 2010). Root, nodal and internodal segments were removed from three months old seedlings as explants. The excised root, nodal and internodal segments were surface sterilized in 70% ethanol for 1 min followed by 4% sodium hypochlorite for 15 min and rinsed thoroughly three times for 3 min each in sterile distilled water. They were then placed individually (horizontal) in culture vessels (200 × 10 mm) containing 5 ml medium. The Murashige and Skoog, 1962 (MS) and Gamborg et al., 1968 (B$_5$) media were adjusted to pH 5.8 and 5.5 with 1 N NaOH or HCl, respectively, before autoclaving at 1.06 kgcm^2 pressure and 121°C temperature for 20 min. All cultures were incubated at 25 ± 2°C with 16/8 h photoperiod under light from cool white fluorescent lamps (125 µmol m^{-2}s^{-1}) and 60 to 70% relative humidity.

Effect of explants type, culture media and plant growth regulators on callus proliferation and shoot regeneration

Root, nodal and internodal segments excised from three-month-old seedlings (Figure 1) were placed on MS medium supplemented with 1 mgL^{-1} thidiazuron (TDZ) to evaluate the effects of different explants on callus proliferation of chestnut as the first experiment. According to the results, nodal segments were the best explants to generate callus. Therefore, the next experiments were done using nodes as initial explants.

For the second experiment, the effects of MS and B$_5$ basal media supplemented with 1 mgL^{-1} TDZ on callus proliferation from nodal segments were evaluated. According to the results, MS medium was better than B$_5$ medium to induce callus formation from nodes. Therefore, the next experiments were performed using nodal segments cultured in MS medium.

The objective of the third experiment was to investigate the effects of growth regulators on callus proliferation and shoot regeneration from chestnut. We used MS medium containing different concentration of TDZ and 6-benzylaminopurine (BA) (0.2, 0.5, 1 and 1.5 mgL^{-1}). Data collected included percentage of explants

Table 1. Effects of different explants types on callus proliferation from *Castanea sativa* Mill.

Explant	Explant forming callus (%)
Node	69.4 ± 0.40^a
Internodal segment	39.1 ± 2.64^b
Roots	14.2 ± 1.36^c

Data was collected after eight weeks of culture. Explants were cultured on MS medium supplemented with 1 mgL⁻¹ TDZ. Values represent the mean ± S.E. Means following the same letter within columns are not significantly different according to Student-Newman-Keuls multiple comparison test (P<0.05). TDZ, Thidiazuron; MS, Murashige and Skoog medium.

Figure 2. Plant regeneration from callus derived from *in vitro* cultured nodal and internodal segment of *Castanea sativa* Mill. (a) Callus formation from node on MS medium with 1 mgL⁻¹ TDZ after four weeks of culture. (b) Callus formation from root segments on MS medium with 1 mgL⁻¹ TDZ after four weeks of culture. (c) Callus formation from internodal stem on MS medium with 1 mgL⁻¹ TDZ after four weeks of culture. TDZ, Thidiazuron; MS, Murashige and Skoog medium.

forming callus, shoot proliferation percent, mean number and length of shoots after eight weeks of culture. Explants were subcultured onto the MS medium with same BA and TDZ concentration at two weeks for further callus and/or shoot development.

Root induction

When shoots developed from initial explants and became 20 to 30 mm in length about eight weeks after culture, they were excised and transferred to the rooting medium. The rooting medium consisted of MS medium supplemented with different concentration of IBA and indole-3-acetic acid (IAA) (0.2, 0.5, 1 and 1.5 mgL⁻¹). The percentage frequency of root formation and its length was calculated after four weeks of culture. The rooted plantlets (four to five weeks old) were then taken out from the culture vessels, washed thoroughly in running tap water to remove any remains of the nutrient-agar medium. Therefore, they were planted into pots (10 cm in diameter), containing a mixture of sterile vermiculite and sand (1:1) and maintained in the growth chamber at a temperature of 25 to 28°C, 16/8 h photoperiod and relative humidity of 80 to 90% covered with plastic bags. Once established in soil, the plants were transferred to the greenhouse at a temperature of 25 to 28°C and 16/8 h photoperiod.

Statistical analysis

The experimental unit was culture vessels. This treatment had five replications with 10 explants plated for each replication. For the comparison between MS and B₅ media, we used T-test. A

completely random design was used for the data analysis. Results were analyzed statistically using the statistical analysis system program (SAS, 2001). The mean values were calculated and compared by Student-Newman-Keuls multiple comparison test (P<0.05).

RESULTS AND DISCUSSION

Effect of explants type, culture media and plant growth regulators on callus proliferation and shoot regeneration

In *C. sativa* Mill., callus formation varied significantly depending on explant type (Table 1). Nodal segments showed the earliest signs of callus formation from the cut edges on MS medium supplemented with 1 mgL⁻¹ TDZ after one week of culture, but root segments and internodal segments started to initiate callus from cut surfaces after two weeks of culture. Figure 2 indicate the callus of different explants after four weeks. The nodal segments showed higher callus formation, while roots and internodal segments exhibited a significantly lower callus induction. The highest callus induction (74.6%) was achieved when MS medium was used in comparison with 69.1% on B₅ medium (Table 2).

Table 2. Effects of different culture media on callus proliferation from node of *Castanea. sativa* Mill.

Medium	Explant forming callus (%)
MS	74.6 ± 1.67^a
B_5	68.1 ± 1.12^b

Data was collected after eight weeks of culture. Different media were supplemented with 1 mgL^{-1} TDZ. Values represent the mean ± S.E. Means following the same letter within columns are not significantly different according to T-test (P<0.05). TDZ, Thidiazuron; MS, Murashige and Skoog medium; B_5, Gamborg's B_5.

Table 3. Effects of different growth regulators on callus proliferation and shoot regeneration from nodal segments of chestnut (*Castanea sativa* Mill.)

Growth regulator (mgL^{-1})		Explant forming callus (%)	Explant regeneration shoot (%)	Mean shoot length (cm)
BA	TDZ			
0.2	0	64.3 ± 0.28^f	24.7 ± 2.24^e	0.5 ± 1.42^e
0.5	0	67.6 ± 0.16^{ef}	31.8 ± 2.73^d	0.6 ± 1.94^e
1	0	72.3 ± 0.43^c	32.6 ± 2.0^e	0.6 ± 2.13^e
1.5	0	76.9 ± 0.67^b	51.5 ± 2.91^b	1.6 ± 1.25^c
0	0.2	85.3 ± 0.68^a	66.9 ± 2.13^a	2.2 ± 1.44^a
0	0.5	71.2 ± 0.22^d	53.3 ± 2.66^b	1.9 ± 1.67^b
0	1	79.4 ± 0.40^e	41.4 ± 2.84^c	1.4 ± 1.25^d
0	1.5	60.8 ± 0.29^e	39.1 ± 2.0^d	0.9 ± 2.85^e

Data was collected after eight weeks of culture. Nodal segments were cultured on MS basal medium. Values represent the mean ± S.E. Means following the same letter within columns are not significantly different according to Student-Newman-Keuls test (P<0.05). BA, 6-benzylaminopurine; TDZ, thidiazuron; MS, Murashige and Skoog medium.

The variations of callus induction on different media may be due to the differences of nitrate/ammonium (NO_3^-/NH_4^+) ratio, an important factor on nitrogen uptake and pH regulation during plant tissue culture (Fracago and Echeverrigaray, 2001). Vieitez et al. (1983) compared several media to obtain the cluster propagation ability in different chestnut clones from the same hybrid group *Castanea sativa* × *Castanea crenata*. The highest number of shoots per culture formed was achieved when the explants were grown on Heller's medium; explants grown on Heller's+$(NH_4)_2SO_4$ medium (Heller, 1953) and on MS medium with the addition of ammonium nitrate (NH_4NO_3) were slightly worse. However, Ballester et al. (2001) indicated that it is very difficult to recommend a mineral medium for general application; nevertheless, half strength MS media appeared to be the most suitable for multiplication through axillary shoot development. Therefore, it is concluded that the best basal medium for the callus induction was MS, although successful callus formation has been achieved on B_5 medium (Table 2). An increase in BA concentration from 0.2 to 1.5 mgL^{-1} increased callus induction from 64.3 to 76.9. The highest frequency of callus formation (85.3%) was obtained on MS medium containing 0.2 mgL^{-1} TDZ.

Shoot regeneration

Calli formed numerous shoots when they were cultured on the same medium after three to four weeks of culture (Table 3). Cytokinin type and concentration also affected the frequency of shoot induction (Maheshwari and Kovalchuk, 2011). In the present experiment, BA could induce shoot regeneration from nodal segments at the percent of 76.9% when cultured on MS medium with 1.5 mgL^{-1} BA. Increasing BA from 0.2 to 1.5 mgL^{-1} resulted in an increase in shoot regeneration ability in callus (Table 3). Vieitez and Vieitez (1980) reported that 6-benzylaminopurine (BAP) showed the most satisfactory effect on promoting the proliferation of axillary shoots, whereas zeatin slightly inhibited the development of axillary shoots but increased the induction rate and caused more vigorous shoots. Tetsumura and Yamashita (2004) achieved similar results with the addition of zeatin also causing the highest proliferation rate. In addition, Arora et al. (2010) found that it is possible to obtain shoot regeneration through indirect organogenesis from nodal segment of *Azadirachta indica* A. Juss. on MS medium supplemented with 0.2 mgL^{-1} BA. Girijashankar (2011) achieved shoot regeneration of *Acacia auriculiformis* from

Table 4. Effect of different concentrations of IBA and IAA on root induction in shoots of chestnut (*Castanea sativa* Mill.)

Growth regulators (mgL⁻¹)		Rooting (%)	Mean root length (cm)
IBA	IAA		
0.2	0	51.2 ± 1.54^d	0.7 ± 1.43^e
0.5	0	52.3 ± 1.94^d	1.7 ± 2.25^c
1	0	68.3 ± 1.11^b	1.6 ± 2.17^c
1.5	0	71.2 ± 1.88^a	2.1 ± 0.73^a
0	0.2	41.5 ± 2.27^e	1.0 ± 1.14^{de}
0	0.5	40.2 ± 1.01^e	1.4 ± 1.20^d
0	1	51.1 ± 1.32^d	1.7 ± 0.94^c
0	1.5	64.2 ± 1.99^c	2.0 ± 1.61^b

Data was collected after four weeks of culture. Nodal segments were cultured on MS basal medium. Values represent the mean ± S.E. Means following the same letter within columns are not significantly different according to Student-Newman-Keuls test (P<0.05). IAA, Indole-3-acetic acid; IBA, indole-3-butyric acid; MS, Murashige and Skoog medium.

nodal stem segments on MS containing BA (2 mgL⁻¹) + naphthaleneacetic acid (NAA) (1 mgL⁻¹). Also single nodal segments of *Castanea mollissima* cv. 'yanshanhong' induced shoot on MS medium with a half concentration of NO_3 supplemented with 0.5 mgL⁻¹ BA (Hou et al., 2010). The result of present study shows that TDZ induced more shoot regeneration. The highest shoot regeneration (66.9%) was obtained on MS medium supplemented with 0.2 mgL⁻¹ TDZ, while further addition of TDZ concentration resulted in decreased shoot regeneration. The lowest percentage of shoot induction was observed on MS medium with 1 mgL⁻¹ BA.

TDZ has been shown to be the most effective in inducing shoot regeneration in woody species and has shown to have a much stronger ability than BA on shoot induction (Huetteman and Preece, 1993). It is a synthetic polyurea which being a cytokinin-like substance, is highly effective for shoot regeneration in tissue culture of recalcitrant plant species (Liu et al., 2003). However, if concentrations of TDZ are high, hyperhydricity or morphological abnormalities could be observed among regenerated shoots (Huetteman and Preece, 1993). Lower concentrations of TDZ are thus preferable for shoot regeneration (Wang et al., 2008). In the current study, the maximum length of shoots was observed from nodal segments in MS supplemented with 0.2 mgL⁻¹ TDZ (2.2 cm) (Table 3). The cytokinins, because of their role in experimentally induced cell division and differentiation, serve as a probe of hormonal involvement in differentiation (Hall, 1976). This study demonstrate the superiority of TDZ over BA as shoot-inducing cytokinin in the *in vitro* induction of adventitious shoots from nodal segments of chestnut. Ballester et al. (2001) also used 0/1 mgL⁻¹ TDZ in MS medium to germinate chestnut embryonic axes.

Root induction from shoots

Rooting of the regenerated shoots did not occur on the

shoot induction medium. Elongated shoots (20 to 30 mm in length) regenerated on MS medium were used fo rooting experiment. Auxins had a significant influence or root formation. Depending on auxin type and concentration, roots were initiated between five to 15 days of culture. The best root formation and maximum mean length (2.1 cm) was observed during regeneration of chestnut on MS basal medium supplemented with 1.5 mgL⁻¹ IBA (the root regeneration percent was 71.2% (Table 4). IAA at 1.5 mgL⁻¹ also induced 64.2% rooting ir regenerated shoots, but the number of roots per shoot was considerably lower. The roots were regenerated within two weeks after transferring onto these media. The roots regenerated on the medium supplemented with 1.5 mgL⁻¹ IBA were longer than others (Table 4).

Among the auxins, IAA and IBA are the most frequently applied chemicals for rooting (Harry and Thrope, 1994) Our results also show that IBA was the best one to be used for root formation. Similar results about response to IBA are observed in *Prosopis ceneria* for root inductior (Kumar and Singh, 2009) and also *Azadirachta indica* A. Juss (Chaturvedi et al., 2004). Figure 3 shows the whole plantlet from nodal segment on MS with 0.2 mgL⁻¹ TDZ and then 1.5 mgL⁻¹ IBA. *In vitro* raised plantlets with wel developed shoots and roots were transferred to pots containing sterile soilrite and acclimated, after which the successfully acclimated plants (80%) were transferred to pots under full sun where they grew well without any detectable phenotypic variation. Ahn et al. (2007) indicated that for plant regeneration from hypocotyls of *Ricinus communis* L., TDZ induced adventitious shoots at a higher rate compared to shoots induced by BA and also IBA was more efficient in root growth and shoot development than NAA.

Conclusion

The limitation of this study lies on the experimental materials, which were from two to three months old

Figure 3 Plantlet from nodal segment via callus on MS with 0.2 mgl⁻¹ TDZ and then 1.5 mgl⁻¹ IBA.

seedlings instead of mature trees with good characters. Seedling materials cannot completely represent the mature ones because of the difference of explants condition (Hou et al., 2010). However, for obtaining the physiology mechanism of adventitious root formation of *C. sativa Mill.* in a comparative short time, this experiment should be the guide of further research. We were successful in plant regeneration in *C. sativa* Mill. using MS medium supplemented with 0.2 mgL⁻¹ concentration of TDZ and 1.5 mgL⁻¹ IBA. This regeneration protocol will be useful not only for further research studies such as genetic cell transformation or protoplast fusion studies, but also for commercial nurseries that could use virus-free plants and agricultural practices to reduce pesticide use and increase yield production.

REFERENCES

Ahn Y, Vang L, McKeon T, Chen G (2007). High-frequency plant regeneration through adventitious shoot formation in castor (*Ricinus communis* L.) *In Vitro* Cell. Dev. Biol. Plant. 43:9-15.

Arora K, Sharma M, Srivastava J, Ranade SA, Sharma A (2010). Rapid *in vitro* cloning of a 40-year-old tree of *Azadirachta indica* A. Juss. (Neem) employing nodal stem segments. Agroforest. Syst. 78:53-63.

Avilés F, Ríos D, González R, Sánchez-Olate M (2009). Effect of culture medium in callogenesis from adult walnut leaves (*Juglans regia* L.). Chilean J. Agric. Res. 69:460-467.

Ballester A, Bourrain L, Corredoira E, Gonçalves JC, Lê CL Miranda ME, San-José MC, Sauer U, Vieitez AM, Wilhelm E (2001). Improving chestnut micropropagation through axillary shoot development and somatic embryogenesis. For. Snow Landscape Res. 76:460-467.

Bandyopadhyay S, Cane K, Rasmussen G, Hamill J (1999). Efficient plant regeneration from seedling explants of two commercially

important temperate eucalypt species-*Eucalyptus nitens* and *E. globulus*. Plant Sci. 140:189-198.

Bounous G (2005). The chestnut: A multipurpose resource for the new millennium. In Proceedings of the Third International Chestnut Congress; Abreu CG, Rosa E, Monteiro AA Eds. Acta Hortic. 693:33-138.

Chandra I, Bhanja P (2002). Study of organogenesis *in vitro* from callus tissue of *Flacurtia jangomonas* (Lour.) Raeush through scanning electron microscopy. Curr. Sci. 83:476-479.

Chaturvedi R, Razdan MK, Bhojwani SS (2004). *In vitro* clonal propagation of an adult tree of neem (*Azadirachta indica* A. Juss.) by forced axillary branching. Plant Sci. 166:501-506.

Dhabhai K, Batra A (2010). Hormonal Regulation Impact on Regeneration of *Acacia nilotica* L. a Nitrogen Fixing Tree. World Appl. Sci. J. 11:1148-1153.

Fereol L, Chovelon V, Causse S, Triaire D, Arnault I, Auger J, Kahane R (2005). Establishment of embryogenic cell suspension cultures of garlic (Allium sativum L.), plant regeneration and biochemical analyses. Plant Cell Rep. 24:319-325.

Fracago F, Echeverrigaray S (2001). Micropropagation of *Cunila galioides*, a popular medicinal plant of south Brazil. Plant Cell Tissue Org. Cult. 64:1-4.

Gamborg OL, Muller RA, Ojima K (1968). Nutrient requirement of suspension cultures of soybean root cells. Exp. Cell Res. 50:151-158.

Girijashankar V (2011). Micropropagation of multipurpose medicinal tree *Acacia auriculiformis*. J. Med. Plant Res. 5:462-466.

Hall RH (1976). Hormonal mechanism for differentiation in plant tissue culture. *In vitro* Cell. Dev. Biol. 12:216-224.

Harry IS, Thrope TA (1994). *In vitro* culture of forest trees. In Plant Cell and Tissue Culture (Vasil IK, Thrope TA eds). Kluwer Acad. Publ. Dodrecht, Netherlands.

Heller R (1953). Reserches sur la nutrition minerale des tissues vegetaux cultives '*in vitro*'. Annales des Sciences Naturelles (Botanique) Biol. Veg. 14:1-223.

Hou JW, Guo SJ, Wang GY (2010). Effects of *in vitro* subculture on the physiological characteristics of adventitious root formation in microshoots of *Castanea mollissima* cv. 'yanshanhong'. J. Forest. Res. 21:155-160.

Huetteman A, Preece EJ (1993). Thidiazuron: a potent cytokinin for woody plant tissue culture. Plant Cell Tissue Org. Cult. 33:105-119.

Kumar S, Singh N (2009). Micropropagation of *Prosopis ceneria* (L.) Druce-A multiple desert tree. Researcher 1:28-32.

Kvaalen H, Gram Daehlen O, Tove Rognstad A, Grønstad B, Egertsdotter U (2005). Somatic embryogenesis for plant production of *Abies lasiocarpa*. Can. J. For. Res. 35:1053-1060.

Liu CZ, Murch SJ, Demerdash MEL, Saxena PK (2003). Regeneration of the Egyptian medicinal plant *Artemisia judaica* L.. Plant Cell Rep. 21:525-530.

Maheshwari P, Kovalchuk I (2011). Efficient shoot regeneration from internodal explants of *Populus angustifolia*, *Populus balsamifera* and *Populus deltoids*. New Biotechnology (in press).

Murashige T, Skoog F (1962). A revised medium for rapid growth and bioassay with tobacco tissue cultures, Physiol. Plant. 15:473-497.

Osterc G, Zavrl Fras M, Vodenik T, Luthar Z (2005). The propagation of chestnut (*Castanea sativa* Mill.) nodal explants. Acta. Agric. Slovenica, 85:411-418.

Rodriguez R (1982). Multiple shoot-bud formation and plantlet regeneration *on Castanea sativa Mill.* seeds in culture, Plant Cell Rep. 1:161-164.

Sánchez MC, San-José MC, Ferro E, Ballester A, Vieitez AM (1997). Improving micropropagation conditions for adult-phase shoots of chestnut. J. Hortic. Sci. 72:433-443.

San-José MC, Ballester A, Vieitez AM (2001). Effect of thidiazuron on multiple shoot induction and plant regeneration from cotyledonary nodes of chestnut. J. Hortic. Sci. Biotech. 76:588-595.

SAS (2001). SAS/STAT User's Guide (8.02) SAS Institute Inc., Cary, NC, USA.

Sauer U, Wilhelm E (2005). Somatic embryogenesis from ovaries, developing ovules and immature zygotic embryos, and improved embryo development of *Castanea sativa*. Biol. Plant. 49:1-6.

Tetsumura T, Yamashita K (2004). Micropropagation of Japanese Chestnut (*Castanea crenata* Sieb. et Zucc.) Seed. Hort. Sci. 39:1684-

1687.

Troch V, Werbrouck S, Geelen D, Van Labeke MC (2010). *In vitro* culture of Chestnut (*Castanea sativa* Mill.): using temporary immersion bioreactors. Acta. Hortic. 885:383-389.

Vieitez AM, Ballester A, Vieitez ML, Vieitez E (1983). *In vitro plantlet* regeneration of mature chestnut. J. Hortic. Sci. 58:457-463.

Vieitez AM, Vieitez ML (1980). Culture of chestnut shoots from buds *in vitro*. J. Horticult. Sci. 55:83-84.

Vieitez FJ, Merkle SA (2005). Castanea spp. Cehstnut. In: Biotechnology of fruit and nut crops (Ed. Litz RE). CABI publishing, Trowbridge. pp. 265-296.

Wang HM, Liu HM, Wang WJ, Zu YG (2008). Effects of Thidiazuror basal medium and light quality on adventitious shoot regeneratio from *in vitro* cultured stem of *Populus alba - P. berolinensis*. J. Fo Res. 19:257-259.

Zheng SJ, Henken B, De Maagd RA, Purwito A, Krens FA, Kik (2005). Two different Bacillus thuringiensis toxin genes confe resistance to beet armyworm (*Spodoptera exigua* Hubner) i transgenic Bt-shallots (*Allium cepa* L.). Trans. Res. 14:261-272.

26

Association of plasma protein C levels and coronary artery disease in men

Olfat A. Khalil[1,2], Kholoud S. Ramadan[1], Amal H. Hamza[1] and Safinaz E. El-Toukhy[1]

[1]Medical Biochemistry Department, Faculty of Science National Research center, Egypt.
[2]Biochemistry Department, Faculty of Medicine (girls)- Al-Azhar University- Cairo Egypt.

Several studies have shown the risk factor causes of coronary heart disease. In this study we tested the hypothesis that plasma protein C level might be used as a biomarker for coronary heart disease and myocardial infarction. The study included 60 men that were classified into 3 groups according to clinical examination; group I set as healthy control group, group II set as patients with ischemic heart disease and group III set as patients suffering from myocardial infarction. Different parameters were measured including, coagulation factor prothrombin time, partial thromboplastin time, fibrinogen and protein C. The activity of the cardiac enzymes (creatine phosphokinase, creatine phosphokinase-MB and lactate dehydrogenase) was also measured. Finally, lipids profile (total lipids, phospholipids, triacylglycerol, total cholesterol, low density lipoprotein cholesterone (LDL-C) and high density lipoprotein (HDL-C) were measured. The results demonstrate significant decrease level of protein C and prothrombin concentration (%) in ischemic heart disease and in myocardial infarction (MI) groups, when compared to the control group. Meanwhile, MI group showed more significant decrease comparing to IHD. Plasma protein C might serve as a marker for coronary artery disease in men. Further studies are warranted to bolster the data and to identify pathogenesis links between innate immune system activation and atherosclerosis.

Key words: Ischemic heart disease, myocardial infarction, protein C, coagulation factor, lipids profile.

INTRODUCTION

Atherosclerosis is a condition in which there is an artery wall thickness as a result of the accumulation of fatty materials such as cholesterol. It is a syndrome affecting arterial blood vessels, a chronic inflammatory response in the walls of arteries, caused largely by the accumulation of macrophage white blood cells and promoted by low density lipoprotein (plasma proteins that carry cholesterol and triglycerides) without adequate removal of fats and cholesterol from the macrophages by functional high density lipoprotein (HDL). It is commonly referred to as a hardening of the arteries. It is caused by the formation of multiple plaques within the arteries (Finn et al., 2010).

Myocardial infarction (MI) commonly known as a heart attack, results from the interruption of blood supply to a part of the heart, causing heart cells to die. This is most commonly due to occlusion of a coronary artery following the rupture of a vulnerable atherosclerotic plaque in the wall of an artery (Didangelos et al., 2009). Moreover, may be a minor event in a lifelong chronic disease, it may even go undetected, but it may also be a major catastrophic event leading to sudden death or severe hemodynamic deterioration. Myocardial infarction may be the first manifestation of coronary artery disease, or it may occur, repeatedly, in patients with established disease (Thygesen et al., 2007). Disturbed lipid profile is one of the most important and potent risk factors in ischemic heart disease (IHD). It has been demonstrated that raised oxidative stress promotes several undesirable pathways including the formation of oxidized low density lipoprotein (O-LDL) and oxidized cholesterol which encourages cholesterol accumulation in arterial tissues (Maharjan et al., 2008)

Excessive consumption of saturated fat and cholesterol has been linked with increased concentration of plasma fibrinogen, a major risk factor for thrombosis that leads to heart attacks and strokes (Avogaro et al., 1988). Abnormalities in the blood coagulation factors regulating thrombosis may also contribute to the risk and extent of thrombosis (Hamker et al., 1991).

Homeostasis is a complex physiologic process involving a promoting factor (procagulants) counter-balance by naturally occurring inhibitors. Derangement of this balance is considered to play an important role in the pathogenesis of thrombosis (Kenneth, 1992). Recently, it was found that heparin coagulation factor (HCII) inhibits thrombin activity by binding to derma tan sulfate and has been shown to be a novel and independent risk factor for atherosclerosis (Huang et al., 2008).

Human protein C is a vitamin K-dependent glycoprotein structurally similar to other vitamin K-dependent proteins affecting blood clotting, such as prothrombin, factor VII, factor IX and factor X. Protein C, also known as autopro-thrombin IIA and blood coagulation factor XIV, is a zymogenic (inactive) protein, the activated form of which plays an important role in regulating blood clotting, inflammation, cell death and maintaining the permeability of blood vessel walls in humans and other animals (Mosnier et al., 2007).

In addition, activated protein C accelerates fibrinolytic activity by raising the level of plasminogen activator or by decreasing the level of plasminogen activator inhibitor (Zateishchikov et al., 1990). The determination of protein C makes it possible to identify patients who are at risk of thrombosis so that preventive measure can be instituted (Sturk et al., 1987). We tested the hypothesis that plasma protein C and different coagulation factors might be able to be used as a biomarker for coronary artery disease (CAD).

MATERIALS AND METHODS

Subjects

Sixty male subjects were selected from Health Insurance Hospital, Al-Azhar University, Cairo Egypt. Their ages ranged between 40-65 years old. They were classified into three groups after complete history taking and through clinical examinations and full investigations by electrocardiogram (ECG), echo, chest X-ray, and laboratory ratory examinations. Group I: Control group, twenty normal health men with no history for disease and drug intake. Group II: Twent cases were suffering from ischemic heart diseases (IHD). Group III Twenty cases were diagnosed as having MI disease. None of the study participants had any of the following disorders, associated with an acute phase reaction, febrile acute infection or acute state of a chronic infection or an inflammatory disease, underlying hematologic or malignant diseases and renal disorders. Current medication and sociodemographic characteristics were also recorded. Participation was voluntary, written informed consent was obtained from each subject upon entry into the study. The study was approved by the ethics committee of the University of Al-Azher

Blood sampling

Ten ml of blood were collected from each subjects, 5 ml blood were added to 3.8% trisodium citrate solution, in the proportion of 9 volumes of blood to one volume of anticoagulant solution and centerifugated at 3000 rpm for 10 min. The plasma was separated for determination of the following parameters: Activated prothrom bin time (PT.second) according to Loeliyer et al. (1985). Activated partial thromboplastin time (PTT.second) was measured according to Munteam et al. (1992). Also plasma fibrinogen g/L and activated protein C% were measured according to Exner and Voasoki (1983).

The other 5 ml of blood were let to clot and the serum was used to determine the following parameters, activity of the enzyme creatine phosphokinase (CPK) according to the study of Szasz (1976) and the activity of the isoenzyme creatine phosphokinase MB (CPK-MB) (Szasz, 1976).

Also the activity of the enzyme lactate dehydrogenase (LDH (Anon, 1977) was measured. Serum total lipid, phospholipids triacylglycerol, cholesterol and HDL-C were also measured according to Knight et al. (1972), Henry (1974), Stavropoulos (1974) and Abell et al. (1952). Low density lipoprotein cholesterone (LDL-C was measured according to the equation:

$$LDL\text{-}C = \text{total cholesterol} - HDL\text{-}C - TG/5$$

Statistical analysis

All results were expressed as mean ± S.E of the mean. Statistica Package for the Social Sciences (SPSS) program, version 11.0 (Chicago, IL, USA) was used to compare significance among three groups. Difference was considered significant when $p \leq 0.05$.

RESULTS

It is clear from Table 1 that there was a highly significant increase in the level of prothrombin time (in seconds) in group II and III when compared to the control group. Also, there was a highly significant decrease in prothrombin concentration percentage in IHD and MI groups compared to the control group.

Moreover, MI group showed a significant decrease compared to IHD group. It could be concluded from this Table that partial thromboplastin time level showed highly significant increase in MI group as compared to IHD group. Plasma fibrinogen increased significantly in IHD and MI groups compared to the control group. Furthermore, IHD group showed lower value of plasma fibrinogen as compared to MI group. Finally, there was a highly

Table 1. Clinical Characteristics and Coagulation Factors in Control, IHD and MI groups.

Parameter	Control group (n=20)	IHD (n=20)	MI (n=20)
Age (yrs)	50.0 ± 11	60 ± 9.5*	56.0 ±10.0
BMI (Kg/m^2)	27.0 ± 3.5	27.8± 3.4	27± 3.5
Current smoker, n (%)	6 (30.0)	15(75.0)	13 (65.0)
Prothrombin time (seconds)	12.4 ± 0.33	14.94± 0.84*	16.17 ± 1.09*§
Prothrombin conc.%	91.88 ± 6.13	57.6 ± 8.41*	49.7 ± 7.3*§
Partial thromboplastin (seconds)	33.74± 2.97	38.64 ± 6.99*	50.47± 6.24*§
Plasma fibrinogen (mg %)	262.8 ± 14.04	306.55 ± 7.33*	346.45 ± 16.63*§
Activity of Protein C (%)	109.53 ± 5.07	81.9 ± 7.77*	62.35 ± 5.13*§

Data are presented as mean ± SD, *p <0.05 vs. control, § p <0.05 significant (IHD) vs. (MI).

Table 2. The activity of cardiac enzymes in control, IHD and MI groups.

Parameter	Control group (n=20)	IHD (n=20)	MI (n=20)
Creatine phosphokinase (U/L)	90.15 ± 19.98	201.8 ± 13.82*	679.9 ± 17.807*§
Creatine phosphokinase –MB (U/L)	10.72 ± 2.62	35.45 ± 6*	107.25 ± 5.487*§
Lactate dehydrogenase (U/L)	114.1 ± 2.81	133.05 ± 3.047*	361.6 ± 10.473*§

Data are presented as mean ± SD, *p <0.05 vs. control, § p <0.05 significant (IHD) vs. (MI).

significant decrease in the levels of protein C in IHD and MI group as compared to control group. Meanwhile, MI group showed significant decrease in protein C level compared to IHD.

It could be seen from Table 2 that there was a highly significant increase in level of CPK, CPK-MB, and lactate dehydrogenase concentration in MI group as compared to IHD group. Also both IHD, and MI groups showed a highly significant increase compared to the control group. The results illustrated in Table 3 showed that lipid profiles (total lipids, total cholesterol, LDL-C) in MI and IHD groups were highly significant increased as compared to the control group, while HDL-C decreased significantly in these groups.

MI groups showed a significant increase in lipids profile as compared to IHD. The present study showed no significant change in phospholipids concen-tration in IHD group while there was highly significant increase in MI group as compared to the control group.

DISCUSSION

Prothrombin time (PT) measured the clotting time of plasma in the presence of an optimal concentration of tissue extract (thromboplastin) and indicates the overall efficiency of the extrinsic clotting system. The test depends on reactions with factors V, VII, and X and on the fibrinogen concentration of plasma (Dacie and Lewisi, 1991). The results of this study are in agreement with that of Gupta et al. (1997) and Folsom et al. (1997). Also, our

results are in accordance with the study of Erbay et al (2004). The elevation of fibrinogen and prothrombin levels acts as a risk factor and may play a causative role in cardiovascular disease. Also, Chambliss et al. (1992) reported highly activated partial thromboplastin time and plasma factor VIII but decrease in value of protein C and antithrombin III activity. Also Kenneth et al. (1992) indicated that plasma fibrinogen concentration factor VII, protein C and antithrmboplastin III levels were signify-cantly higher in early atherosclerosis in carotid arteries which may be a useful marker for identifying individuals at high risk of developing arterial disorders.

In addition, this work showed a highly significant decrease in plasma protein C levels in IHD and MI group as compared to the control group. These results are in agreement with that obtained by Lauribe et al. (1992). The raised fibrinogen and decreased protein C appeared to be risk factor for sudden cardiac death. Gibbs et al. (1992) reported increase in the procoagulants fibrinogen, factor VIII, and decrease in protein C and antithrombin III in cases of myocardial infarction. Treatment with activated protein C significantly improved hemodynamic after ischemia-reperfusion and reduced ischemia-reperfusion-induced myocardial apoptosis in rats (Pirat et al., 2007). Henkens et al. (1993) observed that thromboembolic events occurred in 30% of protein C deficient and in 35% of protein S deficient persons. Also, Dahl back et al. (1993) reported poor anticoagulant response to activated protein C in several families with hereditary tendency to venus thrombosis. Moreover, It was also reported that concentration of protein C level

Table 3. Lipids profile levels in control, IHD and MI groups (Mean ± SD).

Parameter	Control group (n=20)	IHD (n=20)	MI (n=20)
Total lipids (mg %)	628.8 ± 42.81	850.25± 50.43*	982.62 ± 84.38*
Phospholipids (mg %)	207.8 ± 22.84	231.25± 12.84	250.8± 6.44*§
Triacylglycerol (mg %)	125.9 ± 20.67	174.3± 6.891*	185.5 ± 13.4*§
Total cholesterol (mg %)	191.95 ± 15.22	303.6 ± 20.35*	329.35 ± 22.88*
HDL-C (mg %)	55.2 ± 6.12	42.4 ± 4.68*	41 ± 4.21*
LDL-C (mg %)	107.35 ± 7.35	222.5 ± 16.2*	247.75± 19.32*

Data are presented as mean ± SD, *p <0.05 vs. control, § p <0.05 significant (IHD) vs. (MI).

and activity of protein C deficiency has a bearing with pulmonary infarction.

Van-der-Ban et al. (1996) found a reduced response to activated protein C (APC) is associated with an increased risk for cerebrovascular disease but not with an increased risk for myocardial infarction. Goto et al. (1992) reported augmented plasma protein C activity after coronary thrombolysis with urokinase in patients with acute myocardial infarction, thus, it was suggested that urokinase administration for coronary thrombolysis not only causes fibrinolysis, but also induces thrombin activity which may be antagonized by augmented intrinsic protein C activity. The diagnosis of MI is established in patients with chest pain and equivocal electrocardiogram changes by demonstrating a rise in blood levels of creatine kinase MB (CK-MB) and/or an increase in cardiac troponin I (cTnI) or cardiac troponin T (cTnT). Previous studies have shown that levels of CK-MB are increased in the left ventricle of individuals with heart disease Welsh et al, (2002).

While CK-MB as a cardiac marker depended on its relatively high concentration in heart muscle (>20%) compared to typical skeletal muscle (1–2%). There is evidence that higher concentrations of CK-MB in heart may result from ischemic stress. For example, concentrations of CK-MB have been found to be significantly higher in heart muscle of experimental animals and human myocardium with coronary artery disease, aortic stenosis, or heart failure, compared to normals. A number of studies have shown that the concentration of CK-MB is higher in ventricular myocardial tissue in animal models of hypertrophy or ischemia and in humans with a variety of cardiac conditions, compared to controls or young individuals without cardiac disease. In human myocardial biopsy material, concentrations of CK-MB have been reported to be 100-fold greater in hearts from patients with aortic stenosis, coronary artery disease, and coronary artery disease with left ventricular hypertrophy compared to patients without such findings (Welsh et al., 2002).

The study of many enzymes activities are valuable in diagnosis of many disease as the rise in the serum enzyme of CPK and CPK-MB and lactate dehydrogenase are commonly used for diagnosis of coronary heart disease. In the present study, it was found that there was a highly significant increase of serum creatine phosphor-

kinase CPK, CPK-MB and serum lactate dehydrogenase levels in IHD and MI groups as compared to the control group. The discovery of isoenzyme determination has improved the diagnostic value of enzyme tests. The cardio specific isoenzyme of CK (CK-MB) has been used successfully for the detection of myocardial infarction. Our results are in agreement with that of Welsh (2002) and Kato et al. (2006). The European Society of Cardiology (ESC) and American College of Cardiology (ACC) state that any elevation, however small, of a troponin or the creatine kinase MB (muscle, brain) iso enzyme is evidence of myocardial necrosis and that the patient should be classified as having myocardial infarction, however small (Antman et al., 2004).

Hyperlipidemia refers to increased levels of lipids (fats in the blood, including cholesterol and triglycerides. Although hyperlipidemia does not cause you to feel bad, it can significantly increase your risk of developing coronary heart disease, also called coronary artery disease or coronary disease. People with coronary disease develop thickened or hardened arteries in the heart muscle. This can cause chest pain, a heart attack, or both (Saunders 2007). Hyperlipidemia is a disturbance of the lipid transport system that results from abnormalities in the synthesis or degradation of plasma lipoprotein (Brown and Goldstein, 1983). There is strong evidence between abnormalities of lipids metabolism and gradual change of atherosclerosis and coronary heart disease. The substance that gives the atheroma its character is the lipid, chiefly cholesterol esters (Roheim, 1986). Measurements of plasma lipid and lipoprotein levels have been used in diagnostic medicine to assess the risk of coronary artery disease. Cholesterol and triglycerides levels have been recognized as predictors of CAD. HDL-C and LDL-C have been considered the most accurate indicators of CAD. Increased level of LDL cholesterol is associated with increased incidence of CAD.

Our study reveals a high significant increase in serum total lipids in IHD and MI compared to control group. This is in accordance with the results of Brown and Goldstein (1983). They noted the increase in serum total lipid in CAD patients. Also, it was reported that hypertension, smoking and hyperlipidemia are the most important risk factors of IHD. Saturated fat intake has been linked to an increased risk of cardiovascular disease (CVD), and this effect is thought to be mediated primarily by increased concentrations of LDL cholesterol (Patty et al., 2010). The present study shows no significant change in phospholipids concentration in IHD group while there was highly significant increase in MI group as compared to the control group. Natio (1988) proved that the ratio of phospholipids to cholesterol ester level resulted in a corresponding change in phospholipids in similar direction.

Furthermore, triacylglycerol increased significantly in both IHD and MI as compared to the control group. It was suggested by Despres et al. (1990) that triglyceride molecules are not themselves atherogenic. High plasma triacylglycerol level may indirectly represent a cardiovascular risk factor through its effect on lipoprotein composition. Also, the results are in agreement with those of Jensen et al. (1991) who found that triglyceride level were higher in CAD patients and severity of coronary atherosclerosis has been shown to correlate better with serum concentration of triglyceride than of cholesterol.

Sigurdsson et al. (1992) suggested that elevated triglyceride levels are important as a risk factor only when associated with other lipoprotein abnormalities (elevated LDL-C or decreased HDL-C). Also Welin et al. (1991) demonstrated from follow up of incidence of coronary heart disease increased 5 fold from the lowest to the highest of value triacylglycerol. Increased serum triglycerides are a major coronary risk factor in elderly men. Moreover, Assmann (1992) suggested that triglyceridemia is a powerful additional coronary risk factor when excessive triacylglycerol coincide with a high ratio of plasma LDL-C to HDL-C.

In this study, there was a highly significant increase in serum total cholesterol in IHD and MI as compared to control group. The present study is in agreement with Kondreddy et al. (2010) who found that total cholesterol (TC) and LDL-C were significantly increased while HDL-C was significantly decreased among the CHD group. This is in accordance with the results of Bainton et al. (1992). They found that total cholesterol was higher in CAD patients than normal controls. Also, Swedarsen et al. (1991) showed that hypercholesterolemia without associated hypertriglyceridemia was the commonest abnormality. Jensen et al. (1991) observed that plasma cholesterol above the level approximately 270 mg% proportionally increased of CAD. It was concluded by Bainton (1992) that cholesterol was a more important risk factor than HDL-C and was considered to be the most important single lipid risk factor in men. The level of the plasma lipoprotein play an important role in the pathogenesis of atherosclerosis, particularly low level of HDL-C and high level of LDL-C (Roheim, 1986).

In this work the HDL-C levels showed highly significant decrease in both IHD and MI groups as compared to control group. A similar result was found by Duval et al. (1989) who reported that there was s significant decrease

in HDL in cardio vascular disease. Pometta et al. (1987) reported that HDL-C concentration correlated inversely to the development of atherosclerosis; therefore they were considered to be negative risk factor. Finally, there was a highly significant increase of LDL-C in IHD and MI groups comparing to the control group. Similar results were obtained from Henriksen (1984) who found that the lipids in atherosclerotic lesion are derived from plasma LDL. Also Jensen et al. (1991) found that LDL was higher in CAD patients than control. Avogaro et al. (1988) and Badimon et al (1992) studied LDL metabolism and proved that LDL seems to be responsible for transported approximately 70% of the total cholesterol by LDL-C the liver to the tissues.

In conclusion, this study demonstrates the association between protein C and another coagulation factor and other important risk factors as lipid profile and plasma protein C might serves as a marker for coronary artery disease in men.

REFERENCES

Abell LL, Levey BB, Brodie BB (1952). A simplified method for the determination of total cholesterol in serum and determination of its specificity. J. Biol. Chem. 195:357-366.

Anon MY (1977). Method for the estimation of lactate dehydrogenase, J Clin Chem. Clin. Biochem. 15:249.

Antman EM, Armstrong PW, Bates ER, Green LA, Hand M, Hochman JS (2004). ACC/AHA guidelines for the management of patients with ST-elevation myocardial infarction: a report of the American College of Cardiology/American Heart Association Task Force on Practice Guidelines.

Assmann G, Schulte H (1992). Role of triglycerides in coronary artery disease: Lessons from the prospective cardiovascular master study. Am J Cardio, 70(19):10H-13H.

Avogaro P, Bittoto B, Quinici C (1988). Are apolipoprotein better discriminators than lipid for atherosclerosis. Lancet 21(337):1141-1142.

Badimon J, Fuster V, Bedinon L (1992). Role of HDL in the regression of atherosclerosis. Circulation 186:86-94 supplement.

Bainton D, Bolton C, Baker A, Lewis B (1992). Plasma triglycerides and HDL as predictors of ischemic heart disease in British men. Br. Heart J. 68:60-66.

Brown MS, Goldstein JL (1983). Regulation of plasma cholesterol by lipoprotein receptors. Science 212:628-635.

Chambliss LE, McMahon R, Wu K, Folsom A, Finch A, Shen YL (1992). Short term intraindividual variability in homeostasis factors. The ARIC study (atherosclerosis risk in communities) Intraindividual variability study. Ann. Epidemiol. 2(2):723-733.

Dacie JV, Lewisi M (1991). Investigations of acute haemostatic failure. Practical Hematology. 7th edition. Churchill (living stone). p. 279.

Dahl back B, Carlson M, and Sevens son PJ (1993). Familial thrombophilia due to previously unrecognized mechanism characterized by poor anticoagulant response to activated protein C: Prediction of cofactor to activated protein C. Proc Natl. Acad. Sci USA. 90:1004-1008.

Despres JP, Moorjani S, Pouliot ML, Lupien PJ, Bouchard C (1990). Correlates of plasma very low density lipoprotein concentration and composition in premenopausal woman. Metabolism 39(6):577-583.

Didangelos A, Simper D, Monaco C, Mayr M. (2009): Proteomics of acute coronary syndrome. Curr. Atheroscler. Rep. 11(3):188-195.

Duval F, Frommberg K, Drocke T, and Locour B (1989). Influence of end stage renal failure on concentration of free apo AI in serum. Clin. Chem. 335(6):963-966.

Erbay R, Turban A, Aksoy H, Senen Y, Yetkin K (2004). Activation of coagulation system in dilated cardiomyopathy: comparison of patients with and without left ventricular thrombus. Coron. Artery Dis.15 (5):265-268.

Finn AV, Nakano M, Narula J, Kolodgie FD, Virmani R (2010). Concept of vulnerable/unstable plaque. Arterioscler. Thromb. Vasc. Biol. 30(7):1282-1292.

Folsom AR, Wu KK., Rosamond WD, Sharrett AR, Hambles LE (1997). Prospective study of haemostatic factors and incidence of coronary heart disease: The atherosclerosis risk in communities (ARIC) study. Circulation 96(4):1102-1108.

Gibbs NM, Graw Ford GP, And Michalo P (1992). Postoperative changes in coagulant factors following abdominal aortic surgery. J Cardio. Vasc. 6(9):680-685.

Goto S, Hamda S, Kawai Y, Watanabe K, Abe S, Takahashi E, Hori S, Ikeda Y (1992). Augmented plasma protein C activity after coronary thrombolytic with urokinase in patients with acute myocardial infarction. Cardiology 80(3-4):252-256.

Gupta A, Sikka M, Madan N, Dwidedi S, Russia U, Sharama S (1997): Homeostatic function in coronary artery disease. Indian J. Pathol. Microbiol. 40(2):133-137.

Hamker HC, Wieldre S, and Beglum W (1991). The thrombin potential, From: Fraxiprine, the 2nd international world symposium: Recent pharmacological and clinical data. New York, Schattauer. pp. 89-101.

Henkens CM, Van der Mear J, Hillege JL, Vander W and Halie MR (1993). The clinical expression of hereditary protein C and protein S deficiency: a relation to clinical thrombotic risk factors and to levels of protein C and protein S. Thrombos. Hemestas. 4(4):555-62.

Henriksen T (1984). Possible biological mechanisms that make LDL a risk factor in atherogenic. Eur. Surg. Res.16:62-67. Supplement.

Henry RJ (1974): Principles and technology for estimating serum phospholipids. Harper and Row, Clinical chemistry, New York.

Huang SS, Huang PH, Chen YH, Sung SH, Chiang KH, Chen JW, Lin SJ (2008). Plasma heparin cofactor II activity is an independent predictor of future cardiovascular events in patients after acute myocardial infarction. Coron. Artery Dis. 19(8):597-602.

Jensen G, Nyboe J, Schnohr P (1991). Risk factors of acute MI in Copenhagen, II-smoking, alcohol intake, physical activity, obesity, oral contraception, diabetes, lipids and blood pressure. Eur. Heart J. 12(3):298-308.

Kato GJ, McGowan V, Roberto F, Morris C, Nichols J, Wang X, Poljakovic M, Morris M, Gladwin M (2006). Lactate dehydrogenase as a biomarker of hemolysis-associated nitric oxide resistance, priapism, leg ulceration, pulmonary hypertension, and death in patients with sickle cell disease. Blood 107(6):2373.

Kenneth KW, Aaron RF, Geraido H, Maureen GC, Ralph B (1992). Association of coagulation factors and inhibitors with carotid artery atherosclerosis. Ann. Epidemiol. 2:471-480.

Knight JA, Anderson S and Rawle JM (1972). Chemical basis of sulfophosphovanillin reaction for estimating total serum lipids. Clin Chem. 18(3):199.

Kondreddy A, Chenak S, Shakila I, Meka B, Jarari A, Naseb N, Peela J (2010). Correlation of Lipid profile in Coronary Heart Disease patients in Libya. Clinical Chemistry. 56:(6) Supplement.

Lauribe P, Benchimol D, Dartigues J, Dada S, Benchinol H (1992). Biological risk factors for sudden death in patients with coronary artery disease and without heart failure. Int. J. Cardiol. 34(3):307-318.

Loeliyer EA, Van den Besseluar AM, Lewis SM (1985). Reliability and clinical impact of the normalization of the Prothrombin time in oral anticoagulant control. Thromb. Hemost. 53(1):148-154.

Maharjan BR, Jha JC, Adhikari D, Risal S, Alurkar VM, Singh P (2008). Oxidative stress, antioxidant status and lipid profile i ischemic heart disease patients from western region of Nepal. Nepa Med. Coll. J. 10(1):20-24.

Mosnier LO, Zlokovic BV, Griffin JH (2007). The cytoprotective protei C pathway. Blood 109(8):3161-3172.

Natio MK (1988). New diagnostic tests for assessing CAD risk. Recer aspects of diagnosis and treatment of lipoprotein disorders: In put o prevention of atherosclerotic diseases. pp. 49-62. New York.

Pometta D, Janes R and Suenran A (1987). HDL and CAD a familia trend. Adv Exp. Med. Biol. 210:219-24.

Pometta D, Janes R, Suenran A, (1987). HDL and CAD a familial trenc Adv. Exp. Med. Biol. 210:219-24.

Roheim N (1986). Atherosclerosis and lipoprotein metabolism. Role c reverse cholesterol transport. Am. J. Cardiol. 57:3c-10c.

Saunders (2007). Dorland's Medical Dictionary for Health Consumers By, an imprint of Elsevier.

Sigurdsson G, Uggi A and Nikules S (1992). Predictive value o apolipoproteins in a prospective survey of CAD in men. Am. J Cardiol. 69(16):1251-1255.

Siri-Tarino P, Sun Q, Hu F, Krauss R (2010). Saturated fat carbohydrate, and cardiovascular disease. Am. J. Clin. Nutr 91(3):502-509.

Stavropoulos W (1974). A new enzymatic calorimetric procedure fo determination of serum triglycerides. Clin. Chem. 20(8):857.

Sturk A, Morriensalamons WM, Huisman MV, Borm JJ, Buller HR and Tencate Jw (1987). Analytical and clinical evaluation of commercia protein C assays. Clin. Chem. Acta 185:263-270.

Swedarsen M, Becker D and Jialal I (1991). Lipid and lipoprotein abnormalities in South African Indian men with MI. Cardiol 78(4):348-356.

Szasz G. (1976). Creatine kinase in serum: Determination of optimun reaction conditions. Clin. Chem. 22:650-656.

Thygesen K, Alpert JS White HD (2007): Universal definition o myocardial infarction. Eur.Heart J.28 (20):2525-2538.

Van der Ban JG, Bots ML, Haver Kate F, Slagboom PE, Merjer P, de Jong PT, Hofman A, Grobbe DE, Kluft C (1996). Reduced response to activated protein C is associated with increased fo cerebrovascular disease. Ann Intern Med, 15:125(4):265-269.

Welin L, Eriksson H, Larsson N , Ohlson LO (1991). Triglycerides, a major coronary risk factor in elderly men, A study of men born in 1913. Eur Heart J. 12(6):700-4.

Welsh TM, Kukes G, Sandweiss L (2002). Differences of creatine kinase MB and Cardiac Troponin I concentrations in normal and diseased human myocardium. Ann. Clin. Lab. Sci. 32:44 - 49.

Zateishchikov DA, Gratsian Skil NA, Dobrovolskii AB, Metel Skaia V (1990). Plasminogen activator inhibitor and protein C: their relation to plasma lipids and lipo- and apoproteins in ischemic heart diseases o different durations. 30(4):47-50.

27

Toxicity study of the anti-hypertensive agent perindopril on the entomopathogenic fungus *Metarhizium anisopliae* (Metschnikoff) Sorokin assessed by conidia germination speed parameter

Ligia Maria Crubelati Bulla[1], Daniela Andressa Lino Lourenço[2], Sandro Augusto Rhoden[1], Ravely Casarotti Orlandelli[1] and João Alencar Pamphile[1]

[1]Department of Biotechnology, Genetics and Cellular Biology (DBC), Universidade Estadual de Maringá, Brazil.
[2]Department of Zootechny (DZO), Universidade Estadual de Maringá, Brazil.

Hypertension is an important vascular disease to the global public health, since it constitutes the principal cause of death from childhood to adulthood. In order to alleviate its symptoms, the treatment is accomplished by anti-hypertensive drugs, among them, is perindopril, an angiotensin-converting enzyme (ACE) inhibitor. The entomopathogenic filamentous fungus *Metarhizium anisopliae* is widely used for biological control and it has been promising in toxicity studies of substances assessed by conidia germination speed parameter. This study aimed to verify the effect of different concentrations of perindopril on the conidia germination speed of the model fungus *M. anisopliae*, for detecting a possible toxic effect of this medicament in another eukaryote. Conidia of *M. anisopliae* were incubated with perindopril in concentrations of 200 and 20 µg/ml at 28°C for 12 h, sampled at 0, 6, 8, 10 and 12 h and analyzed by light microscopy. The frequency of dormant, embedded, bud and germinated conidia was counted. As a result, perindopril in concentrations of 200 and 20 µg/ml increased the germination speed of *M. anisopliae* conidia, when compared to the negative control (absence of perindopril). It indicates that these two concentrations of perindopril have no toxicity on *M. anisopliae*, considering the Bayesian analysis.

Key words: Angiotensin-converting enzyme inhibitor, vegetative development, model fungus, entomopathogen, Bayesian analysis.

INTRODUCTION

The prevention of vascular diseases associated to hypertension became a priority question in global-health politics (MacMahon et al., 2008). In infancy, the occurrence of arterial hypertension varies between 1-3% and it increases to about 10% in adolescence, especially in obese patients (Dios, 2011). Cardiovascular diseases are the principal cause of death among adults in developed countries and there is an increase of them in developed countries and there is an increase of them in developing countries (Thom et al., 1998).

The estimative is that blood-pressure-related diseases kill around 8 million people per annum (Lawes et al., 2008). In Brazil, cardiovascular diseases are already the principal cause of death, being responsible for 30% of deaths in 2008; especially, the systemic arterial hypertension (SAH) affects more than 30 million Brazilians. In

2010, the number of people affected by SAH in the United States reached 74 million (Giollo et al., 2010).

The antihypertensive pharmacotherapy uses different drug classes, classified as diuretics, adrenergic inhibitors, direct acting vasodilators, calcium channel blockers, angiotensin-converting enzyme (ACE) inhibitors, angiotensin II (AT1 subtype) receptor antagonists and renin inhibitors (Giollo et al., 2010).

ACE inhibitors are widely employed as antihypertensive agents, not only for reducing systemic vascular resistance in hypertension patients but also to prevent and treat cardiovascular, renal and retinal diseases in patients with normal blood pressure (Wang et al., 2012), and it has been suggested that they decrease the risk of cancer (Lever et al., 1998; Yoshiji et al., 2001).

The anti-hypertensive agent perindopril is a long-acting and centrally active ACE inhibitor and according to the Perindopril Protection Against Recurrent Stroke Study (PROGRESS) Collaborative Group (2001) it has also been reported to reduce risks of dementia and cognitive decline in the patients with recurrent strokes. Moreover, a study conducted by Yamada et al. (2010), using a mouse model of Alzheimer's disease, indicates that oral administration of perindopril (0.1, 0.3 or 1 mg/kg/day) significantly ameliorates the cognitive impairment; the dosing regimen of 1 mg/kg/day inhibited the plasma ACE activities by more than 90% and the brain ACE activities by more than 50%.

The asexual filamentous fungus Metarhizium anisopliae (Metschnikoff) Sorokin is capable of infecting more than 300 species of insect-pests. It was used for the first time as a microbial control agent of insects by Elie Metschnikoff, in 1879, to control the wheat grain beetle (Anisoplia austriaca) and afterward, it was used to control the sugar beet curculio (Cleonus punctiventris) (Roberts and St. Leger, 2004). Several products based on Metarhizium spp. strains have been developed in past years (Copping, 2004), both for application in biological control and in the production of drug substances such as antibiotics and immunomodulators (Isaka et al., 2005).

The asexual sporulation is a common mode of reproduction for several filamentous fungi (Osherov and May, 2000). For insect-pathogenic fungi such as M. anisopliae, asexual spores or conidia represent the infective unit and the inoculum source in the field after application in biological control (Alves et al., 2011). Conidia germination represents a crucial step to create enough penetration sites, ensuring the success of these fungi during the infection of their hosts. After a successful germination and penetration, the speed and intensity of the vegetative growth determine the virulence of entomopathogens (Schumacher and Poehling, 2012).

The process of spore germination can be defined as a sequence of events that activates the resting spore (d'Enfert, 1997), what involves water uptake and wall growth (Griffin, 1994). The resting spore is converted into a rapidly growing germ-tube from which the mycelium will be formed by elongation and branching (d'Enfert, 1997).

This process is directly influenced by the incubatio period (Alves et al., 2011) and by environmental factors Water, oxygen, and carbon dioxide are universally require to activate the spore germination (d'Enfert, 1997). More over, optimum conditions such as temperature, humidity pH and nutrient sources are essential for the conidi germination.

To use the germination speed as parameter, conidi are inoculated into a liquid medium and samples ar collected periodically for counting and the number c germinated conidia is determined (Milner et al., 1991) Germination speed of M. anisopliae conidia has bee used as parameter to evaluate the toxicity of substances as employed by Alves et al. (2011) to assess the toxicit of the insect growth regulator lufenuron. Different para meters, including conidia germination, were used b Schumacher and Poehling (2012) to assess the effects c five pesticides on two strains of M. anisopliae.

According to the study of Rangel et al. (2004), th conidial germination parameter is employed to determine whether the substrate on which conidia were produced has influence on the endogenous reserves stored i conidia during conidiogenesis. Therefore, considering th conidia germination of M. anisopliae as a response du to the abiotic factors of environment, this study aimed t verify the effect of different concentrations of perindopr on the conidia germination speed of the model fungus M anisopliae, for detecting a possible toxic effect of thi medicament in a eukaryote.

MATERIALS AND METHODS

Fungal strain and culture media

Mato Grosso (MT) strain of M. anisopliae var. anisopliae wa obtained from the fungal culture collection of Laboratório d Biotecnologia Microbiana from Universidade Estadual de Maring Paraná, Brazil. This fungus was isolated from the insect host Deoi sp. Complete medium (CM) and liquid complete medium (LCM (Pontecorvo et al., 1953) were employed.

Conidia germination speed in the presence of perindopril

The MT strain was grown on Petri dishes containing CM (20 ml) a 28°C in biological oxygen demand (BOD). Conidia were obtained directly from seven to ten days-old sporulating cultures by scraping and suspended in aqueous solution of 0.01% Tween 80 (7 ml). This conidia solution was filtered through a glass funnel containing autoclaved gauze and it was added to saline solution (9 ml), obtaining a solution with a concentration of 1.01×10^7 conidia/ml. Into three glass flasks (10 ml) were inoculated 300 µl of conidia solution Treatment 1 (T1) and treatment 2 (T2) received perindopril solution (perindopril diluted in LCM) in concentration of 200 and 20 µg/ml respectively. The volume of both flasks was completed to 10 ml with LCM. In the control (C), only LCM (9.7 ml) was added.

Samples of C, T1 and T2 were collected in Eppendorf® microtubes and incubated in BOD at 28°C for 12 h. Samples were analyzed in triplicate at 0, 6, 8, 10 and 12 h of incubation. Germinated conidia were counted using Neubauer hemocytometer. A conidium was considered germinated when a germ-tube projected from it (Milner et al., 1991). The percentage germinatior

Table 1. Percentage of germinated *M. anisopliae* conidia in control and treatments throughout the incubation period.

Time/ phase	Dormant conidia			Embedded conidia			Bud conidia			Germinated conidia		
Flask	C	T1	T2	C	T1	T2	C	T1	T2	C	T1	T2
0 h	100	100	100	0	0	0	0	0	0	0	0	0
6 h	22.4	5.1	8.8	27.7	18.7	16.1	31.2	21.5	38.3	18.7	9.1	36.8
8 h	6.8	3.7	2.3	14	6.8	7.2	16.8	11	12.6	62.4	78.4	77.9
10 h	2.0	1.6	0.8	4.5	2.9	1.1	10	15	6.4	83.5	80.2	91.6
12 h	0	1.0	0	3	3.4	0.8	6.2	12.4	6.8	90.7	83.2	92.3

and germination speed were assessed by randomly observing 300 conidia.

Statistical analysis

As the Bayesian statistics is an approach that can handle small data sets and non-normal distributions (Alves et al., 2011), it was considered in order to study the behavior of conidia germination when the perindopril drug is present.

The conidia counting data was analyzed using the statistical package BRugs developed for R software (2008), and the Poisson distribution was assumed for the response variable. For each parameter, 10,000 values were generated in an MCMC (*Monte CarloMarkov Chain*) process, with a burn-in of 1000 initial values, and thinning interval of 10. The multiple comparisons procedure was based on the *a posteriori* samples of the estimates of the parameters. Significant differences were considered at the level of 5% between the treatments and control group if the zero value was not contained in the credibility interval of the desired contrast. A non-informative Gamma distribution was considered *a priori* for means of germinated conidia, that is, $\theta_n \sim G(10^{-3};10^{-3})$, where θ_n is the mean for each n treatment considered.

A model of logistic regression was fitted to study the behavior of conidial germination over time. Data were analyzed using the same package and software described above. The binomial distribution was considered for the data of germination percentage, and the following formula was used:

$\log it(\theta ij) = \beta_0 + \beta_1 time + \beta_2 time^2$,for the control group, treatments 1 and 2.

Where, log *it* is the logistic link function; *θij* is the germination percentage; β_0 is the intercept; β_1 is the linear logistic regression coefficient; β_2 is the quadratic logistic regression coefficient and *time* is the number of hours elapsed since the beginning of incubation.

For each parameter, 50,000 values were generated in an *Monte CarloMarkov Chain* (*MCMC*) process, considering a sample discard period of 5000 initial values, and thinning interval of 10. The significance of logistic regression coefficients was considered at the level of 5% if the zero value was not contained in the credibility interval for the parameter. A non-informative Normal distribution was considered a priori for parameters b_0, b_1 and b_2, that is, b_0, b_1, $b_2 \sim N(0;10^{-6})$.

When a logistic link function is considered, generally the conidia germination percentage is given by:

$$\theta_{ij} = \frac{\exp(\beta_0 + \beta_1 time + \beta_2 time^2)}{1 + \exp(\beta_0 + \beta_1 time + \beta_2 time^2)},$$

With *θij* as the percentage of germinated conidia.

RESULTS AND DISCUSSION

Although fungi were at first time considered to belong to the kingdom of plants, Wainright et al. (1993) and Baldauf and Palmer (1993) pointed that comparative analyses of ribosomal RNA and protein sequences have proved that this organisms are even more closely related to animal cells than was previously known. In 1974, Smith and Rosazza suggested the definition of microbial systems as those that could mimic the biotransformations observed in mammals (Cerniglia, 1997). Also, considering the amenability of fungi to classical and molecular techniques, fungi are considered as model systems for studying fundamental cell biological questions, since basic principles of many cellular processes are conserved between fungi and animals (Steinberg and Perez-Martin, 2008). Therefore, the Mato Grosso strain of *M. anisopliae* was chosen as model system to evaluate the toxic effects of the anti-hypertensive agent perindopril.

The Bayesian analysis of MT conidia germination showed the existence of a statistically significant difference between the germination speed of control and treatments (Tables 1, 2, 3 and 4), mainly for T2 (20 μg/ml of perindopril). According to the study of Alves et al. (2011), this statistical method is an approach that can work on datasets considering the true distribution, being reliable for small groups of data.

At the end of incubation time, it was possible to observe that T2 induced an increase in the number of germinated conidia (92.3%) compared to C (90.7%) and T1 (83.2%) (Table 1). Also, according to the curve of germination speed (Figure 1), the conidia germination started approximately between 2 and 3 h of incubation for T2 and about 3-4 h for C and T1.

The means and ICr for counting of germinated conidia are shown in Table 2. A Bayesian ICr of 95% was considered, in which 95% of samples are contained, and smaller the interval, less dispersed is the parameter. The means of germinated conidia in 12 h were: 85.76 (C), 96.32 (T1) and 113.80 (T2), with a credibility interval formed by 2.5 and 97.5%.

The means of conidia germination in each incubation period sampled throughout the 12 h and the credibility intervals are detailed in Table 3. According to these results, the incubation period of 10 and 12 h were statistically equal, showing a higher germination percentage.

Table 2. Bayesian estimates for the counting of germinated *M. anisopliae* conidia in the presence of different concentrations of perindopril.

Treatment	Mean	Standard error	95% ICr	
			2.50%	97.50%
Control[c]	85.76	0.07292	80.93	90.33
200 µg/ml (T1)[b]	96.32	0.07639	91.17	101.30
20 µg/ml (T2)[a]	113.80	0.08820	108.80	119.30

[a,b,c] Different letters indicate that the means differ.

Table 3. Means and credibility intervals for counting of the germinated *M. anisopliae* conidia throughout the incubation period.

Time (h)	Mean	Standard error	95% ICr	
			2.50%	97.50%
0	1.095e-03[d]	0.0002771	0.000	4.793e-03
6	6.701e+01[c]	0.0586000	61.700	7.253e+01
8	8.952e+01[b]	0.0706200	83.530	9.592e+01
10	1.725e+02[a]	0.0879500	163.700	1.814e+02
12	1.644e+02[a]	0.1044000	155.700	1.729e+02

[a,b,c,d] Different letters indicate that the means differ.

Table 4. Bayesian estimates for the logistic regression coefficients for control and treatments.

Parameter	b0	b1	b2	r²
C	-6.8590	1.4240	-0.04753	0.994821
T1	-1.74e+01	2.512e+00	-1.132e01	0.9987819
T2	-6.4410	1.5640	-0.05422	0.9970927

b0 is the intercept; b1 is the linear coefficient; b2 is the quadratic coefficient and r² is the determination coefficient of regressions.

The periods of 0, 6 and 8 h were statistically different, showing low germination percentage.

The logistic regression adjusted effectively the percentage of germinated conidia over time, with average r² of 0.997. The values of regression coefficients for each treatment, with their respective r² can be seen in Table 4.

Similarly, the germination speed of *M. anisopliae* conidia was employed by Alves et al. (2011) as parameter to assess the toxicity of different concentrations of lufenuron, an insect development inhibitor/ insect growth regulator. Conidia of MT strain were incubated at 28°C and sampled through 12 h. The Bayesian analysis showed that lufenuron do not inhibited the MT conidia in concentrations of 700 /ml and 1 mg/ml, whereas the inhibition occurred with a concentration of 2 mg/ml. These authors concluded that the two low concentrations tested had no toxicity on *M. anisopliae*, suggesting the use of lufenuron in biological-chemical combinations with a low environmental impact to combat insect-pests, maintain viable the fungal inoculums after this application in field.

Among other parameters, the conidia germination speed of two *M. anisopliae* strains was applied recently, by Schumacher and Poehling (2012), to evaluate the effects of pesticides fipronil, permethrin, imidacloprid, amitraz and NeemAzal (in concentrations of 0.32, 1.6, 8.0, 40, and 200 ppm). Also, permethrin and imidacloprid were combined in a ratio of 5:1 and tested in four combinations (1.6 ppm permethrin and 0.32 ppm imidacloprid; 8 ppm permethrin and 1.6 ppm imidacloprid; 40 ppm permethrin and 8 ppm imidacloprid; 200 ppm permethrin and 40 ppm imidacloprid). Analysis of variance showed that the maximum inhibition of germination caused by these pesticides was ≤ 15% and most of the pesticides had no negative influence on the germination. It was concluded that the low of the five pesticides dosages dissolved in 1% DMSO (dimethyl sulfoxide) were compatible with *M. anisopliae* for an integrated pest management approach.

The conidia germination speed of *M. anisopliae* strains has been employed as parameters in studies about solar ultraviolet radiation. Rangel et al. (2004) examined the

Germination speed

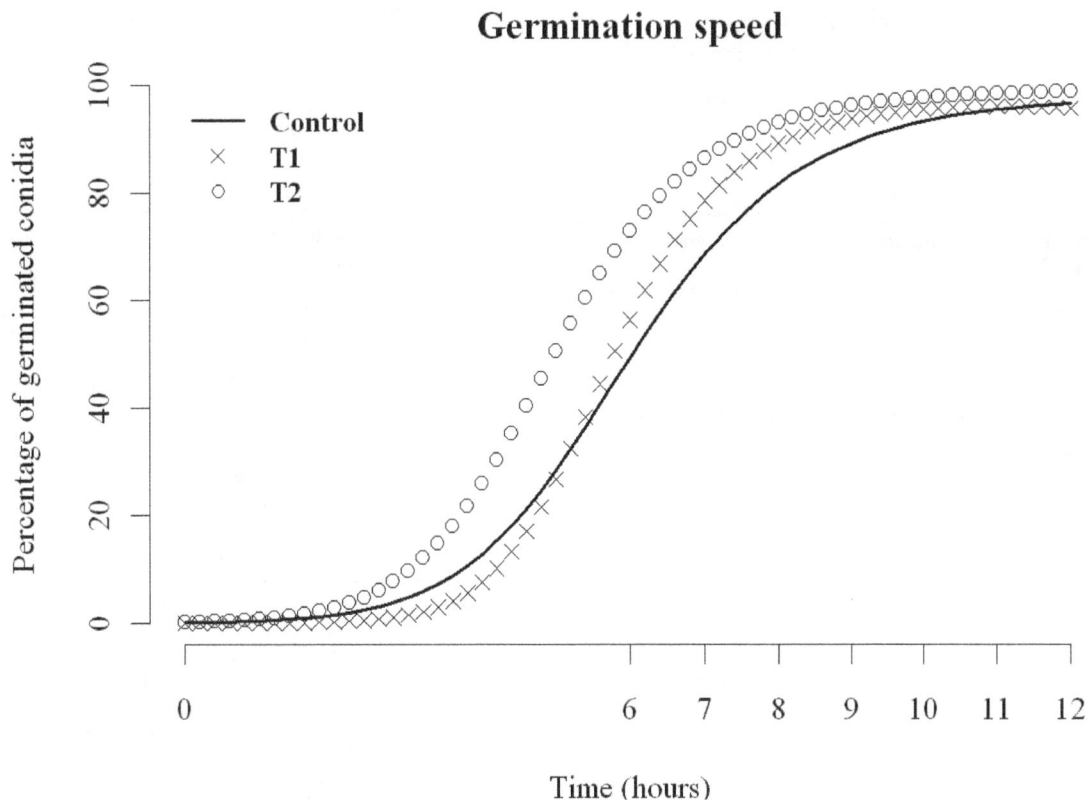

Figure 1. The curve of germination speed of *Metarhizium anisopliae* conidia in the control and treatments.

influence of growth substrate and nutritional environment on the conidial UV-B tolerance of *M. anisopliae* var. *anisopliae*, observing that conidia from insect cadavers germinated slower than those from PDAY culture medium. Rangel et al. (2005) observed that conidia produced on artificial or natural substrates have a similar culturability and tolerance to UV-B radiation, but conidia produced on Czapek's and Emerson's YpSs agar media or rice grains had higher tolerance to UV-B and germinated faster than conidia raised on PDA and PDAY media.

A wild and a transformant strain of *Penicillium roqueforti* were selected by García-Rico et al. (2009) to evaluate the effect of a heterotrimeric G protein α subunit on three parameters: conidia germination, stress res-ponse, and roquefortine C production. Conidia were incubated at 28°C for 14-16 h and sampled at regular intervals every 1-2 h, and then the numbers of conidia and of germinated conidia were counted. As results, at 12 h of incubation, the germination of the parental (wild) strain was very low (2.4% of the total number of conidia), whereas the conidia germination of the transformant strain was about 15% and these differences were main-tained throughout the observation period (6 vs. 36% at 36 h). As conclusion, these authors suggested that G protein-mediated signaling participates in the regulation of these three parameters assessed in *P. roqueforti*.

Pramanik et al. (2010) examined the effects of metabolism of antihistamine drug clemastine on the fungus model *Aspergillus niser* and compared the effect of metabolism with that of human volunteers, using HPLC analysis. As result, it was established that *A. niser* can be a potential model organism for drug metabolism study.

In the present study, the active compound perindopril accelerated the germination speed of conidia (at 6-8 h of incubation). Based on the obtained results, it is possible to conclude that perindopril has no toxicity considering the MT conidia germination speed as parameter. Considering the validity of filamentous fungi as model systems, these results are important data on the toxicity of perindopril and may be associated with other results already obtained for this drug, such as the review of the genotoxicity of marketed pharmaceuticals, published by Snyder and Green (2001), in which perindopril was pointed as negative for tests with following parameters: bacterial mutation, *in vitro* and *in vivo* cytogenetics, mouse lymphoma assay and carcinogenicity.

REFERENCES

Alves MMTA, Orlandelli RC, Lourenço DAL, Pamphile JA (2011). Toxicity of the insect growth regulator lufenuron on the entomopathogenic fungus *Metarhizium anisopliae* (Metschnikoff) Sorokin assessed by conidia germination speed parameter. Afr. J. Biotechnol. 10(47): 9661-9667.

Baldauf SL, Palmer JD (1993). Animals and fungi are each other's closest relatives: congruent evidence from multiple proteins. Proc. Natl. Acad. Sci. USA 90: 11558-11562.

Cerniglia CE (1997). Fungal metabolism of polycyclic aromatic hydrocarbons: past, present and future applications in bioremediation. J. Ind. Microbiol. Biotechnol. 19(5-6): 324-333.

Copping LG (2004). The manual of biocontrol agents. British Crop Protection Council, Farnham, UK.

d'Enfert C (1997). Fungal spore germination: insights from the molecular genetics of Aspergillus nidulans and Neurospora crassa. Fungal Genet. Biol. 21: 163-172.

Dios AM (2011). Empleo de los inhibidores de la enzima convertidora de la angiotensina em el tratamiento de la hipertensión arterial. Rev. Argent. Cardiol. 79(2): 103-105.

García-Rico RO, Chávez R, Fierro F, Martín JF (2009). Effect of a heterotrimeric G protein α subunit on conidia germination, stress response, and roquefortine C production in Penicillium roqueforti. Int. Microbiol. 12:123-129.

Giollo Junior LT, Gomes MAM, Martin JFV (2010). A avaliação da resposta anti-hipertensiva com tonometria de aplanação. Rev. Bras. Hipertens. 17(3): 189-190.

Griffin DH (1994). Fungal physiology. Wiley-Liss, New York, USA.

Isaka M, Kittakoop P, Kirtikara K, Hywell JN, Thebtaranonth Y (2005). Bioactive substances from insect pathogenic fungi. Acc. Chem. Res. 38(10): 813-823.

Lawes CMM, Vander Hoorn S, Rodgers A (2008).Global burden of blood-pressure-related disease, 2001. Lancet 371: 1513-18.

Lever AF, Hole DJ, Gillis CR, McCallum IR, McInnes GT, MacKinnon PL, Meredith PA, Murray LS, Reid JL, Robertson JWK. Do inhibitors of angiotensin-I-converting enzyme protect against risk of cancer? Lancet 352: 179-184.

MacMahon S, Alderman MH, Lindholm LH, Liu L, Sanchez RA, Seedat WK (2008). Blood pressure-related disease is a global health priority. Lancet 371: 1480-2.

Milner RJ, Huppatz RJ, Swaris SC (1991). A new method for assessment of germination of Metarhizium conidia. J. Invert. Pathol. 57: 121-123.

Osherov N, May G (2000). Conidial germination in Aspergillus nidulans requires RAS signaling and protein synthesis. Genetics 155: 647-656.

Pontecorvo G, Roper JA, Hemmons LM, Macdonald KD, Bufton AWJ (1953). The genetics of Aspergillus nidulans. Adv. Genet. 5: 141-238.

Pramanik K, Panda N, Satapathy J, Biswas A (2010). Aspergillus niser for the study of in vitro drug metabolism. Proceedings of 2010 International Conference on Systems in Medicine and Biology 427-431.

Progress Collaborative Group (2001). Randomised trial of a perindopri based blood pressure lowering regimen among 6,105 individuals wit previous stroke or transient ischaemic attack. Lancet 358: 1033 1041.

R Development Core Team (2008). R: A language and environment fc statistical computing. R Foundation for Statistical Computing, Viena Austria.

Rangel DEN, Braga GUL, Flint SD, Anderson AJ, Roberts DW (2004) Variations in UV-B tolerance and germination speed of Metarhiziur anisopliae conidia produced on insect and artificial substrates. J Invert. Pathol. 87: 77-83.

Rangel DEN, Braga GUL, Anderson AJ, Roberts DW (2005). Influenc of growth environment on tolerance to UV-B radiation, germinatio speed, and morphology of Metarhizium anisopliae var. acridur conidia. J. Invert. Pathol. 90: 55-58.

Schumacher V, Poehling HM (2012). In vitro effect of pesticides on th germination, vegetative growth, and conidial production of two strain of Metarhizium anisopliae. Fungal Biol. 116(1): 121-32.

Snyder RD, Green JW (2001). A review of the genotoxicity of markete pharmaceuticals. Mutat. Res. 488: 151-169.

Steinberg G, Perez-Martin J (2008). Ustilago maydis, a new funga model system for cell biology. Trends Cell Biol. 18(2): 61-67.

Thom TJ, Kannel WB, Sibershatzi H, D'Agostino RB (1998). Incidence prevalence, and mortality of cardiovascular diseases in the Unite States. In: Alexander RW, Schlant RC, Fuster V. Hurst's the hear McGraw-Hill, New York, USA.

Wainright PO, Hinkle G, Sogin ML, Stickel SK (1993). Monophyleti origins of the Metazoa: an evolutionary link with fungi. Science 260 340-342.

Wang N, Zheng Z, Jin HY, Xu X (2012). Treatment effects of captopril o non-proliferative diabetic retinopathy. Chin. Med. J. 125(2): 287-292.

Yamada K, Uchida S, Takahashi S, Takayama M, Nagata Y, Suzuki N Shirakura S, Kanda T (2010). Effect of a centrally active angiotensin converting enzyme inhibitor, perindopril, on cognitive performance ir a mouse model of Alzheimer's disease. Brain Res. 1352: 176-186.

Yoshiji H, Kuriyama S, Kawata, Yoshii J, Ikenaka Y, Noguchi R Nakatani T, Tsujinoue H, Fukui H (2011). The angiotensin-I-con verting enzyme inhibitor perindopril suppresses tumor growth an angiogenesis: possible role of the vascular endothelial growth factor Clin. Cancer Res. 7: 1073-1078.

Isolation, characterization and antimicrobial activity of *Streptomyces* strains from hot spring areas in the northern part of Jordan

M. J. Abussaud[1]*, L. Alanagreh[1] and K. Abu-Elteen[2]

[1]Department of Biological Sciences, Yarmouk University, Irbid- Jordan.
[2]Department of Biological Sciences, Alhashemia University, Zarqa.

A total of 30 *Streptomyces* isolates (28 from soil and 2 from water) were isolated and purified from hot-springs areas in the northern part of Jordan. Four strains were thermopile. They grew at 45 and 55°C but not at 28°C. Strains were described morphologically on four different media: on glycerol yeast extract, oatmeal, yeast malt-extract and starch casein agar. White and grey color series were the most frequent series on all media. The results showed that glycerol yeast extract and starch casein were the best media for sporulation. And yeast malt-extract was the best medium for the production of soluble pigment. Physiological and biochemical tests showed that the highest number of *Streptomyces* isolates were able to hydrolyze tyrosine was 26 (87%). This was followed by 25 (83%) for starch, 24 (80%) for urea, 21 (70%) for casein and 10 (33%) for gelatin. Twenty two (73%) strains showed the ability to reduce nitrate and 8 (27%) strains produced melanin. Carbon source utilization showed that 26 (87%) strains were able to utilize L- arabinose, 25 (83%) strains were able to utilize meso-inositol, 8 (27%) strains were able to utilize D-sorbitol, 18 (60%) strains were able to utilize D-mannitol, 28 (93%) strains were able to utilize L-rhamnose and all isolates exhibited the ability to utilize D-fructose and D- glucose. The ability to exhibit antibacterial activity against *Escherichia coli* and *Staphylococcus aureus* was detected among 20 and 26% of the isolates, respectively, while the ability to exhibit antifungal activity against *Candida albicans* was detected among 23% of the isolates. Molecular identification of the 8 antibiotics producers was carried out by PCR technique using two sets of primers specific to *Streptomyces* 16S rDNA gene sequences; strepB/strepE and strepB/strepF which amplified 520 and 1070 bp, respectively. All these antibiotic producer isolates showed positive results for the genus *Streptomyces* specific primers.

Key words: Characterization, streptomyces, antimicrobial activity, hot springs, thermophile, PCR.

INTRODUCTION

Since the discovery of penicillin from the filamentous fungus, *Penicillium notatum,* by Fleming in 1929 and the observation of the broad therapeutic use of this agent in the 1940s, the so-called "Golden Age of Antibiotics", many countries around the world have developed intensive programs to increase the number of described antibiotics or to find new one's (Abussaud, 2000; Cragg and Newman, 2005).

In spite of the large number of antibiotics that have been discovered since that time, a large number of patho-genic bacteria have became resistant to antibiotics in common use (Mellouli et al., 2003; Cirz et al., 2005). As a result of the increasing prevalence of these antibiotic-resistant pathogens and the pharmacological limitations of the present antibiotics, searching for new antibiotics or modification of the present types has become an urgent focus for many researches (Rintala, 2001; Sahin and Ugur, 2003).

Filamentous bacteria belonging to the genus *Streptomyces* are well-known as the largest antibiotic-producing genus in the microbial world discovered so far (Taddei et al., 2006; Jayapal et al., 2007). Most *Streptomyces* and other Actinomycetes produce a diverse array of anti-biotics including aminoglycosides, anthracyclins, glycol-peptides, β-lactams, macrolides, nucleosides, peptides,

*Corresponding author. E-mail: m.abussaud@gmail.com, abussaud@yu.edu.jo.

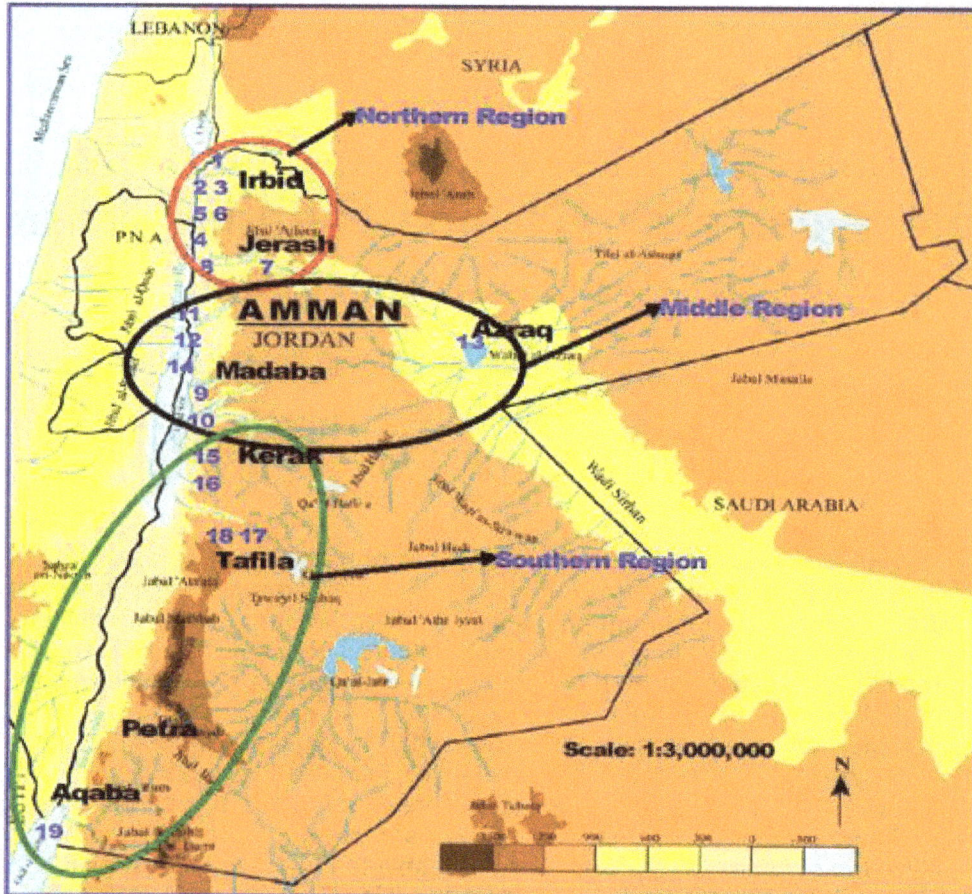

Figure 1. Distribution of hot-springs in in northern region of Jordan. 1: Al Hammah; 2: Ashouneh; 3: Abu Dablah; 4: Waggas; 5: Al Mansheyyah; 6: Abu Ziad; 7: Jerash; 8: Deir Alla.

polyenes, polyethers and tetra-cyclines. They produce about 75% of commercially and medically useful antibiotics (Mellouli et al., 2003; Sahin and Ugur, 2003).

The genus *Streptomyces* proposed by Waksman and Henrici in 1943, are a Gram-positive, aerobic, filamentous soil bacteria, produce an extensive branching substrate and aerial mycelium bearing chains of arthrospores. The substrate mycelium and spores could be pigmented, but also diffusible pigments could be produced. *Streptomyces* have high G+C (69 - 78%) content in their DNA and their cell wall is characterized as Type I (Lechevalier and Lechevalier, 1970; Williams et al., 1989; Rintala, 2001).

In the course of screening for new antibiotics, attention has primarily been concentrated to isolate *Streptomyces* from soil. Most recently, attention has been focused on greater diversity of organisms, those which are considered "rare", those which are difficult to isolate and/or culture and those which grow under extreme conditions such as thermophiles, acidophiles, halophiles etc (Yallop et al., 1997; Thakur et al., 2007).

Mesophilic *Streptomyces* are usually cultivated at temperature from 22-37°C while thermophilic *Streptomyces*

grow between 25 and 55°C, they grow quite well at 50°C (Kim et al., 1999; Rintala, 2001). These organisms are useful as producer of antibiotics, enzymes and other bioactive metabolites because of their rapid autolysis of mycelium (Xu et al., 1998).

In continuing our screening program for *Streptomyces* flora in Jordan (Abussaud, 1996; Abussaud, 2000), we tend our attention to isolate *Streptomyces* strains from new locations and conditions such as hot-springs areas and test their capability to produce antimicrobial substances in order to look for the possibility of finding novel antibiotics. We started with 4 locations in the northern part of Jordan: Alshouneh, Waggas, Almansheyyah and Deir Alla springs (Figure 1).

MATERIALS AND METHODS

Collection of sample

A total of 12 soil samples and 12 water samples were collected from four different hot spring areas (Figure 1) Alshouneh, Waggas, Al-Mansheyah and Deir Alla. Six samples (3 soil samples and 3 water samples) from each location were collected as follows.

Water samples from the spring water column

About 2 l of water have been collected in sterile container, closed immediately and stored in ice box, in the presence of ice pads until shipped to the laboratory for analysis.

Soil samples

Soil samples were taken from sites along water streams at a depth of 10 cm, after removing approximately 3 cm of the soil surface. Samples were placed in polyethylene bags, closed tightly and stored in ice box as previously described. Physical factors such as temperature of water were directly measured at the sampling site by using a thermometer (Brannan co. England). The pH was also measured by using pH indicator paper (Whattman co. England).

Isolation of *Streptomyces* strains

Soil samples

Soil samples were analyzed following a modification of the procedure of Abussaud and Saadoun (1991): one gram of soil was suspended in 100 ml sterile distilled water, shaked in a reciprocal shaker at 190 rpm for 30 min, and then allowed settling. Serial dilutions (10^{-1} to 10^{-6}) were made. A 0.1 ml of each dilution was pipetted and spread evenly over the surface of Starch Casein (SCM) agar plates (g/l): (starch 10, casein 0.3, $NaNO_3$ 2, K_2HPO_4 2, NaCl 2, $MgSO_4.7H_2O$ 0.05, $CaCO_3$ 0.02, $FeSO_4.7H_2O$ 0.01 and agar powder 20, pH = 7.2) supplemented with cyclohexamide (50 ug/ml) and filter-sterilized rifampicin (0.5 ug/ml) using a sterile L-shaped glass rod. These plates were incubated at 28 and 55°C until good growth occurred. Dilutions that gave about 100 colonies per plate were chosen for the isolation of *Streptomyces* isolates.

Water samples

(a) 100 ml of each water sample was filtered through a Millipore membrane (0.22 - 0.45 μm pores, Sartorious. Germany), after that, the membranes were transferred to the surface of Starch Casein (SCM) plates and incubated at 55 and 28°C for 7 - 14 days.
(b) 10 ml of each water sample was inoculated into 90 ml starch casein (SCM) broth and Tryptone-yeast extracts broth (TYE): (g/l) Tryptone 5, yeast extract 3, pH = 7.2) in 250 ml flask, then incubated for 24 h with shaking; (225 rpm, HT shaker. Germany), at 55 and 28°C, after that about 200 μl were transferred to starch casein (SCM) plates. These plates were incubated following the previous procedure of incubation.
(c) A combination between the previous two steps was done to ensure our results, in details: 100 ml of each water sample was filtered through a Millipore membrane; the membranes were transferred to 100 ml starch casein (SCM) broth in 250 ml flask, incubated at 55 and 28°C for 24 h with shaking at 225 rpm. Then aliquots (0.2 ml) of 10^{-2} to 10^{-5} ten-fold serial dilutions were spread over the surface of dried SCM agar plates.
    The plates were incubated as described previously at 55 and 28°C for 7 - 14 days.
    Streptomyces colonies were then picked up and transferred to yeast malt-extract agar (g/l) (yeast-extract): (3, malt-extract 3, peptone 5, glycerol 10 ml/l, agar 20, pH = 7.0 ± 0.2), starch casein agar plates, glycerol yeast-extract agar plates (g/l): yeast-extract 2, K_2HPO_4 1, glycerol 5 ml/l, agar 20, pH = 7.2 and oatmeal agar for further purification.

Maintenance media

After purification, *Streptomyces* isolates were maintained as suspend-sions of spores and mycelia fragments in 20% glycerol (v/v) at -20°C.

Morphological characterization

Morphological characterization of *Streptomyces* isolates were done according to the ISP recommendations (Shirling and Gottlieb, 1966). A pure culture of each isolate was picked up and transferred to grow on four different media: Yeast Malt-Extract (YME) agar (ISP2), Starch Casein (SCM) agar, Glycerol Yeast-Extract (GYE) agar and Oatmeal agar (ISP3) for 5 - 10 days at 55°C for themophilic *Streptomyces* and at 28°C for mesophilic *Streptomyces*. Then, colors of the aerial and substrate mycelium and those of the soluble pigments were examined.

Cultural, physiological and biochemical tests

Determination of the cultural, biochemical and physiological characteristics was carried out according to Williams et al. (1983), Brown et al. (1999) and Babcock (1979).

Growth temperature

Also the ability of the isolates to grow at different temperatures (28, 37, 45 and 55°C) was studied.

Carbon source

The ability of the strains to use different carbon sources was determined according to the ISP recommendation (Shirling and Gottlieb, 1966); seven sugars were used as a carbon sources: L-arabinose, Meso-inositol, D-sorbitol, D-mannitol, L-rhamnose, D-fructose and D- glucose, the results were determined after 14 days incubation at optimum temperatures.

Antimicrobial activity

Antimicrobial activity on agar media

Antimicrobial activity was tested by growing *Streptomyces* strains on agar plates until good and thick growth occurred. Then agar block from these plates were transferred to plates previously seeded with the indicator organisms (*Escherichia coli, Staphylococcus aureus* and *Candida albicans*). The plates were incubated at appropriate temperature for each indicator. Activity was measured as inhibition zone in millimeters around the agar block (Abussaud and Saadoun, 1991).

Antimicrobial activity in broth media (antibiotics fermentation)

Spores from each 10 days-old culture of isolates grown on glycerol yeast-extract agar plates were used to inoculate 100 ml of glycerol yeast-extract broth into 250-ml Erlenmeyer flasks. These cultures were grown in a rotary shaker at 150 rpm, 28°C, for eight days. Glycerol Yeast-Extract medium was preliminarily tested and was found to be suitable for antibiotic production by our isolates.
    During fermentation, 2 ml sample of each culture were collected in eppendorf tube every 2 days and tested for antibiotic activity against *E. coli, S. aureus* and *C. albicans*. Activity was tested by

Table 1. Primers used for detection of *Streptomyces* in PCR reactions.

Primer	Sequence (5′ 3′)	Primer target sequence	Amplicon length (bp)	Reference
strepB(F) /strepE(R)	ACAAGCCCTGGAAACGGGGT CACCAGGAATTCCGATCT	16S rDNA	520	Rintala et al., 2002
strepB(F) /strepF(R)	ACAAGCCCTGGAAACGGGGT ACGTGTGCAGCCCAAGACA	16S rDNA	1070	Rintala et al., 2002

Table 2. Physical properties of hot springs.

Location	Temperature (°C)	pH
Alshouneh (Sh)	54	6.5
Waggas (Wg)	47	8
Almansheyyah (Mn)	48	8
Deir Alla (DA)	36	7

preparing wells into agar plates previously seeded with the indicator organism and transferring 100 microliters from the centrifuged fermentation broth into these wells.

Molecular identification of antibiotics producer isolates

Extraction of genomic DNA from pure culture

Genomic DNA was isolated from the isolates using a bacterial genomic DNA isolation kit (Biobasic Inc. Canada). One separate colony from each bacterial isolate was inoculated into 10 ml nutrient broth and incubated overnight at 28°C. Then 1 ml was taken and centrifuged at 14000 rpm for 15 min at room temperature, the pellets were suspended in 200 µl cold TE (10 mM tris base, 1 mM EDTA, pH 8.0) buffer and 400 µl digestion solution and mixed well, then a 3 µl of Proteinase K were added and incubated at 55°C for 2 h.

After incubation, 260 µl of 100% ethanol were added to the solution, and then the whole mixture was applied into 2 ml EZ-10 column provided with the Kit and centrifuged at 8,000 rpm for 1 min. The pellets were resuspended again with 500 µl of wash solution and centrifuged at 8,000 rpm for 1 min, this step was repeated again.

After that, the column was placed in a clean microfuge tube and a 50 µl of elution buffer were added to the center of the column, incubated at 37°C for 2 min and finally centrifuged at 10,000 rpm for 1 min to elute the DNA.

Polymerase chain reaction (PCR)

PCR amplification of 16S rDNA was carried out in 50 µl volumes containing: 25 µl of Econo Taq PLUS GREEN 2X Master Mix, 0.25 µl (100 pmol) of each primer, 2 µl (10 ng) of DNA template and 22.5 µl DNAse free water, each primer pair has its program that will be mentioned later.

For each PCR reaction a negative PCR reaction tube was performed where no DNA template was added (not shown in figures) and all PCR reactions were performed in a Perkin Elmer DNA thermal cycler (Perkin Elmer 480).

Identification of bacterial isolates using StrepB/StrepE primer pair specific to genus Streptomyces 16S rDNA gene sequences

The primer pairs StrepB/StrepE (sequences listed in Table 1)

amplified 520 bp fragments, nucleotides 139 - 657. The PCR wa programmed as follows: after the initial denaturation for 5 min a 98°C, 30 cycles of denaturation (1 min at 95°C), primer annealin 40 s at 54°C and primer extension (2 min at 72°C) were performed A final extension at 72°C for 10 min had followed.

Identification of bacteria isolates using StrepB/StrepF prime pair specific to genus Streptomyces 16S rDNA gene sequences

The primer pairs StrepB/StrepF and StrepB/StrepE amplified 107(and 520 bp fragments, nucleotides 139 -1212. The PCR were programmed as follows: after the initial denaturation for 5 min a 98°C, 30 cycles of denaturation (1 min at 95°C), primer annealing 40 s at 58°C and primer extension (2 min at 72°C) were performed A final extension at 72°C for 10 min followed.

Gel electrophoresis and photography

The PCR amplified products were separated on 1% w/v ultra-pure agarose powder in 1X TBE buffer (pH 8.3) at 100 V for 60 - 70 mi using mini-gel set (Bio Rad,). Gels were stained with ethidiun bromide (0.5 µg/ml) and analyzed using BioDocAnalyze (Biometra Germany). A 250 base pair (bp) molecular weight marker was included on every gel.

RESULTS

Physical properties of hot springs

Physical properties such as temperature and acidity degree (pH) for each sampling location (water sample are indicated in Table 2.

Isolation of *Streptomyces*

Thirty different bacterial isolates were isolated during this study. Six isolates were isolated from Deir Alla, only one from water (DA1-DA6); 8 isolates were isolated from Almansheyyah (Mn1-Mn8); 9 isolates were isolated from Alshouneh, only one from water (Sh1-Sh9) (two of these isolates were thermophilic they grow at_55°C) and 7 isolates were isolated from Waggas (Wg1 - Wg7); also two of these isolates were thermophilic; (Wg6 and Wg7).

Morphological characterization

The aerial mycelium color of the isolates ranged from red

Table 3. Physiological and biochemical characteristics of *Streptomyces* isolate (number of positive/total in each site).

Location	Starch hydrolysis	Urea hydrolysis	Casein hydrolysis	L-tyrosine hydrolysis	Gelatin lequification	Nitrate reduction	Melanin formation
DA	4/6	6/6	6/6	6/6	1/6	4/6	1/6
Mn	7/8	6/8	7/8	7/8	3/8	6/8	3/8
Sh	9/9	7/9	6/9	7/9	4/9	6/0	3/9
Wg	5/7	5/7	2/7	6/7	2/7	6/7	1/7

Table 4. Ability of *Streptomyces* isolates to grow at different temperatures.

Location	28°C	37°C	45°C	55°C
DA	6/6	6/6	3/6	0
Mn	8/8	8/8	6/8	0
Sh	7/9	7/9	5/9	2/9
Wg	5/7	6/7	4/7	2/7

white, blue and green, grey to purple. While the substrate mycelium color shows narrower diversity. On the other hand, the isolates produced different pigments: yellow, pink and grey. These differences might reflect the diversity among the isolates.

On GYE medium, out of 30 isolates, 14 were white, 10 grey, 2 red, 2 green, 1 blue and 1 purple. On the other hand, 9 isolates were found to be producers of diffusible pigment, 7 of them produced brown color while the other produced a yellow pigment.

On Oatmeal medium, out of 30 isolates, 12 isolates were white, 9 grey, 2 pink, 1 brown, 1 purple and 1 red and 4 isolates failed to sporulate on this medium. 10 isolates produced diffusible pigment, 5 of them brown, 4 yellow and one isolates produced pink pigment.

On YME media, out of 30 isolates, 10 isolates were white, 7 grey, 1 green, 1 brown, 1 yellow and 10 of them were not able to sporulate. 12 isolates were able to produce diffusible pigment, 7 produced brown pigment and the rest produced yellow pigment.

On STC media, out of 30 isolates, 13 isolates were white, 10 grey, 3 pink, 2 purple, 1 green and 1 brown, only 3 isolates were able to produce diffusible pigment, 2 produced brown color and 1 produced red color.

Physiological and biochemical characteristics of *Streptomyces* isolates

As indicated in Table 3, 25 (83%), 24 (80%), 21 (70%), 26 (87%), 10 (33%), 22 (73%) and 8 strains (27%) showed the ability to degrade starch, hydrolyze urea, degrade casein, hydrolyze tyrosine, liquefy gelatin, reduce nitrate and produce melanin, respectively.

Also the ability of the isolates to grow at different temperatures (27, 37, 45 and 55°C) was examined. The

results depicted in Table 4 showed that, out of 30 isolates only 4 (Thermophilic) isolates were able to grow at 55°C, 18 isolates included thermophilic isolates were able to grow at 45°C, 27 at 37°C and 26 at 27°C.

Carbon source utilization

Table 5 represents the ability of *Streptomyces* to utilize different carbon sources (7 different sugars were used). Out of 30 isolates, 26 (87%) isolates were able to utilize L- arabinose, 25 (83.%) isolates were able to utilize Meso-inositol, 8 (27%) isolates were able to utilize D-sorbitol, 18 (60%) isolates were able to utilize D-mannitol, 28 (93%) isolates were able to utilize L-rhamnose and all isolates exhibited the ability to utilize D-fructose and D-glucose.

Antimicrobial activity on agar media

Antimicrobial activity was tested against *E. coli*, *S. aureus* and *C. albicans*. Out of 30 isolates 8 were antibiotics producers. These isolates were assigned: Mn1, Mn3, Mn8, Sh1, Sh3, Sh6, Wg4 and Wg5. Six of them showed antibacterial activity against G-ve (represented by *E. coli*), all of them were active against G+ve bacteria (represented by *S. aureus*) and 7 of them were able to produce antifungal activity against *C. albicans*.

Table 6 represents the antimicrobial activity during a period of 8 incubation days. After 2 days, only 1 (12.5%) isolate was active against all tested microorganisms. After 4 days, 3 (37.5%) of the isolates were active against all tested microorganisms. After 6 days, 5 (62.5%) isolates showed activity, 40% of them were active against *E. coli*, all of them exhibited activity against *S. aureus* and 80% of them exhibited activity against *C. albicans*. After 8 days, 4 (50%) of the isolates exhibited the activity, 75% of them were active against *E. coli* and all of the rest were active against *S. aureus* and *C. albicans*.

Molecular identification of the bacterial isolates

Genomic DNA was isolated from the 8 antibiotic producers (Mn1, Mn3, Mn8, Sh1, Sh3, Sh6, Wg4 and Wg5) and from the positive control (*Streptomyces halstedii* ATCC

Table 5. Utilization of carbon sources (positive/total).

Location	Arabinose	Inositol	Sorbitol	Mannitol	Rhamnose	Fructose	Glucose
DA	5/6	5/6	3/6	4/6	5/6	6/6	6/6
Mn	8/8	5/8	1/8	5/8	8/8	8/8	8/8
Sh	7/9	8/9	2/9	6/9	9/9	9/9	9/9
Wg	6/7	7/7	2/7	3/7	6/7	6/7	7/7

Table 6. Antimicrobial activity *of *Streptomyces* isolates (on agar media).

Location	E. coli	S. aureus	C. albicans
DA	0	0	0
Mn	3/8	3/8	2/8
Sh	3/9	3/9	3/9
Wg	0	2/7	2/7

*Number of active isolates/total number.

10897). A large amount and good quality of genomic DNA was obtained from each bacterial isolates.

Two primer pairs were used in the PCR reactions to identify the bacterial isolates as a *Streptomyces* isolates (StrepB/StrepE and StrepB/StrepF). Using 16S rDNA StrepB/StrepE primer pair, all antibiotic producers showed positive results with 520 bp PCR amplification products (Figure 2)

Using 16S rDNA StrepB/StrepF primer pair all antibiotic producers and the control *S. halstedii* showed positive results with 1070 bp PCR amplification product (Figure 3).

DISCUSSION

Streptomyces represent an important source of biologically active compounds. They are used extensively in industry as producers of antibiotics, enzymes, enzyme inhibitors and antitumour agents. However, it is important to continue the screening for novel bioactive compounds as the number of microorganisms resistant to the existing antibiotics is growing every year. However, it is becoming increasingly difficult to discover new commercially useful secondary metabolites from common streptomycetes, thereby emphasizing the need to isolate, characterize and screen novel members of the genus *Streptomyces*. *Streptomyces* from under explored habitats are proving to be a rich source of new bioactive compounds, including antibiotics (Berdy, 2005; Okoro et al., 2009). Therefore, we decided to isolate *Streptomyces* strains from Jordanian hot springs and study their capability to produce antibiotic activity.

A total of 30 different bacterial isolates were recovered during this study from non-cultivated hot spring areas in Jordan. All bacterial isolates were typically *Streptomyces*; they grew on a range of agar media showing morphology typical of *Streptomyces*.

The majority of the isolates showed good growth or SCM agar medium. This medium seems to be specific and sensitive for *Streptomyces*, since it contains starch that most *Streptomyces* use as a carbon source and the basic minerals that are needed for good growth. In addition its transparency facilitates colony observation. Earlier studies have shown the importance of the constituents of the screening media under which the producing micro organisms were cultivated (Williams et al., 1989; Rintala 2001).

The number of *Streptomyces* isolates that were isolated from soil samples (28 isolates) was highly greater than that isolated from water samples (2 isolates), this may by due to the presence of organic matter that make *Streptomyces* abundant in soil, it is the dominant genus in the soil that gives it it's odor (Kutzner, 1986 ; Rintala 2001).

Streptomyces have been isolated from fresh water as well as marine environments (Delabre et al., 1998) although, it has been a subject of debate. Many scientists considered *Streptomyces* to be part of the marine ecosystem, while many others failed to isolate *Streptomyces* from water samples and did not consider *Streptomyces* to be indigenous to the marine environments (Okazaki, 2006) and the debate point was whether they are indigenous, or have been washed off from the surrounding soils, so such studies could explain the presence of few *Streptomyces* strains in hot-springs water samples, since its water coming from under ground surrounded by rocks not soils (Goodfellow and Simpson 1987; Rintala, 2001).

All of these isolates fitted the genus description as reported by several studies (Shirling and Gottlieb, 1966; Kutzner, 1986; Williams et al., 1989). The color of the substrate mycelium and aerial spore mass was varied which reflect the diversity of *Streptomyces* isolates.

Msameh (1992) in his study on distribution and antibiotic activity of *Streptomyces* flora in Jordan reported that the white and grey color series showed the highest percentage of occurrence (43.6 and 28.3%, respectively). In the present study, 50% of the isolated *Streptomyces* isolates were from Alshoneh and Deir Alla (30 and 20%, respectively). White and grey color series had also the percentage (46 and 33.3%, respectively).

The comparison of the physiological and biochemical characteristics of the presented isolates with the actionmycetes as described in Bergey's Manual of Determinative

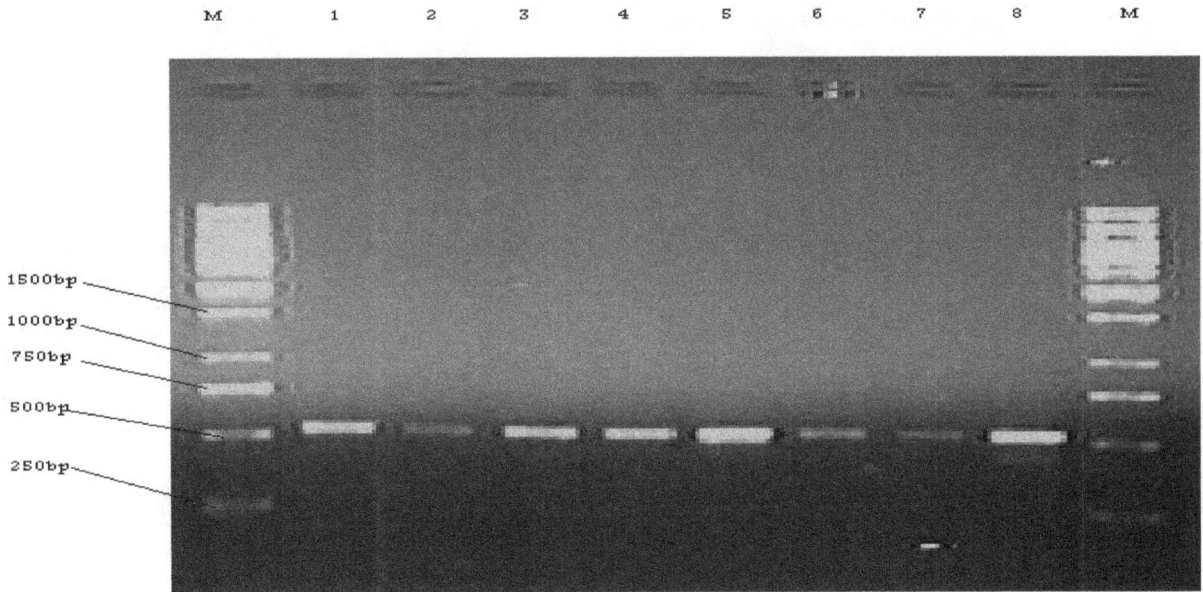

Figure 2. Agarose gel electrophoresis of PCR amplification products of genomic DNA isolated from *Streptomyces* pure culture using strepB(F)/strepE(R) of 16S rDNA gene. Lane M, 1 kb DNA ladder; lane 1, Mn8; lane 2, Sh3; lane 3, Mn1; lane 4, Sh6; lane 5, Wg5; lane 6, Wg4; lane 7, Sh1; lane 8, Mn3.

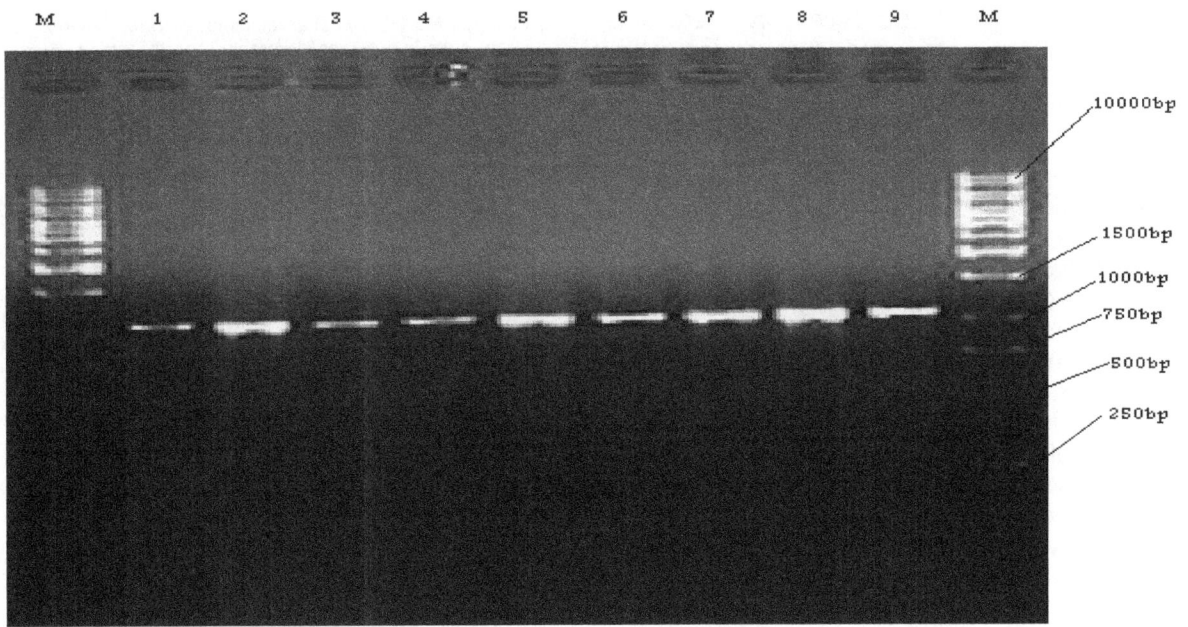

Figure 3. Agarose gel electrophoresis of PCR amplification products of genomic DNA isolated from *Streptomyces* pure culture using strepB(F)/strepF(R) of 16S rDNA gene. Lane M, 1 kb DNA ladder; lane 1, Mn8; lane 2, Sh3; lane 3, Mn1; lane 4, Sh6; lane 5, Wg5; lane 6, Wg4; lane 7, Sh1; lane 8, Mn3) and lane 9, *S. halstedii*.

Bacteriology determined that these isolates belongs to the genus *Streptomyces*.

Antibacterial activity and antifungal activity was observed in 8 (27%) and 7 isolates (23%), respectively. In former studies, it was shown that the isolation rate of *Streptomyces* with antimicrobial activity was higher than 40%

(Lemriss et al., 2003) and in others less than 10% (Jiang and Xu 1996). This variation may be due to many factors example, soil type, climate, strain type and isolation methods. We found the best percentage (37.5%) of antibacterial and antifungal activity among Almansheya strains, followed by those from Alshouneh and Waggas

Table 7. Antimicrobial activity of antibiotics producers (in broth media).

Strain	2 days against			4 days Against			6 days against			7 days against		
	E. coli	S. a	C. alb	E. coli	S. a	C. alb	E. coli	S. a	C. alb	E. coli	S. a	C. alb
Mn1	-	-	-	-	-	-	-	-	-	-	-	-
Mn3	+	+	+	+	+	+	+	+	+	+	+	+
Mn8	-	-	-	-	-	-	-	+	-	-	-	-
Sh1	-	-	-	-	-	-	ND	+	+	+	+	+
Sh3	-	-	-	+	+	+	+	+	+	-	-	ND
Sh6	-	-	-	+	+	+	ND	+	+	+	+	+
Wg4	-	-	-	-	-	-	-	-	-	-	+	+
Wg5	-	-	-	-	-	-	-	-	-	-	-	-

E. coli: *Escherichia coli*, *S. a*: *Staphylococcus aureus*, *C. alb*: *Candida albicans*. + = Active, - = not active, ND = not determined.

areas 33.3 and 28.5%, respectively. No antibiotics producers were isolated from Deir Alla area.

The highest percentage of activity was recorded against Gram-positive bacteria followed by Yeast and Gram-negative bacteria. Some isolates did not show activity in liquid media. Out of the 8 active isolates on agar medium, only 5 (62.5%) isolates were found to exhibit antibacterial activity in liquid media (Table 7).

During the screening of the secondary metabolite, *Streptomyces* isolates were often encountered which show antimicrobial activity on agar but not in liquid culture (Thakur et al., 2007).

Molecular identification

Molecular identification was performed using polymerase chain reaction (PCR) which is currently used as a sensitive and specific detection method for micro-organisms (Rintala et al., 2002). The 16S rDNA gene was chosen as the target gene for the PCR primers in the PCR assay, aiming at the detection of the 8 antibiotic producers of the *Streptomyces* isolates. In this study two sets of primers strepB/strepE and strepB/strepF specific to 16S rDNA gene fragment were used to identify bacterial isolates; positive results were recorded for all bacterial isolates with amplification and corresponding to 520 and 1070 bp and thus, confirm that all antibiotics produce bacterial isolates belong to *Streptomyces* species.

In order to detect the presence of *Streptomyces* isolates in the water samples, the same two sets of primers were used to amplify the 16S rDNA gene collected from water samples; all samples exhibited negative results which indicated the inexistence of *Streptomyces* in these water samples.

In comparison between cultural and molecular methods for identification of *Streptomyces* isolates, we could say that molecular methods are more sensitive, rapid and not laborious in opposite to cultural methods that are laborious and time consuming.

ACKNOWLEDGMENT

This work was funded by the Deanship for Academi Research and Higher Studies at Yarmouk University in Irbid-Jordan.

REFERENCES

Abussaud MJ (1996). Characteristics of *Streptomyces* isolates isolate from soils in two landfills areas in north Jordan. Acta Microbiologica et Immunologica Hungarica. 43:47-53.

Abussaud MJ (2000). The antibiotic activity of a *Streptomyces* strai and its cultural and orphological characteristics. (Dirasat) Bas. Sc Eng. 9:179-190.

Abussaud MJ, Saadoun IM (1991). *Streptomyces* flora of some Jorda valley soils, characteristics and seasonal distribution. Dirasat. 18:66 75.

Babcock JB (1979). Tyrosine Degradation in Presumptive Identification of Peptostreptococcus anaerobius. J. Clin. Microb. 9:358-361.

Berdy J (2005). Bioactive microbial metabolites. J. Antibiotics 58:1-26

Brown JM (Indicate Initials), McNeil MM(Indicate Initials), Desmond EF (1999). In Murray, Baron, Pfaller, Tenover and Yolken (ed.), Manua of clinical microbiology, 7th ed. Washington, D.C. Am. Soc. Microbiol p.370.

Cirz RT, Chin JK, Andes DR, Vale´ Rie De C-L, Craig WA, Romesberg FE (2005). Inhibition of mutation and combating the evolution o antibiotic resistance. PLoS Biol. 3(6):1024.

Cragg GM, Newman DJ (2005). Biodiversity: A continuing source o novel drug leads. Pure Appl. Chem. 77(1):7-24.

Delabre K, Cervantes P, Lahoussine V, Roubin MRD (1998). Detection of viable pathogenic bacteria from water samples by PCR. OECD Workshop on Molecular Methods for Safe Drinking Water. France.

Goodfellow M, Simpson KE (1987). Ecology of *Streptomyces*. Frnts Appl. Microbiol. 2:97-125.
including the description of *Streptomyces thermoalcalitolerans* sp nov. Int. J. Syst. Bacteriol. 49:7-17.

Jayapal KP, Lian W, Glod F, Sherman DH, Hu WS (2007). Comparative genomic hybridizations reveal absence of large *Streptomyces coelicolor* genomic islands in *Streptomyces lividans*. BMC Genomics 8:229

Jiang CL, Xu LH (1996). Diversity of aquatic actinomycetes in lakes o the middle plateau, Yunnan, China. Appl. Environ. Microbiol. 62:249 253.

Kim B, Sahin N, Minnikin DE, Zakrzewska-Czerwinska J, Mordarski M Goodfellow M (1999). Classification of thermophilic *Streptomyces*,

Korn-Wendisch F, Kutzner HJ (1992). The family Streptomycetaceae. In The Prokaryotes, Edited by Balows A, Truper HG, Dworkin M, Harde W, Schleifer KH. N.Y., Springer. pp. 921-995.

Kutzner KJ (1986). The family *Streptomycetaceae*. In: Starr MP, Stolp H, Tr_Per HG, Balows A, Schlegel HG (eds) The prokaryotes, A Handbook on Habitats, Isolation, and Identification of Bacteria, Springer-Verlag, N.Y. 2:2028-2090.

Lechevalier MP, Lechevalier HA (1970). Chemical composition as a criterion in the classification of aerobic actinomycetes. Int. J. Syst. Bacteriol. 20:435-443.

Lemriss S, Laurent F, Couble A, Casoli E, Lancelin JM, Saintpierre-Bonaccio D (2003). Screening of nonpolyenic antifungal metabolites produced by clinical isolates of actinomycetes. Can. J. Microbiol. 49:669-674.

Mellouli L, Ameur-Mehdi R, Sioud S, Salem M, Bejar S (2003). Isolation, purification and partial characterization of antibacterial activities produced by a newly isolated *Streptomyces* sp. US24 strain. Res. Microbiol. 154:345-352.

Msameh YM (1992). *Streptomyces* in Jordan, distribution and antibiotic activity. MS thesis supervised by Dr. Abussaud, Department of Biological Sciences, Yarmouk Univesity, Irbid, Jordan.

Okazaki T (2006). Intrigued by actinomycete diversity. Actinomycetology 20:15-22.

Okoro CK, Brown R, Jones AL, Andrews BA, Asenjo JA, Goodfellow M, Bull AT (2009). Diversity of culturable actinomycetes in hyper-arid soils of the Atacama Desert, Chile. Antonie van Leeuwenhoek. 95:121-133

Rintala H (2001). *Streptomyces* in Indoor Environments- PCR based detection and diversity. Department of Environmental Health. National Public Health Institute. Kuopio, Finland.

Rintala H, Nevalainen A, Suutari M (2002). Diversity of *Streptomyces* in water-damaged building materials based on 16S rDNA sequences. Lett. Appl. Microbiol. 34:439-443.

Sahin N, Ugur A (2003). Investigation of the antimicrobial activity of some *streptomyces* isolates. Turk. J. Biol. 27:79-84.

Shirling E, Gottlieb D (1966). Methods for characterization of *Streptomyces* species. Int. J. Syst. Bacteriol. 16:313-334.

Taddei A, Rodrı´guez MJ, Ma´rquez-Vilchez E, Castelli C (2006). Isolation and identification of *Streptomyces* spp. from Venezuelan soils: Morphological and biochemical studies. Microbiol. Res. 161:222-231.

Thakur D, Yadav A, Gogoi BK, Bora TC (2007). Isolation and screening of *Streptomyces* in soil of protected forest areas from the states of Assam and Tripura, India, for antimicrobial metabolites. J. de Mycologie Médicale. 17:242-249.

Williams ST, Goodfellow M, Alderson G (1989). Genus *Streptomyces* (Waksman & Hanrici 1943) 339AL. In: Bergey's Manual of Systematic Bacteriology, Edited by Williams ST, Sharpe ME, Holt JG. Williams and Wilkins. Baltimore, 4:2452-2492.

Williams ST, Goodfellow M, Alderson G, Wellington EMH, Sneath PHA, Sackin MJ (1983). Numerical classification of *Streptomyces* and related genera. J. Gen. Microb. 129:1743-1813.

Xu LH, Yong-Qian Tiang, Yun-Feng Zhang, Li-Xing Zhao, Cheng-Lin Jiang (1998). *Streptomyces thermogriseus*, a new species of the genus *Streptomyces* from soil, lake and hot-spring. Int. J. Syst. Bacteriol. 48:1089-1093.

Yallop CA, Edwards C, Williams ST (1997). Isolation and growth physiology of novel thermoactinomycetes. J. Appl. Microbiol. 83:685-692.

29

Helminth parasites of *Synodontis nigrita* at lower Niger (IDAH), Nigeria

O. O. Toluhi and S. O. Adeyemi

Department of Biological Sciences, Kogi State University, Anyigba, Kogi State, Nigeria.

The isolation and identification of helminth parasites of *Synodontis nigrita* and length–weight relationship of the fish in the lower Niger (Idah), Kogi State, Nigeria were carried out in order to describe the pattern of occurrence of the helminth and to establish the well-being of the host fish. A total of 102 randomly sampled fish were studied and three genera of helminths were recovered; 39.1, 48.7 and 12.2% respectively, and were harboured in the fish' intestine. The three genera of helminths isolates identified include two nematodes (*Capillaria* and *Contracaecum* species), Acanthocephala (*Acanthocephalus* species) and Trematode (*Posthodiplostomum* spp.). Of the 102 fish studied, 16 were infected with 80 helminth parasite giving a prevalence rate of 15.7%. The overall mean intensity and mean abundance of helminth parasite occurrence for the sampled fish were 13.6 and 1.5, respectively. The mean standard length of the fish was 7.45 ± 2.59 cm. The need for fish seeds from the wild to be examined for the helminth parasites during culture practice and the socio-economic and human health implications of eating infected fish is also recommended.

Key words: Acanthocephala, nematode, isolation, trematode, intestine.

INTRODUCTION

Catfish is a common name for about 2,200 species of fishes that make up the order Siluriformes and class Actinopterygii (ray-finned fishes). These two families of the order Ariidae and Plotosidae are primarily marine, while all other families are freshwater dwellers. Catfishes are a collection of scaleless, tenacious fish mostly nocturnal scavengers that have adapted to life in a variety of environments with some living near the bottom in shallow waters (Gunder and Fink, 2004).

Fish, like all living organisms, are susceptible to infections with various parasites (Hilderbrand et al., 2003). Chiefly among the parasites afflicting fish are the helminths. Helminths comprising nematodes, trematodes, cestodes and Acanthocephala commonly parasitize both wild and cultured fish with the former constituting heavier parasitic burden (Merck, 2006). Direct association of wild species with cultured fish farms has been established as a way of contaminating cultured fish by parasites (Okaeme and Olufemi, 1997).

The wellbeing, robustness and degree of fatness of fish is a measure of its condition factor with respect to the same specie taken from other water bodies or to other species of fish taken from the same water body (Pauly, 1983). It is expressed by relating length of fish to its weight. A plump or fat fish will give a higher condition factor than a lean and thin fish. Lower value means that the fish are in poor condition which may be a reflection of either over population or outbreak of diseases (Gupta and Gupta, 2006).

MATERIALS AND METHODS

Study area

The study area is the lower Niger (Idah), Kogi State, Nigeria. It is located on latitude 7° 06'N and longitude 6° 43'E of the Greenwich Meridian in the Guinea Savannah vegetation zones of Nigeria (Areola et al., 1992). A total of 102 fish samples of *Synodontis*

Table 1. Pattern of helminth parasites occurrence in *S. nigrita* in relation to standard length of fish in the lower Niger (Idah), Nigeria.

Standard Length group (cm)	Total number examined	Total number infected	Total parasites recovered	Prevalence rate (%)	95% C I	Mean intensity (No)	Mean abundance (No)
0.0 - 5.4	1	1	18	100.0	-	18.0	18.0
5.4 - 19.0	20	3	12	19.0	18.96 - 19.04	2.6	5.0
5.8 - 19.0	36	2	26	33.0	32.99 - 33.00	3.1	1.0
5.4 - 26.1	22	2	14	28.9	28.78 - 29.02	4.1	1.2
5.8 - 25.1	13	4	8	6.7	6.42 - 6.98	2.0	1.3
19.0 - 26.1	10	4	2	3.7	3.31 - 4.09	53.0	26.5
Total	102	16	80	15.7	26.37 – 26.43	5.0	0.8

CI = Confidence interval.

Table 2. Identity and pattern of helminth parasites recovered from *S. nigrita* in the lower Niger (Idah), Nigeria in relation to predilection sites.

Helminth parasites		Predilection sites					Sub total	
Taxonomic group	Species	Skin	Gill	Fin	Anus	Small intestine	Number	%
Nematoda	*Capillaria* sp.	-	-	-	-	4	4	16.5
	Contraceacum sp.	-	-	-	-	11	11	22.6
Acanthocephala	*Acanthocephalus* sp.	-	-	-	-	62	62	48.7
Trematoda	*Posthodiplostomum* sp.	-	-	-	-	3	3	12.2
Grand total		-	-	-	-	80	80	100.0
Percentage (%)		-	-	-	-	100	100.0	

nigrita of different sizes caught by fishermen using gillnets, cast nest and hook and line at Idah area of the lower Niger were identified, bought and transported alive to the Biological Science laboratory, Kogi State University, Anyigba between January and August 2008 for the study.

The skin, fins, eyes, anus, intestinal organs, buccal and opercula cavities of fish were cut open and placed in 0.9% physiological saline and examined under a dissecting microscope. Helminths recovered were counted and placed in saline solution (0.9%) which was kept overnight in refrigerator to enable them stretch and relax. They were later fixed and preserved in 70% alcohol. The helminths were stained over night with weak Erlich's haematoxylin; and dehydrated in graduated alcohol (30, 50, 70, 90% and absolute) for 45 min, cleared in methyl-salicylate and mounted on a slide in Canada balsam. The occurrence (prevalence, mean intensity and abundance) of the helminths on the *Synodontis* fish hosts was determined by standard procedures described by Khalil and Polling (1997).

With the aid of a measuring board and sensitive Mettler weighing balance, the total length (cm), standard length (cm) and weight (g) of each fish sample was measured fresh to the nearest 0.1 cm and 0.1 g, respectively. The length weight relationships were estimated from the allometric formula, $W = aL^b$, where W is total body weight (g), L the total length (cm), a and b are the coefficients of the functional regression between W and L (Ricker, 1973).

RESULTS

From Table 1, grouping of the fish into 6 categories according to standard length (SL) showed that every group had helminth infection but the group with highest parasitism is the 0.0-5.4 cm group (n = 1) as it had prevalence rate of 100%; this group had mean intensity and mean abundance of 18.0. Mean intensity (number of parasite per fish) was highest for the SL group 5.8- 26. 1 cm.

A total of 80 helminth parasites comprising nematode, Acanthocephalan and trematode were isolated from one predilection site of the fish sampled, the small intestine. The types and pattern of helminth parasites isolated from *S. nigrita* species were nematodes, 15 (30.3%); Acanthocephala, 62 (55.4%) and trematode, 3 (14.3%) (Table 2).

The pattern of helminth parasites occurrence in relation to season of the year is as shown in Table 4. Sixty one (61) fish were sampled for wet season, out of which 5 (21.1%) were infected, mean intensity 11.6 and mean abundance 1.0 of helminth parasites occurrence were recorded. In dry season, forty one (41) fishes were sampled, 11 (78.9%) had infection with mean intensity of 2.0 and mean abundance of 0.5, respectively (Table 3).

The standard lengths (SL) of the sampled fish ranged from 5.4 to 26.1 cm, 5.8 to 19.0 cm and 5.4 to 26.1 cm. The mean SL for the fish were 7.29 ± 2.82, 7.66 ± 2.25 and 7.45 ± 2.59. The weights of the fish ranged from 2.6 - 379.9, 3.4-206.6 and 2.6-379.9 g with the overall mean

Table 3. Pattern of helminth parasites occurrence recovered in *S. nigrita* in relation to season in the lower Niger (Idah) Nigeria.

Season	Fish species	Total fish examined	Total fish infected	Total parasites recovered	Prevalence rate (%)	95% CI	Mean intensity (No)	Mean abundance (No)
Wet season	*S .nigrita*	61	5	58	7.5	7.11 - 7.14	11.6	1.0
Dry season	*S .nigrita*	41	11	22	8.2	8.52 - 8.63	2.0	0.5
Total		102	16	80	15.7	15.63 -15.77	13.6	1.5

CI = Confidence interval

Table 4. Length - weight relationship of *S. nigrita* in the lower Niger (Idah) Nigeria.

Fish species	Sex	Standard length (cm)			Weight (g)			n	a	b	r
		Min.	Max.	Mean ± SD	Min.	Max.	Mean ± SD				
	Males	5.4	26.1	7.29 ± 2.82	2.6	379.9	13.90 ± 49.41	58	0.0197	2.9948	0.9697
S. nigrita	Females	5.8	19.0	7.66 ± 2.25	3.4	206.6	13.31 ± 30.53	44	0.0164	3.0887	0.9694
	Combined sex	5.4	26.1	7.45 ± 2.59	2.6	379.9	13.65 ± 42.13	102	0.0182	3.0364	0.9701

n = Number of fish examined; a = intercept; b = slope; r = correlation coefficient of determination.

weight of 13.65 ± 42.13 g (Table 4).

DISCUSSION

The result of this study reveals the occurrence of four helminth parasites in *S. nigrita* in the lower Niger (Idah), Kogi State, Nigeria. The four parasites belonged to nematode, Acanthocephala and trematoda. The large number of helminths infection recorded indicated that helminths were considerable parasites of the *Synodontis* species studied. This agrees with the study of Boomker (1994), Akinsanya et al. (2008) and Owolabi (2008). The parasites recovered from the studied fish were *Capillaria* spp., *Contraceacum* spp., *Acanthocephalus* spp. and *Posthodiplostomum* species. This conforms with the studies of Boomker (1994), Khalil (1969) and Boomker (1994).

According to the host parasite checklist on African freshwater fishes of Khalil and Polling (1997) and other relevant studies, the present work is the first scientific record of *Contraceacum* spp. and *Posthodiplostomum* spp. in the *Synodontis* species examined in lower Niger (Idah) Nigeria. The high proportion of Acanthocephala (48.7%) than nematode (39.1%) and trematodes (12.2%) showed that *Acanthocephalus* spp were the commonest infection of this genus in lower Niger (Idah).

The standard lengths (SL) of the sampled fish ranged from 5.4 to 26.1 cm, 5.8 to 19.0 cm and 5.4 to 26.1cm. The mean SL for the fish were (7.29 ± 2.82), (7.66 ± 2.25) and (7.45 ± 2.59). The weights of the fish ranged from 2.6 - 379.9 g, 3.4 - 206.6 g and 2.6 - 379.9 g, while the mean was 13.65 ± 42.13 g. The result of the length-weight relationship of the fish showed that the fish exhibits isometric growth in the water body.

It is therefore recommended that fish seeds from the wild should be examined for the presence of helminth parasites prior to use and periodically during culture practice. Awareness should also be created on the socio economic and human health implications of eating infected fish among the fisher folks and the general

REFERENCES

Areola O, Irueghe O, Ahmed K, Adeleke B, Leong GC (1992). Certificate Physical and Human Geography for Secondary Schools. University Press Plc, Ibadan. p. 406.

Boomker J (1994). Parasites of South African freshwater fish VI Nematode parasites of some fish species in the Kruger National Park. Onderstepoort J. Vet. Res. 61(1):35-43.

Gunder H, Fink W (2004). *Clarias gariepinus* (on-line), Animals Diversityweb.http://animaldiversity.ummz.unmich.edu/site/accounts/in formation/clariasgareipinus.htm. Accessed on June 15, 2007.

Gupta K, Gupta PC (2006). General and Applied Ichthyology (Fish and Fisheries). First Edition S. Chand and Company Ltd. 7361, Ram Nagar, New Delhi-110055. ISBN, 81-219-7 code 03339. p. 1133.

Hilderbrand KS, Price RJ, Olson RE (2003). Parasites in Marine Fishes Questions and Answers for Seafood Retailers Oregon State University, USA.seagrant.oregonstate.edu/index/html

Khalil LF (1969). Studies on the helminth parasites of freshwater fishes of the Sudan. J. Zool. London, 158:143-170.

Khalil LF, Polling L (1997). Checklist of the helminth parasites of African freshwater fishes. University of the North Republic of South Africa, River Printers, Pieturburg, South Africa. p. 185.

Okaeme AN, Olufemi BE (1997). Fungi associated with *Tilapia* culture ponds in Nigeria. J. Aquat. Trop. 12:267-274.

Owolabi OD (2008). Endoparasitic helminths of the upside-down catfish *Synodontis membranaceus* (Geoffroy Saint Hillaire) in the Jebba Lake, Nigeria. International J. Zool. Res. 4 (3):188-188.

Pauly D (1983). Some simple methods for the assessment of tropical fish stocks, Food and Agriculture Organization (FAO). Fish. Technol Paper 234. FAO, Rome.

Ricker WE (1973). Linear regressions in fishery research. J. Fish. Res Board Can. 30:409-439.

30

In vitro antibacterial activity of alkaloid extracts from green, red and brown macroalgae from western coast of Libya

Rabia Alghazeer[1] , Fauzi Whida[2], Entesar Abduelrhman[3], Fatiem Gammoudi[4] and Mahboba Naili[1]

[1]Chemistry department, Faculty of Sciences, Tripoli University, Tripoli, Libya.
[2]Botany department, Faculty of Sciences, Tripoli University, Tripoli, Libya.
[3]Biology department, Faculty of Education, Azzawiya University, Azzawiya, Libya.
[4]Microbiology and Parasitology department, Faculty of Veterinary Medicine, Tripoli University, Tripoli, Libya.

Marine organisms and microorganisms are known to be a rich source of alkaloids with unique chemical feature and interesting biological activities. The current study presents the antibacterial effect of the alkaloid extracts of some green, red and brown algae were collected from western coast of Libya, against, *Escherichia coli*, *Salmonella typhi*, *Klebsiella* spp., and *Pseudomonas aeruginosa*, *Staphylococcus aureus*, *Bacillus subtilis*, *Bacillus* spp. and *Staphylococcus epidermidis* were investigated. Although alkaloid extracts of green algae inhibited all tested bacteria, maximum effect was exhibited by brown and red algae species. Thus, *Cystoseira barbata* alkaloid extract showed remarkable inhibition of human pathogen *Klebsiella* spp. *Dictyopteris membranacea* alkaloid extract also demonstrated similar considerable effect against *S. typhi* with MIC value 1.56 mg/ml. The pronounced antibacterial activity of *C. barbata* and *D. membranacea* can be attributed to their high alkaloid contents. These results suggest that red and brown algae secondary metabolites are important sources that could produce potential chemotherapeutic agents.

Key words: Macroalgae, alkaloids, antibacterial activity.

INTRODUCTION

Algae are a large and diverse group of organisms from which a wide range of secondary metabolites have been isolated. A number of these compounds possess biological activities such as toxicity, antibacterial, antifungal, antiviral, antitumour and other specific activity (Cannell, 1993). These bioactive compounds include alkaloids (Guven et al., 2010), polyphenols) Pereira et al., 2002), terpenoids (Cen-Pacheco et al., 2010), flavonoids (Stafford, 1991), tannins (Serrano et al., 2009) and acetogenins (Narkowicz and Blackman, 2006) which are applicable for antioxidant (Rocha et al., 2007), antimicrobial

(Li, 2009; Saidani et al., 2011), antiviral (Romanos et al., 2002; Mayer et al., 2009), anti-inflam-matory and anticancer activities (Jaswir and Monsur, 2011; Bhakuni and Rawat, 2005; Vasanthi et al., 2004; Natarajan and Kathiresan, 2010). Nevertheless, in Libya this kind of study has not been well explored, despite the wealth of Libyan marine flora. Therefore, the present study is to investigate the alkaloid compounds extracted from macro-algae from the Libyan coast.

Alkaloids are heterocyclic nitrogen compounds, naturally occurring in plants, microbes, animals and marine organisms. The first medically useful example of an alkaloid was morphine, isolated in 1805 from opium *papaver somniferum* (Fessenden and Fessenden, 1982). Although, alkaloids have been extensively studied in plants and few studies in marine algae are reported due to the fact that alkaloids of marine algae are relatively rare compared with terrestrial plant alkaloids (Guven et al., 2010). The alkaloids found in marine algae can be classified into three groups: Phenylethylamine alkaloids, indole and halogenated indole alkaloids, and other alkaloids. Structurally, phenylethylamine and indole groups are the most alkaloids isolated from marine algae. Biological activities of halogenated and non-halogenated forms have been reported as bioactive compounds and as biological probes for physiological studies (Kasim et al., 2010). In addition, *Caulerpin* isolated from macroalgae, was the only indole alkaloid from marine sources which has been reported to have anti-inflammatory potentials (Carolina et al., 2011; Everton et al., 2009). In addition, there are two derivatives: lophocladine A and lophocladine B which have been isolated from a red alga *Lophocladia* spp., collected from Fijian Island, New Zealand (Gross et al., 2006) and their anticancer activity has been proved successfully in various cancer cell lines (Patricia et al., 2010).

Previous studies revealed that seaweed extracts, especially polyphenols have antioxidant activity (Chandini et al., 2008; Ganesan et al., 2008); whereas alkaloids are commonly found to have antimicrobial properties against both Gram-positive and Gram-negative bacteria (Guven et al., 2010) such as halogenated indole alkaloids which are isolated from red algae (Ayyad and Badria, 1994). These compounds have been approved for their antibacterial activity (Guella et al., 2006). The biological activity of marine indole alkaloids is clearly a product of the unique functionality and elements involved in the biosynthesis of marine natural products which increase the biological activity of seaweeds. For instance, bromination of many natural products has the potential to increase biological activity significantly (Gul and Hamann, 2005). The current study was undertaken to investigate the antibacterial effect of alkaloid extracts of 6 species of marine algae (two Chlorophyceae, three Phaeophyceae and one Rhodophyceae) collected from the Libyan western coast, against pathogenic Gram-negative bacteria: *Escherichia coli*, *Salmonella typhi*, *Klebsiella* spp., and *Pseudomonas aeruginosa* as well as Gram positive bacteria, *Staphylococcus aureus*, *Bacillus subtilis*, *Bacillus* spp., and *Staphylococcus epidermidis*.

MATERIALS AND METHODS

Sample collection

Ulva lactuca, *Codium tomentosum* (Chlorophyta), *Cystoseira barbata*, *Sargassum vulgare*, *Dictyopteris membranacea* (Phaeophyta) and *Gelidium latifolium* (Rhodophyta) were collected from western coast of Libya between February and spring, 2009.

The algal samples were taxonomically identified at Botany Department, Faculty of Science, Tripoli University. Algae were washed properly with distilled water, then they were shade dried at room temperature, after which they were crushed in an electric mill until a fine powder was obtained (Chiheb et al., 2009).

Bacterial strains

Eight bacterial strains (Gram positive and negative) were selected for the study. The Gram positive species were: *S. aureus* (*S. aur*) was obtained from the Clinical Microbiology Laboratory, Azzawiya Medical Center (Azzawiya, Libya). *B. subtilis* (*B. sub*) were kindly provided by Mohamed Elghazali, Department of Microbiology Biotechnology Research Center (Twaisha, Libya), while *Bacillus* spp. (*B. spp.*) and *S. epidermidis* (*S. epi*) were obtained from the Department of Microbiology, Faculty of Veterinary Medicine, Tripoli University. The Gram negative species *S. typhi* (*S. typhi*), *E. coli* (*E. coli*), *P. aeruginosa* (*P. aer*) and *klebsiella* spp. (*K. spp.*) were obtained from the Department of Microbiology, Faculty of Veterinary Medicine, Tripoli University, Libya.

Alkaloids extraction

Powdered algae materials (50 g) were extracted several times with methanol (300 ml). Methanol extraction was continued until the plant material gave a negative result for alkaloids (Mayer's test). The obtained methanolic extract was evaporated under reduced pressure at 40°C, to minimize any possible thermal degradation of the alkaloids and other thermo labile compounds. The crude alkaloid mixture was then separated from neutral and acidic materials and water soluble ingredients by extraction with aqueous acetic acid, followed by dichloromethane extraction, then basification of the aqueous solution which was subjected to further dichloromethane extraction thereafter (Hadi and Bremner, 2001).

Thin layer chromatography (TLC)

Identification of alkaloids was further carried out by TLC using pre coated silica gel 60 F264 plates (Wagner and Bladt, 2004). Different screening systems were used to obtain better resolution of the components. Dragendorff's reagent was used as a locating reagent (Harborne, 1992). R_f value of each spot was calculated as R_f = Distance travelled by the solute/Distance travelled by the solvent.

Determination of antibacterial activity

The antimicrobial activity test of algal crude extracts was performed *in vitro* using the "hole-plate diffusion method" (Sarvanakumar et al., 2009). Each test organism was maintained on nutrient agar slant and was recovered for testing by growth in nutrient broth (Biolab, Difco) for 14 h at 37°C before streaking. Cultures were routinely adjusted to a suspension of 1×10^6 to 2×10^6 CFU/ml using pre-made calibration curve representing viable cell count ($X \times 10^6$) against OD 660 nm (Y). The plates with bacteria were incubated at 37°C for 24 h. After incubation, the inhibition zones formed around the holes were measured. Methanol (100%) without seaweed extract was used as negative control and ciprofloxacin disc (30 µg) was used as the positive control.

Determination of minimum inhibitory concentration (MIC)

The MICs were determined by the agar dilution method (Daud and

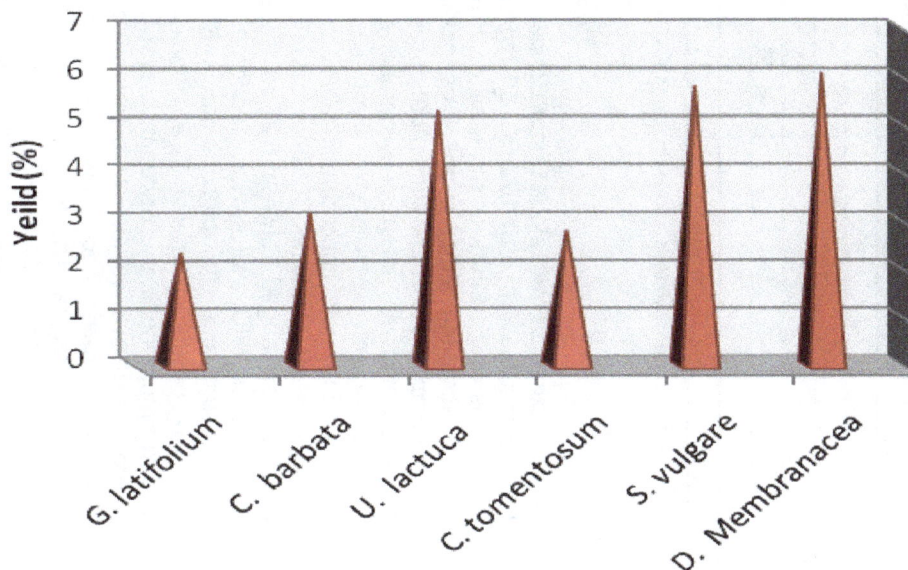

Figure 1. Percentage of alkaloid yield (mg alkaloid/g dry weigh) extracted from *U. lactuca, C. tomentosum, C. barbata, S. vulgare, D. membranacea* and *G. latifolium*.

Sanchaz, 2005). Two-fold serial dilutions of the original algae extract (100 mg/ml) were prepared in nutrient broth to obtain concentration from 100 to 1.56 mg/ml solvent. The plates incubated at 27°C for 18 h.

Bioautography method

Ten microliter (10 µl) of solutions corresponding to 1000 µg of alkaloids extract were applied to precoated Silica-gel TLC plates, developed with $CHCl_3$/ MeOH/ Na_2CO_3 (3:8:1, v/v) for each extract, and was evaporated to complete dryness. The dried plates were overlaid with nutrient agar medium seeded with *E. coli* (10^6 to 10^7 CFU/ml) and then incubated overnight at 37°C.

Statistical analyses

All assays were done in triplicate. All data are expressed as means ± S.D. Data were analyzed by an analysis of variance (P < 0.05) and the means separated by one-way ANOVA with Turkey's b test using SPSS version 20.0.

RESULTS AND DISCUSSION

The qualitative phytochemical analysis for *U. lactuca, C. barbata, D. membranacea, C. tomentosum, S. vulgare* and *G. latifolium* showed the presence of alkaloids according to previous finding (Alghazeer et al., 2013). The present study was performed in order to extract alkaloids from the same species and then assess their antibacterial activity. Figure 1 shows alkaloid contents (% mg/g) extracted from green, red and brown algae species. The highest content was recorded for *D. Membranacea* (6.11%), *S. vulgare* (5.84%), *U. lactuca* (5.33%), (6.11%); whereas *C. barbata* and *C. tomentosum* showed moderate content of alkaloid (3.2 and 2.84%, respectively),

while the lowest alkaloid content was obtained from *G. latifolium* (2.37%). Alkaloids present special interest because of their pharmacological activities. In fact, many reports revealed the presence of alkaloids in marine algae and some of them have been investigated for their biological activity (Guven et al., 2010; Kasım et al., 2010). Antimicrobial activities of alkaloid extracts from six seaweeds species represented by three Phaeophyta (*S. vulgare, D. membranacea* and *C. barbata*), two Chlorophyta (*U. lactuca* and *C. tomentosum*) and one Rhodophyta (*G. latifolium*) were examined against eight test bacteria (*Bacillus* spp., *B. subtilis, S. aureus, S. epidermidis, E. coli, kleb.* spp., *P. aeruginosa* and *Salmonella typhi*). The inhibition zones of brown, green and red algae extracts against Gram positive and Gram negative bacteria ranged between 13 to 35, 12 to 29 and 15 to 34 mm, respectively, all values are shown in Table 1.

The alkaloid extract of *U. lactuca* showed a relatively high mean zone of inhibition (21 ± 0.11 mm) against the Gram positive *S. aureus, S. epidermis* then *Bacillus* spp. (17 ± 0.3 mm), *E. coli* (16 ± 0.12 mm) and *B. subtilis* (14 ± 0.10 mm). While the alkaloid extract of *C. tomentosum* showed a remarkable high inhibition zone against *S. epidermis* (29 ± 0.35 mm) then *Bacillus* spp. (20 ± 0.3 mm), *S. aureus* (16 ± 11) and *B. subtilis* (13 ± 0.09 mm). For Gram negative bacteria, maximum zone of inhibition was recorded with alkaloid extract of *U. lactuca* against *kleb* spp. (18 ± 0.15 mm) and *S. typhi* (17 ± 0.11 mm). Also, maximum inhibition zones were recorded by *C. tomentosum* alkaloid extract against *klebsiella* spp. (27 ± 0.35 mm) then *E. coli* (23 ± 0.11 mm), *S. typhi* (21 ± 0.23 mm) and *P. aeruginosa* (12 ± 0.09 mm) (Table 1). The alkaloid extract of *S. vulgare* and *C. barbata* showed highest mean zone of inhibition against the Gram positive

Table 1. *In vitro* antimicrobial activity of the algal alkaloids extracts (100 mg/ml) against gram positive and gram negative bacteria.

Algal species	U. lactuca	C. tomentosum	G. latifolium	S. vulgare	D. membranacea	C. barbata	Ciprofloxacin[a]	Chloramphenicol[a]	Neomycin[a]
Test organism	DIZ (mm)	DIZ (mm)	DIZ (mm)	DIZ (mm)	DIZ (mm)	DIZ (mm)	DIZ (mm)	DIZ (mm)	DIZ(mm)
E. coli	16±0.12	23±0.11	29±0.54*	23±0.11	30±0.31*	22±0.22	23	19	-
S. typhi	17±0.11	21±0.23	18±0.11	25±0.6	35±0.74*	26±0.45	26	21	-
Kleb. sp	18±0.15	27±0.35*	15±0.11	24±0.09	28±0.6*	35±0.54*	24	18	-
P. aer	15±0.12	12±0.09	ND	ND	ND	ND	20	-	-
B. sub	14±0.10	13±0.09	ND	ND	23±0.6	ND	29	-	21
B. sp	17±0.11	20±0.35	24±0.11	19±0.21	ND	31±0.15*	24	-	26
S. aur	21±0.15	16±0.11	15±0.12	13±0.09	16±0.11	22±0.15	25	-	20
S. epi	21±0.6	29±0.35*	34±0.6*	18±0.11	23±0.15	20±0.11	23	-	23

Data are expressed as the mean ± standard deviation (SD) of three replicates. * represent the statistical comparisons between alkaloid extracts and positive control by using ANOVA followed by post hoc Tukey's b test (p<0.05). ND: not detectable.

Bacillus spp. (19 ± 0.21 mm and 31 ± 0.15 mm, respectively) then against *S. epidermis* (18 ± 0.11; 20 ± 0.11 mm, respectively), and *S. aureus* (13 ± 0.09; 22 ± 0.15). However, *D. membranacea* alkaloid extract had no effect on *Bacillus* spp., but showed high inhibition zones against *B. subtilis* and *S. epidermis* (23 ± 0.6 and 23 ± 0.15 mm). Maximum inhibition zone of Gram negative bacteria was recorded for alkaloid extract of *S. vulgare* and *D. membranacea* against *S. typhi* (25 ± 0.6 mm; 35 ± 0.74 mm) while the highest effect by the alkaloid extract of *C. barbata* was observed against *klebsiella* spp. (35 ± 0.54 mm). Whereas, *P. aeruginosa* was not susceptible to the alkaloid extracts of *S. vulgare*, *D. membranacea* and *C. barbata*.

The antibacterial activity of the alkaloid extract of *D. membranacea* and *C. barbata* were significantly high (P<0.05) compared with the positive control (Ciprofloxacin and Chloramphenicol) against Gram negative bacteria (Table 1), the antibacterial activity of green, brown and red algae is well documented (Del Val et al., 2001) as well as their isolated alkaloids (Masuda et al., 1997; Kasim et

al., 2010). The alkaloid extract of *G. latifolium* showed significant high mean zone of inhibition against the Gram positive *S. epidermis* (34 ± 0.6 mm) compared with positive control (Ciprofloxacin and Neomycin) (P<0.05) which is in consistent with earlier finding where alkaloid isolated from red algae exhibited different modes of bioactivity (Sato et al., 1998; Gross et al., 2006). The recorded inhibition zone against *Bacillus* spp. and *S. aureus* were 24 ± 0.11 and 15 ± 0.12 mm respectively, while no inhibition was observed against *B. subtilis*. For Gram negative bacteria, maximum zone of inhibition was recorded with alkaloid extract of *G. latifolium* against *E. coli* (29 ± 0.54 mm), *S. typhi* (18 ± 0.11 mm) and *Kleb.* sp. (15 ± 0.11 mm), while no inhibition was observed against *P. aeruginosa* (Table 1). The activity of red, green and brown algae against both Gram positive and Gram negative bacteria may be indicative of presence of broad spectrum antibiotic compounds or simply the content of pharmacological active constituents like alkaloids (Omulokoli et al., 1997; Phang et al., 1994).

Minimum inhibitory concentrations (MICs) of the

alkaloid extracted from algae (Figure 2) were found to be within the range of 100 to 1.56 mg/ml. The high levels of the MIC's of some alkaloid extracts can be attributed either to the presence of the active components in low concentrations, or to the presence of some antagonistic components that serve as growth promoters for the bacteria. The minimum inhibitory concentration (MIC) value of green algae (*U. lactuca, C. tomentosum*) against bacteria was ranged between 6.25 to 100 mg/ml. The lowest MIC value was recorded for *C. tomentosum* and *U. lactuca* extracts (6.25, 25 mg/ml respectively) against *S. epidermidis* while The minimum inhibitory concentration (MIC) value of brown algae (*S. vulgare, D. membranacea* and *C. barbata*) against bacteria was ranged between 6.25 to 100 mg/ml. The lowest MIC (1.56, 6.25 and 12.5 mg/ml) values were recorded for *C. barbata*, *D. membranacea* and *S. vulgare* extracts, respectively against *klebsiella* spp. (Figure 1). For alkaloid extracted from red algae (*G. latifolium*), the minimum inhibitory concentration (MIC) values of brown algae (*S. vulgare*, *D. membranacea* and *C. barbata*) against bacteria were

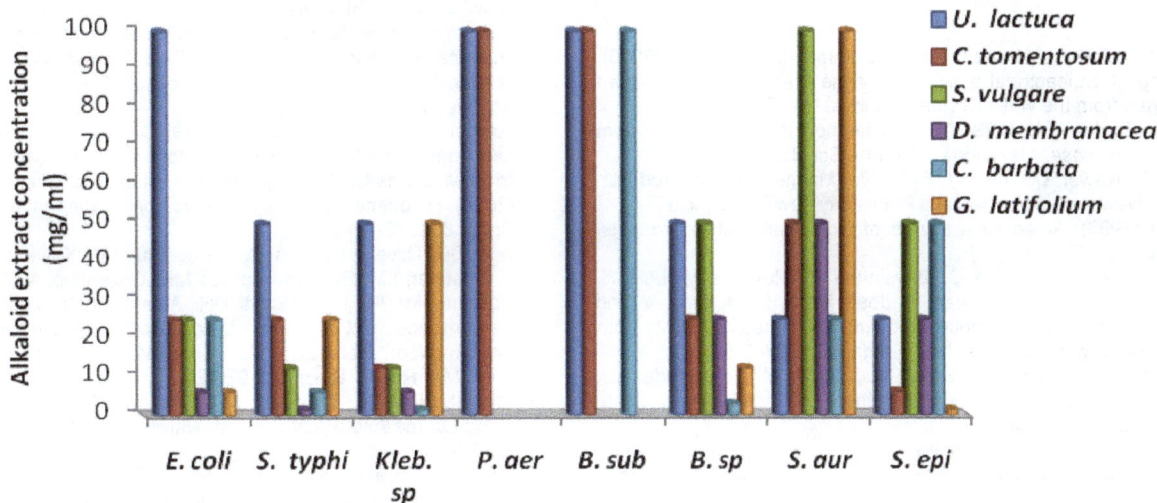

Figure 2. The *in vitro* antimicrobial activity of alkaloids extracts of tested algae expressed as minimum inhibitory concentration (MIC) (mg/ml) against some bacteria. S. *aur*: *Staphylococcus aureus*, B. *sub*: *Bacillus subtilis*, E. *coli*: *Escherichia coli*, P. *aer*: *Pseudomonas aeruginosa*, B. *sp*: *Bacillus* spp., S. *typhi*: *Salmonella typhi*, S. *epi*: *Staphylococcus epidermidis*, Kleb. sp: *klebsiella* spp.

Table 2. Location and prominence of zones of inhibition at different R_f values of alkaloid extracts against *E. coli*.

Algal species	Spot Number					
	S1		S2		S3	
	R_f	*E. coli*	R_f	*E. coli*	R_f	*E. coli*
U. lactuca	0.57	+	0.68	+	-	-
C. tomentosum	0.59	-	0.72	+	-	-
G. latifolium	0.57	++	0.61	-	0.68	+
S. vulgare	0.55	+	0.76	+	-	-
D. membranacea	0.54	++	0.64	-	0.75	++
C. barbata	0.52	+	0.74	++	-	-

Degree of inhibition: ++ = Prominent; + = moderate; - = Nil, *E. coli*: *Escherichia coli*; Mobile phase system is: chloroform: methanol: sodium carbonate (3:8:1, v/v).

ranged between 1.56 to 100 mg/ml. The lowest MIC (1.56 mg/ml) value was recorded against *kleb*. sp (Figure 1).

Preliminary TLC and the bioautographic assay tests were carried out on extracts to separate the compounds that were responsible for the inhibition tested the bacteria hence one spot may contain more than one compound. The results showed that the antibacterial assay can be attributed to the compounds observed at the various R_f values on the TLC separation. Although, sometimes the activity of compounds is not easily detected by this assay, if the compound does not diffuse through the agar, then the activity could be masked. The bioautoghraphy method was applied to the extracts using *E. coli* isolate to which all extracts exhibited antimicrobial activity (Table 1). The results showed good activity against *E. coli*, with prominent inhibition zones for the alkaloid of *D. membranacea*, *C. barbata*, *S. vulgare*, *G. latifolium* and *U. lactuca* extracts had two zones of inhibition, whereas,

the alkaloid of *C. tomentosum* extract exhibited one zone of inhibition (Table 2).

Conclusions

The results of this work indicate the presence of alkaloids in tested algae that play an indispensable role in antibacterial activity, however further studies to identify and characterize the specific active compounds, as well as the evaluation of the toxic aspects are recommended.

ACKNOWLEDGMENTS

The authors wish to thank Dr Hussein and Dr Asma AlNajar for their constant encouragement and consultation.

REFERENCES

Alghazeer R, Whida F, Abduelrhman E, Gammoudi F, Azwai S (2013). Screening of antibacterial activity in marine green, Red and brown macroalgae from the western coast of Libya. Nat. Sci. 1:7-14.

Ayyad SN, Badria FA (1994). An antitumor indole alkaloid from *Caulerpa racemosa*. Alexandria J. Pharm. Sci. 8: 217–219.

Bhakuni DS, Rawat DS (2005). Bioactive Marine Natural Products. Springer: New York and Anamaya Publisher, New Delhi, India

Cannell RJ (1993). Algae as a source of biologically active products. Pestic Sci. 39:147-153.

Carolina B, Everton T, Aline C, Daysianne P, Morgana V, Luiz H, Cavalcante-S, George E, Joao X, José Maria B, Bárbara V and Magna S (2011). Antinociceptive and Anti-Inflammatory Activity from Algae of the Genus Caulerpa. Mar. Drugs 9(3): 307-318.

Cen-Pacheco F, Nordstrom L, Souto ML, Martin MN, Fernandez JJ, Daranas AH (2010). Studies on polyethers produced by red algae. Mar Drugs 8: 1178-1188 .

Chandini SK, Ganesan P, Bhaskar N (2008). In vitro antioxidant activities of three selected brown seaweeds of India. Food Chem. 107: 707-713.

Chiheb I, Riadi H, Martinez-Lopez J, Dominguez S, Gomez V J, Bouziane H , Kadiri M (2009). Screening of antibacterial activity in marine green and brown macroalgae from the coast of Morocco. Afr. J. Biotechnol. 8: 1258-1262.

Del Val AG, Platas G, Basilio A, Cabello A, Gorrochateui J, Suay I, Vicente F, Portillo E, DeRio MJ, Reina GG and Pelaez F (2001). Screening of antimicrobial activities in red, green and brown macroalgae from Gran Canaria (Canary Islands, Spain). Int. J. Microbiol. 4: 35-40.

Fessenden RJ, Fessenden JS (1982). Organic Chemistry. 2nd eidition. Willard Grand press. Boston. Mass.

Ganesan P, Chandini SK, Bhaskar N (2008). Antioxidant properties of methanol extract and its solvent fractions obtained from selected Indian red seaweeds. Bioresour. Technol. 99: 2717-2723

Gross H, Goeger DE, Hills M, Ballantine DL, MurrayT F, Valeriote F A, Gerwick WH (2006). *Lophocladines*, bioactive alkaloids from the red alga *Lophocladia sp.* J. Nat. Prod. 69:640–644.

Guella G, N'Diaye I, Fofana M, Mancini I (2006). Isolation synthesis and photochemical properties of almazolone, a new indole alkaloid from a red alga of Senegal. Tetrahedron 62:1165–1170

Gul W, Hamann MT (2005). Indole alkaloid marine natural products: An established source of cancer drug leads with considerable promise for the control of parasitic, neurological and other diseases. Life Sci. 78: 442–453.

Guven KC, Percot A, Sezik E (2010). Alkaloids in marine algae. Mar. Drugs 8: 269-284.

Hadi S, Bremner B (2001). Initial studies on alkaloids from Lombok medicinal plants. Molecules 6:117- 129.4

Harborne JB (1992). Phytochemical methods. Chapman and Hall Publications, London. pp. 7-8.

Jaswir I, Monsur H (2011). Anti-inflammatory compounds of macro algae. J. Med. Plant Res. 5(33): 7146-7154.

Kasım CG, Aline P, Ekrem S (2010). Alkaloids in Marine Algae. Mar. Drugs 8: 269-284.

Li ZY (2009). Advances in marine microbial symbionts in the China Sea and related pharmaceutical metabolites Mar. Drugs 7: 113-129.

Masuda M, Abe T, Sato S, Suzuki T and Suzuki M (1997). Diversity of halogenated secondary metabolites in the red alga Laurencia nipponica (Rhodomelaceae, Ceramiales). J. Phycol. 33:196-208.

Mayer AMS, Rodriguez AD, Berlinck RS, Hamann MT (2009). Marine pharmacology in 2005–6: Marine compounds with anthelmintic, antibacterial, anticoagulant, antifungal, anti-inflammatory, antimalarial, antiprotozoal, antituberculosis, and antiviral activities; affecting the cardiovascular, immune and nervous systems, and other miscellaneous mechanisms of action. Biochim Biophys Acta 1790(5): 283-308.

Narkowicz C.K, Blackman AJ (2006). Further acetogenins from Tasmanian collections of Caulocystis cephalornithos demonstrating chemical variability. Biochem. Syst. Ecol. 34: 635-641.

Natarajan S, Kathiresan K (2010). Anticancer Drugs from Marine Flora: An Overview. J. Oncol. 2010: 1-18.

Omulokoli E, Khan B, Chhabra SC (1997).). Antiplasmodial activity of four Kenyan medicinal plants. J Ethnopharmacol. 56:133-7.

Patricia M, Lucas S, Caio G, Ademar A, Marcio L, Wilson R and Ana F (2010). Halogenated Indole Alkaloids from Marine Invertebrates. Mar Drugs 8: 1526-1549.

Pereira SB, Oliveira-Carvalho MF, Angeiras JAP, Oliveira NMB, Torres J, Gestinari LM, Badeira-Pedrosa ME, Cocentino ALM, Santos MD, Nascimento PRF, Cavalcanti DR. Algas bentônicas do Estado de Pernambuco (2002). In Diagnstico da Biodiversidade de Pernambuco; Tabarelli, M., SilvaJ.M.C., Eds.; Massagana & SECTMA: Recife, Brazil. pp. 97-124.

Phang SM, Lee YK, Browitzka MA, Whiltow BA (1994). Algal biotech nology in the Asia-Pacific region: University of Malaya, Kualalumpur pp. 75-81.

Rocha FD, Pereira RC, Kaplan MAC, Teixeira VL (2007). Natural products from marine seaweeds and their antioxidant potential. Braz J. Pharmacogn. 17: 631-639

Romanos MV, Andrada-Serpa MJ, Santos MGM, Ribeiro ACF, Yoneshiguevalentin Y, Costa SS, Wigg MD (2002). Inhibitory effect of extracts of Brazilian marine algae on human T-cell lymphotropic virus type 1 (HTLV-1) induced syncytium formation in vitro. Cancer Invest 20: 46-54 .

Saidani K, Bedjou F, Benabdesselam F, Touati N (2011). Antifungal activity of methanolic extracts of four Algerian marine algae species. Afr. J. Biotechnol. 11(39):9496-9500

Sato H, Tsuda M, Watanabe K, Kobayashi J, Rhopaladins A–D (1998). New indole alkaloids from marine tunicate Rhopalaea sp Tetrahedron 54: 8687–90.

Serrano J, Puupponen-Pimia R, Dauer A, AuraA.M, Saura-Calixto F (2009). Tannins: Current knowledge of food sources, intake bioavailability and biological effects. Mol. Nutr. Food Res. 53: S310 S329.

Stafford HA (1991). Flavonoid evolution: an enzymic approach. Plant Physiol. 96:680-685 .

Vasanthi H, Rajamanickam G, Saraswathy A (2004). Tumoricidal effect of the red algae Acanthophora spicifera on Ehrlich's ascites carcinoma in mice. Seaweed Res. Util. 217–224.

Wagner H, Bladt S (2004). Plant drug analysis-A thin layer chromatography atlas 2nd edition. New Delhi: Thompson Press Ltd.

Everton TS, Daysianne PL, Aline CQ, Diogo JCS, Anansa BA, Eliane ACM, Vitor PL, George ECM, Joao XAJ, Maria COC, José MBF, Petrônio FA, Barbara VOS, Magna SA (2009). The Antinociceptive and Anti-Inflammatory Activities of Caulerpin, a Bisindole Alkaloid Isolated from Seaweeds of the Genus Caulerpa. Mar. Drugs. 7: 689-704.

Permissions

All chapters in this book were first published in AJB, by Academic Journals; hereby published with permission under the Creative Commons Attribution License or equivalent. Every chapter published in this book has been scrutinized by our experts. Their significance has been extensively debated. The topics covered herein carry significant findings which will fuel the growth of the discipline. They may even be implemented as practical applications or may be referred to as a beginning point for another development.

The contributors of this book come from diverse backgrounds, making this book a truly international effort. This book will bring forth new frontiers with its revolutionizing research information and detailed analysis of the nascent developments around the world.

We would like to thank all the contributing authors for lending their expertise to make the book truly unique. They have played a crucial role in the development of this book. Without their invaluable contributions this book wouldn't have been possible. They have made vital efforts to compile up to date information on the varied aspects of this subject to make this book a valuable addition to the collection of many professionals and students.

This book was conceptualized with the vision of imparting up-to-date information and advanced data in this field. To ensure the same, a matchless editorial board was set up. Every individual on the board went through rigorous rounds of assessment to prove their worth. After which they invested a large part of their time researching and compiling the most relevant data for our readers.

The editorial board has been involved in producing this book since its inception. They have spent rigorous hours researching and exploring the diverse topics which have resulted in the successful publishing of this book. They have passed on their knowledge of decades through this book. To expedite this challenging task, the publisher supported the team at every step. A small team of assistant editors was also appointed to further simplify the editing procedure and attain best results for the readers.

Apart from the editorial board, the designing team has also invested a significant amount of their time in understanding the subject and creating the most relevant covers. They scrutinized every image to scout for the most suitable representation of the subject and create an appropriate cover for the book.

The publishing team has been an ardent support to the editorial, designing and production team. Their endless efforts to recruit the best for this project, has resulted in the accomplishment of this book. They are a veteran in the field of academics and their pool of knowledge is as vast as their experience in printing. Their expertise and guidance has proved useful at every step. Their uncompromising quality standards have made this book an exceptional effort. Their encouragement from time to time has been an inspiration for everyone.

The publisher and the editorial board hope that this book will prove to be a valuable piece of knowledge for researchers, students, practitioners and scholars across the globe.

List of Contributors

W. Chahrour
Laboratory of Applied Microbiology, Department of Biology, Faculty of Science, Oran University, BP 16, Es-Senia, 31100, Oran, Algeria

Y. Merzouk
Laboratory of Applied Microbiology, Department of Biology, Faculty of Science, Oran University, BP 16, Es-Senia, 31100, Oran, Algeria

J. E. Henni
Laboratory of Applied Microbiology, Department of Biology, Faculty of Science, Oran University, BP 16, Es-Senia, 31100, Oran, Algeria

M. Haddaji
Laboratory of Applied Microbiology, Department of Biology, Faculty of Science, Oran University, BP 16, Es-Senia, 31100, Oran, Algeria

M. Kihal
Laboratory of Applied Microbiology, Department of Biology, Faculty of Science, Oran University, BP 16, Es-Senia, 31100, Oran, Algeria

N. Ogbonna Christiana
South-East Zonal Biotechnology Centre, University of Nigeria, Nsukka, Enugu State, Nigeria
ASAMA CHEMICAL Co., Ltd. 20-3, Nihonbashi-Kodenmacho, Chuo-ku, Tokyo, 103-0001, Japan
Department of Plant Science and Biotechnology, University of Nigeria, Nsukka, Enugu state, Nigeria

K. Nozaki
ASAMA CHEMICAL Co., Ltd. 20-3, Nihonbashi-Kodenmacho, Chuo-ku, Tokyo, 103-0001, Japan

H. Yajima
ASAMA CHEMICAL Co., Ltd. 20-3, Nihonbashi-Kodenmacho, Chuo-ku, Tokyo, 103-0001, Japan

Y. Quiñones-Gutiérrez
School of Biological Sciences, UANL Ciudad Universitaria, AP 46-F, CP 66451, San Nicolas de los Garza, N. L., Mexico

M. J. Verde-Star
School of Biological Sciences, UANL Ciudad Universitaria, AP 46-F, CP 66451, San Nicolas de los Garza, N. L., Mexico

C. Rivas-Morales
School of Biological Sciences, UANL Ciudad Universitaria, AP 46-F, CP 66451, San Nicolas de los Garza, N. L., Mexico

A. Oranday-Cárdenas
School of Biological Sciences, UANL Ciudad Universitaria, AP 46-F, CP 66451, San Nicolas de los Garza, N. L., Mexico

R. Mercado-Hernández
School of Biological Sciences, UANL Ciudad Universitaria, AP 46-F, CP 66451, San Nicolas de los Garza, N. L., Mexico

A. Chávez-Montes
School of Biological Sciences, UANL Ciudad Universitaria, AP 46-F, CP 66451, San Nicolas de los Garza, N. L., Mexico

M. P. Barrón-González
School of Biological Sciences, UANL Ciudad Universitaria, AP 46-F, CP 66451, San Nicolas de los Garza, N. L., Mexico

KOUASSI Kan Modeste
Laboratoire Central de Biotechnologies, Centre National de Recherche Agronomique, 01 P. O. Box 1740 Abidjan 01, Côte d'Ivoire

KOFFI Kouablan Edmond
Laboratoire Central de Biotechnologies, Centre National de Recherche Agronomique, 01 P. O. Box 1740 Abidjan 01, Côte d'Ivoire

KONKON N'dri Gilles
Université de Cocody, UFR Sciences Biosciences, Laboratoire de Botanique, 22 BP 1414 Abidjan 22, Côte d'Ivoire

GNAGNE Michel
Station de recherche de Bimbresso, Centre National de Recherche Agronomique, 01 P. O. Box 1740 Abidjan 01, Côte d'Ivoire

KONÉ Mongomaké
Université d'Abobo Adjamé, UFR Sciences de la Nature, Laboratoire de Biologie et Amélioration des Productions Végétales, 02 BP 801 Abidjan 02, Côte d'Ivoire

KOUAKOU Tanoh Hilaire
Université d'Abobo Adjamé, UFR Sciences de la Nature, Laboratoire de Biologie et Amélioration des Productions Végétales, 02 BP 801 Abidjan 02, Côte d'Ivoire

Yun Hee Choi
Department of Pharmacy, College of Pharmacy, Chosun University, Gwangju 501-759, Korea

Seung Sik Cho
Department of Pharmacy, College of Pharmacy, Mokpo National University, Muan, Jeonnam, 534-729, Korea

Jaya Ram Simkhada
Department of Pharmacy, College of Pharmacy, Chosun University, Gwangju 501-759, Korea

Chi Nam Seong
Department of Biology, College of Natural Sciences, Sunchon National University, Sunchon, Jeonnam, 540-742, Korea

Hyo Jeong Lee
Department of Alternative Medicine, Gwangju University, Gwangju 503-703, Republic of Korea

Hong Seop Moon
Department of Pharmacy, College of Pharmacy, Mokpo National University, Muan, Jeonnam, 534-729, Korea

Jin Cheol Yoo
Department of Pharmacy, College of Pharmacy, Chosun University, Gwangju 501-759, Korea

Dahlia M. El Maghraby
Department of Botany and Microbiology, Faculty of Science, Alexandria University, 21511 Alexandria, Egypt

Ashraf El-Sayed
Cairo University Research Park (CURP), Faculty of Agriculture, Cairo University, 12613 Giza, Egypt
Department of Animal Production, Faculty of Agriculture Cairo University, 12613 Giza, Egypt

Salem M. Salem
Department of Animal Production, Faculty of Agriculture Cairo University, 12613 Giza, Egypt

Amany A. El-Garhy
Department of Pharmacology, National Organization for Drug Control and Research, Giza, Egypt

Zeinab A. Rahman
Department of Pharmacology, National Organization for Drug Control and Research, Giza, Egypt

Asmaa M. Kandil
Department of Pharmacology, National Organization for Drug Control and Research, Giza, Egypt

Jamylla Mirck Guerra de Oliveira
Medicinal Plants Research Center, Federal University of Piauí, Teresina, Piauí, Brazil

Denise Barbosa Santos
Medicinal Plants Research Center, Federal University of Piauí, Teresina, Piauí, Brazil

Francimarne Sousa Cardoso
Departament of Veterinary Morphophysiology, Federal University of Piauí, School of Agrarians Sciences, Teresina, Piauí, Brazil

Márcia de Sousa Silva
Departament of Veterinary Morphophysiology, Federal University of Piauí, School of Agrarians Sciences, Teresina, Piauí, Brazil

Yatta Linhares Boakari
Departament of Veterinary Morphophysiology, Federal University of Piauí, School of Agrarians Sciences, Teresina, Piauí, Brazil

Silvéria Regina de Sousa Lira
Departament of Veterinary Morphophysiology, Federal University of Piauí, School of Agrarians Sciences, Teresina, Piauí, Brazil

Amilton Paulo Raposo Costa
Medicinal Plants Research Center, Federal University of Piauí, Teresina, Piauí, Brazil
Departament of Veterinary Morphophysiology, Federal University of Piauí, School of Agrarians Sciences, Teresina, Piauí, Brazil

Thiriloshani Padayachee
Department of Biosciences, Vaal University of Technology, Vanderbijlpark South Africa

Bharti Odhav
Department of Biotechnology and Food Technology, Durban University of Technology, South Africa

Dilafroza Jan
Centre of Research for Development, University of Kashmir, Srinagar-190006, Jammu and Kashmir, India

Ashok K. Pandit
Centre of Research for Development, University of Kashmir, Srinagar-190006, Jammu and Kashmir, India

Azra N. Kamili
Centre of Research for Development, University of Kashmir, Srinagar-190006, Jammu and Kashmir, India

Hossein Salehizadeh
Department of Civil Engineering, University of Ottawa, Ottawa, ON, K1N 6N5, Canada
Chemical Engineering Group, Faculty of Engineering, University of Isfahan, Isfahan, Iran

Aida Ranjbar
Chemical Engineering Group, Faculty of Engineering, University of Isfahan, Isfahan, Iran

Kevin Kennedy
Department of Civil Engineering, University of Ottawa, Ottawa, ON, K1N 6N5, Canada

Bhabesh Mili
Division of Physiology and Climatology Indian Veterinary Research Institute, Izatnagar-243122 (U.P.) India

Sujata Pandita
Dairy Cattle Physiology, National Dairy Research Institute, Karnal, Haryana, India

B. S. Bharath kumar
Dairy Cattle Physiology, National Dairy Research Institute, Karnal, Haryana, India

Anil Kumar Singh
Dairy Cattle Physiology, National Dairy Research Institute, Karnal, Haryana, India

Madhu Mohini
Division of Dairy Cattle Nutrition, National Dairy Research Institute, Karnal, Haryana, India

Manju Ashutosh
Dairy Cattle Physiology, National Dairy Research Institute, Karnal, Haryana, India

Eshaq A. Mohmid
Department of Microbiology, Medical Laboratory, Ahmadi Hospital, Kuwait Oil Company, Ahmadi, Kuwait

El-Sayed A. El-Sayed
Department of Botany, Faculty of Science, Zagzig University, Zagzig, Egypt

Mahmoud F. Abdel El-Haliem
Department of Botany, Faculty of Science, Zagzig University, Zagzig, Egypt

Jissa G. Krishna
Microbial Technology Laboratory, Department of Biotechnology, Cochin University of Science and Technology, Cochin 682 022, Kerala, India
National Centre for Biological Sciences, Bangalore 560 065, India

Ansu Jacob
Department of Polymer Science and Rubber Technology, Cochin University of Science and Technology, Cochin-682 022, Kerala, India

Philip Kurian
Department of Polymer Science and Rubber Technology, Cochin University of Science and Technology, Cochin-682 022, Kerala, India

KK Elyas
Microbial Technology Laboratory, Department of Biotechnology, Cochin University of Science and Technology, Cochin 682 022, Kerala, India
Department of Biotechnology, Calicut University, Kerala, India

M. Chandrasekaran
Microbial Technology Laboratory, Department of Biotechnology, Cochin University of Science and Technology, Cochin 682 022, Kerala, India

M. E. Balogun
Department of Physiology, Faculty of Basic Medical Sciences, College of Health Sciences, Ebonyi State University, Abakaliki, Nigeria

J. O. Oji
Department of Physiology, Faculty of Basic Medical Sciences, College of Health Sciences, Ebonyi State University, Abakaliki, Nigeria

E. E. Besong
Department of Physiology, Faculty of Basic Medical Sciences, College of Health Sciences, Ebonyi State University, Abakaliki, Nigeria

A. A. Ajah
Department of Physiology, Faculty of Basic Medical Sciences, College of Health Sciences, University of Port Harcourt, Choba, Port Harcourt, Nigeria

E. M. Michael
Department of Anatomy, Faculty of Basic Medical Sciences, College of Health Sciences, Ebonyi State University, Abakaliki, Nigeria

Ding Ting
School of Plant Protection, Anhui Agricultural University, Hefei 230036, People's Republic of China

Sun Wei-Wei
School of Plant Protection, Anhui Agricultural University, Hefei 230036, People's Republic of China

Qi Yong- Xia
School of Plant Protection, Anhui Agricultural University, Hefei 230036, People's Republic of China

Jiang Hai-Yang
Anhui Provincial Key Laboratory of Crop Biology, Hefei 230036, People's Republic of China

Shabana Maqsood
Department of Microbiology, Quaid –e- Azam University, Islamabad, Pakistan

Fariha Hasan
Department of Microbiology, Quaid –e- Azam University, Islamabad, Pakistan

Tariq Masud
Department of Food, Technology, PMAS-Arid Agriculture University, Rawalpindi, Pakistan

Nashwa A. Ezzeldeen
Department of Microbiology, Faculty of Veterinary Medicine, Cairo University, Giza, Egypt

Khaled F. Al-Amary
Department of Microbiology, Faculty of Veterinary Medicine, Cairo University, Giza, Egypt

Mohamed M. Abdalla
Department of Microbiology, Central Laboratory of Residue Analysis of Pesticides and Heavy Metals in Foods (QCAP), Dokki, Giza, Egypt

Sherein I Abd El-Moez
Department of Microbiology and Immunology, National Research Center (NRC), Giza, Egypt
Food Risk Analysis Group- Center of Excellence for Advanced Sciences, NRC, Giza, Egypt

A. Muthukumar
Department of Plant Pathology, Faculty of Agriculture, Annamalai University, Annamalainagar-608 002, Chidambaram, Tamil Nadu, India

A. Venkatesh
Department of Plant Pathology, Faculty of Agriculture, Annamalai University, Annamalainagar-608 002, Chidambaram, Tamil Nadu, India

Mona A. Aldamegh
Departmet of Biology, College of Sciences and Arts at Onaizah, Qassim University, Saudi Arabia

Emad M. Abdallah
Department of Laboratory Sciences, College of Sciences and Arts at Al-Rass, Qassim University, P. O. Box 53, Saudi Arabia

Anis Ben Hsouna
Department of Laboratory Sciences, College of Sciences and Arts at Al-Rass, Qassim University, P. O. Box 53, Saudi Arabia

M. Radha Krishna REDDY
Department of Biotechnology, Krishna University, Machilipatnam-521001, Andhra Pradesh, India

S. A. MASTAN
P.G Department of Biotechnology, PG Courses, Research Center, DNR College, Bhimavaram- 534202, Andhra Pradesh, India

Regildo Márcio Gonçalves da Silva
Universidade Estadual Paulista (UNESP), Departamento de Ciências Biológicas - Laboratório de Fitoterápicos, Faculdade de Ciências e Letras de Assis, Avenida Dom Antônio 2100, CEP: 19806-900, Assis, São Paulo, Brasil

Eni Aparecida do Amaral
Centro Universitário de Patos de Minas (UNIPAM), Laboratório de Química Instrumental e Central Analítica

Vanessa Marques de Oliveira Moraes
Universidade Estadual Paulista (UNESP), Departamento de Ciências Biológicas - Laboratório de Fitoterápicos, Faculdade de Ciências e Letras de Assis, Avenida Dom Antônio 2100, CEP: 19806-900, Assis, São Paulo, Brasil

Luciana Pereira Silva
Universidade Estadual Paulista (UNESP), Departamento de Ciências Biológicas - Laboratório de Fitoterápicos, Faculdade de Ciências e Letras de Assis, Avenida Dom Antônio 2100, CEP: 19806-900, Assis, São Paulo, Brasil

Mansoor Saljooghianpour
Islamic Azad University, Iranshahr Branch, Iranshahr, Iran

Taiebeh Askari Javaran
Islamic Azad University, Iranshahr Branch, Iranshahr, Iran

Mohamed E. H. Osman
Botany Department, Faculty of Science, Tanta University, Tanta, Egypt

Atef M. Aboshady
Botany Department, Faculty of Science, Tanta University, Tanta, Egypt

Mostafa E. Elshobary
Botany Department, Faculty of Science, Tanta University, Tanta, Egypt

Mehrcedeh Tafazoli
Department of Forestry, Faculty of Natural Resources, Sari Agricultural Science and Natural Resources University, Sari, Iran

Seyed Mohammad Hosseini Nasr
Department of Forestry, Faculty of Natural Resources, Sari Agricultural Science and Natural Resources University, Sari, Iran

Hamid Jalilvand
Department of Forestry, Faculty of Natural Resources, Sari Agricultural Science and Natural Resources University, Sari, Iran

Dariush Bayat
Deputy of Humid and Semi-humid Areas, Forests Ranges and Watershed Organization, Chalus, Iran

Olfat A. Khalil
Medical Biochemistry Department, Faculty of Science National Research center, Egypt
Biochemistry Department, Faculty of Medicine (girls)- Al-Azhar University- Cairo Egypt

Kholoud S. Ramadan
Medical Biochemistry Department, Faculty of Science National Research center, Egypt

Amal H. Hamza
Medical Biochemistry Department, Faculty of Science National Research center, Egypt

Safinaz E. El-Toukhy
Medical Biochemistry Department, Faculty of Science National Research center, Egypt

Ligia Maria Crubelati Bulla
Department of Biotechnology, Genetics and Cellular Biology (DBC), Universidade Estadual de Maringá, Brazil

Daniela Andressa Lino Lourenço
Department of Zootechny (DZO), Universidade Estadual de Maringá, Brazil

Sandro Augusto Rhoden
Department of Biotechnology, Genetics and Cellular Biology (DBC), Universidade Estadual de Maringá, Brazil

Ravely Casarotti Orlandelli
Department of Biotechnology, Genetics and Cellular Biology (DBC), Universidade Estadual de Maringá, Brazil

João Alencar Pamphile
Department of Biotechnology, Genetics and Cellular Biology (DBC), Universidade Estadual de Maringá, Brazil

M. J. Abussaud
Department of Biological Sciences, Yarmouk University, Irbid- Jordan

L. Alanagreh
Department of Biological Sciences, Yarmouk University, Irbid- Jordan

K. Abu-Elteen
Department of Biological Sciences, Alhashemia University, Zarqa

O. O. Toluhi
Department of Biological Sciences, Kogi State University, Anyigba, Kogi State, Nigeria

S. O. Adeyemi
Department of Biological Sciences, Kogi State University, Anyigba, Kogi State, Nigeria

Rabia Alghazeer
Chemistry department, Faculty of Sciences, Tripoli University, Tripoli, Libya

Fauzi Whida
Botany department, Faculty of Sciences, Tripoli University, Tripoli, Libya

Entesar Abduelrhman
Biology department, Faculty of Education, Azzawiya University, Azzawiya, Libya

Fatiem Gammoudi
Microbiology and Parasitology department, Faculty of Veterinary Medicine, Tripoli University, Tripoli, Libya

Mahboba Naili
Chemistry department, Faculty of Sciences, Tripoli University, Tripoli, Libya

www.ingramcontent.com/pod-product-compliance
Lightning Source LLC
Chambersburg PA
CBHW080655200326
41458CB00013B/4861